AF794228

Spintronics for Next Generation Innovative Devices

Wiley Series in Materials for Electronic and Optoelectronic Applications

www.wiley.com/go/meoa

Series Editors

Professor Arthur Willoughby, *University of Southampton, Southampton, UK*
Dr Peter Capper, *Formerly of Selex ES, Southampton, UK*
Professor Safa Kasap, *University of Saskatchewan, Saskatoon, Canada*

Published Titles

Bulk Crystal Growth of Electronic, Optical and Optoelectronic Materials, Edited by P. Capper
Properties of Group-IV, III–V and II–VI Semiconductors, S. Adachi
Charge Transport in Disordered Solids with Applications in Electronics, Edited by S. Baranovski
Optical Properties of Condensed Matter and Applications, Edited by J. Singh
Thin Film Solar Cells: Fabrication, Characterization and Applications, Edited by J. Poortmans and V. Arkhipov
Dielectric Films for Advanced Microelectronics, Edited by M. R. Baklanov, M. Green and K. Maex
Liquid Phase Epitaxy of Electronic, Optical and Optoelectronic Materials, Edited by P. Capper and M. Mauk
Molecular Electronics: From Principles to Practice, M. Petty
CVD Diamond for Electronic Devices and Sensors, Edited by R. S. Sussmann
Properties of Semiconductor Alloys: Group-IV, III–V and II–VI Semiconductors, S. Adachi
Mercury Cadmium Telluride, Edited by P. Capper and J. Garland
Zinc Oxide Materials for Electronic and Optoelectronic Device Applications, Edited by C. Litton, D. C. Reynolds and T. C. Collins
Lead-Free Solders: Materials Reliability for Electronics, Edited by K. N. Subramanian
Silicon Photonics: Fundamentals and Devices, M. Jamal Deen and P. K. Basu
Nanostructured and Subwavelength Waveguides: Fundamentals and Applications, M. Skorobogatiy
Photovoltaic Materials: From Crystalline Silicon to Third-Generation Approaches, G. Conibeer and A. Willoughby
Glancing Angle Deposition of Thin Films: Engineering the Nanoscale, Matthew M. Hawkeye, Michael T. Taschuk and Michael J. Brett

Spintronics for Next Generation Innovative Devices

Edited by

KATSUAKI SATO
Program Officer PRESTO Project, Japan Science and Technology Agency (JST), Tokyo, Japan

EIJI SAITOH
Institute for Materials Research and WPI Advanced Institute for Materials Research, Tohoku University, Sendai, Japan

WILEY

This edition first published 2015
© 2015 John Wiley & Sons, Ltd

Registered office
John Wiley & Sons Ltd, The Atrium, Southern Gate, Chichester, West Sussex, PO19 8SQ, United Kingdom

For details of our global editorial offices, for customer services and for information about how to apply for permission to reuse the copyright material in this book please see our website at www.wiley.com.

The right of the author to be identified as the author of this work has been asserted in accordance with the Copyright, Designs and Patents Act 1988.

All rights reserved. No part of this publication may be reproduced, stored in a retrieval system, or transmitted, in any form or by any means, electronic, mechanical, photocopying, recording or otherwise, except as permitted by the UK Copyright, Designs and Patents Act 1988, without the prior permission of the publisher.

Wiley also publishes its books in a variety of electronic formats. Some content that appears in print may not be available in electronic books.

Designations used by companies to distinguish their products are often claimed as trademarks. All brand names and product names used in this book are trade names, service marks, trademarks or registered trademarks of their respective owners. The publisher is not associated with any product or vendor mentioned in this book.

Limit of Liability/Disclaimer of Warranty: While the publisher and author have used their best efforts in preparing this book, they make no representations or warranties with respect to the accuracy or completeness of the contents of this book and specifically disclaim any implied warranties of merchantability or fitness for a particular purpose. It is sold on the understanding that the publisher is not engaged in rendering professional services and neither the publisher nor the author shall be liable for damages arising herefrom. If professional advice or other expert assistance is required, the services of a competent professional should be sought

The advice and strategies contained herein may not be suitable for every situation. In view of ongoing research, equipment modifications, changes in governmental regulations, and the constant flow of information relating to the use of experimental reagents, equipment, and devices, the reader is urged to review and evaluate the information provided in the package insert or instructions for each chemical, piece of equipment, reagent, or device for, among other things, any changes in the instructions or indication of usage and for added warnings and precautions. The fact that an organization or Website is referred to in this work as a citation and/or a potential source of further information does not mean that the author or the publisher endorses the information the organization or Website may provide or recommendations it may make. Further, readers should be aware that Internet Websites listed in this work may have changed or disappeared between when this work was written and when it is read. No warranty may be created or extended by any promotional statements for this work. Neither the publisher nor the author shall be liable for any damages arising herefrom.

Library of Congress Cataloging-in-Publication Data applied for.

A catalogue record for this book is available from the British Library.

ISBN: 9781118751916

Set in 10/12pt TimesLTStd by SPi Global, Chennai, India

1 2015

Contents

List of Contributors

Matthias Althammer, Walther-Meißner-Institut für Tieftemperaturforschung, Bayerische Akademie der Wissenschaften, Germany

Tomoteru Fukumura, Department of Chemistry, Graduate School of Science, University of Tokyo, Japan

Sebastian Goennenwein, Walther-Meißner-Institut für Tieftemperaturforschung, Bayerische Akademie der Wissenschaften, Germany

Kohei Hamaya, Department of Electronics, Faculty of Information Science and Electrical Engineering, Graduate School of Information Science, Kyushu University, Japan

Hans Huebl, Walther-Meißner-Institut für Tieftemperaturforschung, Bayerische Akademie der Wissenschaften, Germany

Norikazu Mizuochi, Graduate School of Engineering Science, Osaka University, Japan

Shuichi Murakami, Department of Physics, Tokyo Institute of Technology, Japan

Theo Rasing, Institute for Molecules and Materials, Radboud University Nijmegen, the Netherlands

Eiji Saitoh, Institute for Materials Research and WPI Advanced Institute for Materials Research, Tohoku University, Japan

Masashi Shiraishi, Solid State Electronics, Department of Systems Innovation, Graduate School of Engineering Science, Osaka University, Japan

Yukiko Takahashi, Magnetic Materials Group, Magnetic Materials Center, National Institute for Materials Science, Japan

Koki Takanashi, Magnetic Materials Lab, Institute for Materials Research, Tohoku University, Japan

Tomoyasu Taniyama, Materials and Structures Laboratory, Division of Novel Functional Ceramics, Tokyo Institute of Technology, Nagatsuta, Japan

Arata Tsukamoto, Department of Electronic Engineering, College of Science and Technology, Nihon University, Japan

Ken-ichi Uchida, Institute for Materials Research, Tohoku University, Japan

Mathias Weiler, Walther-Meißner-Institut für Tieftemperaturforschung, Bayerische Akademie der Wissenschaften, Germany; and National Institute of Standards and Technology, CO, USA

Jiang Xiao, Department of Physics, Fudan University, China

Akinobu Yamaguchi, Laboratory of Advanced Science and Technology for Industry, University of Hyogo, Japan

Series Preface

Wiley Series in Materials for Electronic and Optoelectronic Applications

This book series is devoted to the rapidly developing class of materials used for electronic and optoelectronic applications. It is designed to provide much-needed information on the fundamental scientific principles of these materials, together with how these are employed in technological applications. The books are aimed at (postgraduate) students, researchers and technologists, engaged in research, development, and the study of materials in electronics and photonics, and industrial scientists developing new materials, devices and circuits for the electronic, optoelectronic, and communications industries.

The development of new electronic and optoelectronic materials depends not only on materials engineering at a practical level, but also on a clear understanding of the properties of materials, and the fundamental science behind these properties. It is the properties of a material that eventually determine its usefulness in an application. The series therefore also includes such titles as electrical conduction in solids, optical properties, thermal properties, and so on, all with applications and examples of materials in electronics and optoelectronics. The characterization of materials is also covered within the series in as much as it is impossible to develop new materials without the proper characterization of their structure and properties. Structure–property relationships have always been fundamentally and intrinsically important to materials science and engineering.

Materials science is well known for being one of the most interdisciplinary sciences. It is the interdisciplinary aspect of materials science that has led to many exciting discoveries, new materials, and new applications. It is not unusual to find scientists with a chemical engineering background working on materials projects with applications in electronics. In selecting titles for the series, we have tried to maintain the interdisciplinary aspect of the field, and hence its excitement to researchers in this field.

Arthur Willoughby
Peter Capper
Safa Kasap

Preface

Semiconductor manufacturing technologies are indivisibly related to nanotechnology, since they become more and more sophisticated as exemplified by the fact that the manufacturing accuracy of the CMOS microprocessing plunges into the 10 nm range. Consequently the technology is opposing a limitation due to (1) consumption of energy and (2) transmission delay in wiring. International Technology Roadmap for Semiconductors (ITRS) published a roadmap to overcome the limit, which suggested three directions of research and development to overcome the limit: (1) More Moore: extension of the limit by invention of novel technologies, (2) More than Moore: addition of higher functionalities by integration of different technologies, and (3) Beyond CMOS: development of devices based on new concept. Among these three directions, the "beyond CMOS" direction should be most eagerly pursued because of its innovative nature.

Diverse materials and processes have been proposed for the beyond CMOS technology, such as low-dimensional materials (nano-mechanical memory, nanotube, nanowire, grapheme, etc.), macromolecules (molecular memory, molecular devices, resists, imprint polymers, etc.), self-assembled materials (sublithographic patterns, selective etch, etc.), spintronics materials (MRAM by spin injection, semiconductor spin transport, ferro-magnetic semiconductors, etc.), complex metal oxides (multiferroics), and interfaces and hetero-interfaces (electrical and spin contacts).

Majority of the contents of this book are based on the achievements of researchers and advisors affiliated with the PRESTO (Precursory Research for Embryonic Science and Technology) project titled "Materials and Processes for Next Generation Innovative Devices," which was promoted by Japan Science and Technology Agency (JST) between FY 2007 and 2013. The PRESTO project for which Prof. Katsuaki Sato, one of co-editors of this book, dedicated himself as a Research Supervisor covered wide area of materials including spintronics materials, materials of strongly correlated system such as high-temperature superconductors, high-mobility wide-gap semiconductors quantum dots, nano-carbons, and organics. All these fields have provided successful achievements; some of which are highly evaluated and already in the second stage of R&D involving industries. Among them, the most prominent innovative achievements were obtained in spintronics, since it has opened up the most advanced areas of physics, leading to completely new concepts for electronic devices. For this field Prof. Eiji Satoh, another co-editor, dedicated himself as a strong promoter of this field.

This book is planned to give comprehensive insight into spintronics with a particular reference to materials and processes for the next-generation innovative devices. We hope this book is helpful to not only scientists who started researches in spintronics, but also engineers who are pioneering the next-generation devices beyond CMOS.

Sincere gratitude is due to Prof. Koki Takanashi, an advisor of the PRESTO project, and all the PRESTO members who joined the hot discussions on all the topics of the project, as well as other authors who cooperated to publish this book. Also, thanks to the editorial office of Wiley Publishing Company.

Katsuaki Sato and Eiji Saitoh
May 2015
Tokyo

Introduction

Spintronics or electronics using spin-related phenomena has been attracting attention because of its potential applicability to new functional devices combining the charge transport and the spin properties. Magnetic semiconductors and ferromagnetic/nonmagnetic hybrid structures are now the most important topics of investigation in the field of new functional semiconductor devices. There is a long history of research on this category of materials. The first-generation materials were europium chalcogenides and chalcogenides of chromium with spinel-type crystal structures, which were studied intensively in the late 1960s and early 1970s. Interesting physical properties of magnetic semiconductors, such as magnetic red shift of the absorption edge and huge negative magnetoresistance (MR) around the Curie temperature, were discovered at that time. However, researchers lost interest in these materials because their Curie temperatures were far below the room temperature and because growth of good-quality single crystals was very difficult. The second-generation materials are II–VI-based diluted magnetic semiconductors (DMSs), among which $Cd_{1-x}Mn_xTe$ was the focus of attention due to its capability to accommodate a high percentage of Mn atoms (as high as 77%) and its appropriate energy gap for optical application. The magnetic properties of most of these materials are either paramagnetic or spinglass. Although the controllability of transport properties is relatively poor, the material shows a good optical property that led to its application to optical isolators.

The third-generation materials are III–V-based diluted semiconductors, in which magnetic properties have been found to be strongly dependent on the carrier concentration in the material. This series of materials can only be produced by using an MBE technique with very low substrate temperatures. Since III–V compound semiconductors are widely used in electronic devices, the III–V-based DMSs are inherently capable of device integration. Despite these efforts, magnetic semiconductors have not been put into practical use due to their low Curie temperature. Recently, a room-temperature ferromagnetic semiconductor FET has been realized by Fukumura using a transition metal-doped oxide semiconductor, which enabled the control of magnetic properties by the gate voltage.

Another direction of spintronics research is a spin FET proposed by Datta and Das, in which the spin polarization of an electron injected from a ferromagnetic source electrode into a semiconductor is modulated by application of the gate voltage using the Rashba effect, and analyzed by a ferromagnetic drain electrode. However, the Datta–Das device has not been realized yet in the original setups.

The innovative development of spintronics had its gates opened in 1988, with the giant magnetoresistance (GMR) effect of the magnetic metal/nonmagnetic metal hybrid structure by Fert and Grünberg, *et al.* Over a few years, the GMR contributed to the increase in the density of the magnetic storages as a spin valve, and humanity obtained a means to efficiently convert magnetic information into electric signals without using a coil.

After that, Miyazaki *et al.* and Moodera *et al.* discovered the tunnel magnetoresistance (TMR) effect at room temperature, providing an opportunity for the emergence of non-volatile solid-state memory devices, magnetoresistive random-access memory (MRAM). Furthermore, the TMR was significantly improved and it greatly advanced due to the study by Yuasa *et al.* and Parkin *et al.*, in which MgO was employed as the tunnel barrier.

The next innovative development was brought on by a theoretical prediction and an empirical validation of the magnetization reversal phenomena that uses spin-transfer torque (STT). The STT-MRAM that utilizes this phenomenon does not require any electric current wires for generation of a magnetic field, allowing for a high-density integration that is more than DRAM, and finally, a sample was able to be released to the market. At last, humanity had obtained a way to convert electric signals into magnetic information without a coil.

Up to the point, the spin current had been something that accompanied the electrical charge current, but then a theoretical prediction of the existence of a pure spin current that was not accompanied by the electric charge current was made, and it has been empirically validated in the last 10 years. Using the pure flow of spin allows information transmission without having Joule heat, so the expectation is that this can solve the issues of energy dissipation from the metal wire of high-density minute integrated circuits.

Theory and its empirical validation of the spin Hall effect (SHE) greatly contributed to the generation and detection of the spin current. These effects are based on the concept of Berry's phase and it is assumed that a cosmology can be established in solid matter. Furthermore, theories that are the basis of spintronics, such as the discovery of the spin Seebeck effect with the thermal spin current, the spin current carried by a spin valve, the interaction between the spin current and phonons, and topological insulators, are experiencing a great leap forward. In addition, recently, a path to control the magnetic properties by using an electric voltage instead of an electric current using a principle that is different from the STT, is about to open up.

In the following, topics included in this book are briefly reviewed as a guide to those readers who are not familiar with the field of spintronics. First part is devoted to materials for spintronics and the rest to various functions associated with spintronics.

MATERIALS FOR SPINTRONICS

For practical applications of spintronics, research and development of materials for both spin injection and spin guideline are important.

Spin Injector Materials with High Polarization

As spin-injector electrodes, full-Heusler compounds have been intensively studied. The compounds are predicted by electronic structure calculations as half-metals, in which at Fermi level, a half of the spin-polarized band has a finite value of density of state while the other half has zero density of state, leading to a 100% degree of spin polarization in ideal cases. Dr. Yukiko Takahashi gives a comprehensive review of the Heusler compounds and describes methods how to measure the degree of spin polarization, as well as how to increase the degree of spin polarization by appropriate combination of elements.

Carbon Spintronics

As a guideline for spin current, the spin diffusion lengths are rather short in metals such as Cu and Al which have predominantly been used. On the other hand, longer spin diffusion length is expected in lighter atoms which are subjected to a reduced spin–orbit coupling due to smaller relativistic effect. In this context, nano-carbon materials such as carbon nanotubes (CNT), graphene sheets, and organic compounds are promising, for which Prof. Seiji Shiraishi provides comprehensive descriptions.

Silicon Spintronics

Silicon is also a very important candidate for a spin current guideline, since silicon is a dominantly used material in the semiconductor device technology so that spintronics can easily be combined to conventional electronics circuits. Until recently, a tunnel contact with heavily doped "degenerate" silicon materials has been used for spin-injection experiments. However, for the purpose of electrical control of spin currents by application of the gate-voltage (spin-FET operation), nondegenerate materials should be pursued. For this purpose, improvement of spin injection efficiency is necessary. Prof. Kohei Hamaya describes how he succeeded in realization of a spin-MOSFET using high-quality Schottky tunnel contact.

SPINTRONICS FUNCTIONS

Spin-Dependent Transport

The first epoch-making issue in the field of spintronics function was introduced by the Nobel-Prize winners Albert Fert and Peter Grünberg, who independently observed prominent spin-dependent transport phenomena in superstructures consisting of magnetic and nonmagnetic layers, which was named "giant magnetoresistance (GMR)." Next important issue was an observation of spin-dependent tunneling conductance, "tunnel magnetoresistance (TMR)," at room temperature by Terunobu Miyazaki. Astonishingly it took no longer than 5 years before these discoveries are applied to practical devices in both GMR and TMR; the former being used as magnetic sensors for hard disk drives, while the latter being used as MRAM. Anomalous magnetoresistance (AMR) effects, even though small, have been investigated for long time, which have been providing an important background for present-day spintronics. Physical and technical aspects for magnetoresistance phenomena including AMR, GMR, and TMR are described in detail by Prof. Koki Takanashi.

Spin Current, Spin Hall Effect, and Spin Pumping

Dramatic reduction of energy consumption in the LSI wiring is expected by using the spin current instead of the charge current. The spin current is defined as a difference between a flow of up-spin electrons and that of down-spin electrons. If the charge current direction of up-spins is opposite to that of down-spins, net charge current vanishes and only the spin current exists; that is a pure spin current. Not only the conventional spin current, but also the spin-wave spin current and the topological spin current are recently proposed and experimentally confirmed. In addition, new paradigm of spin-related physics has been opened up such as the concept of spin Hall effect (SHE) and inverse spin Hall effect (ISHE), which can be used for mutual conversion between the spin current and the charge current. The latter is used to detect the spin injection, and other spin current-related phenomena. Detailed description of diverse types of spin current is provided by Prof. Eiji Saitoh, one of the editors of this book. Theoretical backgrounds for concept of the SHE and ISHE and topological insulators are given by Prof. Shuichi Murakami. Concept of spin pumping is also a significant issue for injection of spin-wave spin current to insulators and semiconductors, for which Prof. Sevastian Goennenwein gives physical and technical backgrounds.

Spin Torque

Since the spin is a quantum of the angular momentum, a spin-polarized electron transfers a torque when injected into magnetized materials as a current flow. The torque is called "spin-transfer torque (STT)." The torque can be used to invert the direction of magnetization in MRAMs instead of the current-induced magnetic field used in conventional MRAMs, which has been an obstacle to miniaturization of the device due to increase in the current density. MRAM making use of spin-transfer torque, STT-MRAM, is considered to be a promising candidate for a nonvolatile memory component, required in realization of "normally off" computing technology. Not only the STT-MRAM, but also the spin torque technology is expected to be applicable to domain-wall drive memories, spin-torque oscillators, etc. Theoretical and experimental descriptions of spin torque phenomena are given by Prof. Akinobu Yamaguchi.

Spin Seebeck Effect

It is recently observed that temperature gradient in a slab of magnetic materials induces a pure spin current without a flow of the charge current. By the use of ISHE, the spin current can be converted to the electric current, which in turn can be converted to voltage. The effect is named "spin Seebeck effect" and is promising for energy harvesting, which converts an environmental heat to an electric energy. Dr. J. Xiao gives theoretical and experimental details for the spin Seebeck effect.

Electric Control of Spin Phenomena

Tunable Spin Sources

In conventional spintronics devices, spin polarization of an injector electrode is solely dependent on magnetization of the material and cannot be controlled externally. A few ideas have been proposed to obtain tunable spin sources using some sort of magnetic phase change. One idea is to use an antiferromagnetic (AF)/ferromagnetic (FM) phase transition induced by spin injection and another is to use temperature-induced phase transition in magnetite electrode for LED. Prof. Tomoyasu Taniyama describes experimental observation of AF/FM transition in FeRh alloy by spin injection from Co electrodes and of spin injection to GaAs/GaAlAs quantum well from the Fe_3O_4 electrode which undergoes the famous Verway transition around 120 K.

Enhancement of Rashba Effect

In spin-transistors proposed by Datta and Das, the spin polarization direction is modulated by using the Rashba effect, in which a spin–orbit coupling is controlled by an application of the gate voltage. Since the effect is so small that considerably long channel length is necessary to invert electron spins flowing between source and drain electrodes. Therefore, the Datta–Das device requires a long spin diffusion length, which is not the case in conventional semiconductors. To overcome the problem, Dr. Hiroyuki Nakamura proposed a method to enhance the spin–orbit interaction by means of dielectric materials. He succeeded to realize an enhancement of spin–orbit interaction by using the third-order Rashba effect in $SrTiO_3$ dielectric.

Ferromagnetic Semiconductor

Electrical control of magnetism has been a long-lasting wish of researchers from the age of the first-generation magnetic semiconductors. Although dramatic voltage-controlled change in the magnetization curves was reported in InMnAs semiconductors, the operation temperature was as low as order of 10 K due to low ferromagnetic transition temperature. Quite recently, the electric field-induced ferromagnetism was observed in ferromagnetic oxide semiconductors at room temperature. Prof. Tomoteru Fukumura describes how he succeeded in control of magnetism by an electric field in TiO_2:Co room temperature ferromagnetic semiconductor with an FET structure. He also refers to some of recent developments in voltage-controlled spin states in metals and other materials.

Spin Photonics

Spin Quantum Devices using Diamond

Room-temperature operation of quantum information processing has been a dream of researchers of this field. The diamond NV^- center consisting of a carbon vacancy (V) and a nitrogen atom (N) substituting a carbon site has been attracting attention as a promising quantum information device operating at room temperature, since the center forms a "deep" center due to wide-gap nature of the material. By using optically controlled magnetic resonance technique, the coupling of an electron spin and a nuclear spin becomes available, which in turn enables optical initialization of the quantum system. Professor Norikazu Mizuochi describes how the NV^- centers can be coupled to enable 5 qubit operations, as well as how he managed to realize a current-operated single photon source by using a specially designed LED composed of a diamond p–i–n junction.

Ultrafast Light-Induced Spin Reversal

Magnetic devices are facing a limitation of data-transfer rate determined by the magnetization-reversal time associated with the ferromagnetic resonance, which is of the order of GHz. Recently, optically induced ultrafast spin-reversal phenomenon of subpicosecond range has been observed in a few materials. The most prominent achievement was the realization of the ultrafast light-induced magnetization reversal in amorphous rare earth-transition metal alloy films around the angular moment compensation temperature. Prof. Arata Tsukamoto describes the details of the phenomenon and gives a physical background of dynamical behavior of spins in nonequilibrium conditions.

Finally, we sincerely hope that this book may provide a perspective guide to spintronics for young researchers who are pioneering the science and technology for next-generation innovative devices.

1

Fundamentals of Magnetoresistance Effects

Koki Takanashi
Institute for Materials Research, Tohoku University, Aoba-ku, Sendai, Japan

1.1 Giant Magnetoresistance (GMR) Effect

Giant magnetoresistance (GMR) effect is the most fundamental phenomenon in the field of spintronics. In general, the electric resistance of a material is changed when a magnetic field is applied. This phenomenon is known as a magnetoresistance (MR) effect. Although there are many types of MR effects, the MR effect showing a particularly large resistance change among them is called the GMR effect. However, GMR is different from the conventional MR effects that had been known before, not only quantitatively, but also qualitatively. In 1988, GMR was first reported in the experiment of Fe/Cr superlattices by Fert and his collaborators [1]. That was followed by an enormous amount of studies concerning GMR in nanometer-scaled layered structures containing various kinds of ferromagnetic metals, and GMR was practically used for a read head of a hard disk drive (HDD) in only 10 years after the discovery of GMR. The GMR-based read head had dramatically improved the storage density of HDD until GMR was replaced by tunnel magnetoresistance (TMR) later.

Before the discovery of GMR by Fert *et al.*, the antiferromagnetic coupling of Fe layers through a Cr interlayer had been reported in a trilayer structure consisting of Fe/Cr/Fe by Grünberg and coworkers in 1986 [2]. The antiferromagnetic interlayer exchange coupling is closely related to GMR as more details are shown later. In fact, the MR effect equivalent

Spintronics for Next Generation Innovative Devices, First Edition. Edited by Katsuaki Sato and Eiji Saitoh.
© 2015 John Wiley & Sons, Ltd. Published 2015 by John Wiley & Sons, Ltd.

Picture 1.1 *Dr. A. Fert (left) and Dr. P. Günberg shared their joy at the award lecture of 2007 Nobel Prize in Physics*

to GMR, which was smaller than GMR reported by Fert *et al.*, was also observed for the Fe/Cr/Fe trilayer prepared by Grünberg *et al.* [3]. Therefore, both Fert and Grünberg were regarded as the discoverers of GMR, and jointly awarded the Nobel prize in physics in 2007 (Picture 1.1). According to the words by the Nobel Foundation, GMR was considered "the first major application of nanotechnology." From not only the technological but also physical point of view, as mentioned above, GMR is completely different from the MR known before, and it has brought out a new physical concept, i.e., spin-dependent transport. GMR is the origin for spintronics that is a very popular field now, and without the discovery of GMR no development of spintronics afterward would have appeared.

Section 1.1 starts with a brief explanation of the conventional MR in ferromagnetic materials that had been known before the discovery of GMR. Then the phenomenon, the mechanism and the application of GMR are given. Here, it is noted that the word of GMR means a giant MR effect, and also contains a giant TMR effect[1] (once called the tunnel-type GMR), a CMR (colossal magnetoresistance) effect observed in magnetic oxides, and further different types of giant MR effects in a broad sense. However, GMR normally concerns a giant MR effect observed in "metallic systems" composed of ferromagnetic and nonferromagnetic metals in a nanometer scale. Section 1.1 focuses on the "metallic systems."

1.1.1 Magnetoresistance Effects in Ferromagnetic Materials

All conductive materials, whichever are ferromagnetic or not, exhibit a MR effect. This MR effect is called an ordinary magnetoresistance effect. The electric resistance of a material increases as the applied magnetic field is increased, i.e., the sign of MR is positive. The resistance increase is caused by the Lorentz force that affects the motion of conduction electrons. On the other hand, ferromagnetic materials (including ferrimagnetic materials) with spontaneous magnetization show a characteristic MR effect, which depends on the spontaneous magnetization, and this MR effect is called an anomalous magnetoresistance effect. The anomalous magnetoresistance effect is classified into the following two kinds: an anisotropic magnetoresistance (AMR) effect and a forced effect. In a magnetic field region where magnetic moments are not saturated, the AMR effect appears depending on the relative orientation between magnetization and electric current. The forced effect appears under the application of high magnetic field after saturating the magnetization, and a slight increase in the spontaneous magnetization leads to the corresponding decrease in electric resistance.

The AMR effect, which is well known as a MR in ferromagnetic materials, is usually expressed as

$$\begin{aligned} \rho &= \rho_l \cos^2\theta + \rho_t \cos^2\theta \\ &= \rho_t + (\rho_l - \rho_t)\cos^2\theta, \end{aligned} \tag{1.1}$$

where ρ represents the electric resistivity and θ represents the relative angle between the magnetization vector and the electric current. ρ_l and ρ_t are the resistances when the magnetization is parallel ($\theta = 0$) and perpendicular ($\theta = \pi/2$) to the electric current, respectively. Generally ρ_l is not equal to ρ_t ($\rho_l \neq \rho_t$), leading to the appearance of AMR. The origin for AMR is known to be caused by spin–orbit interaction [4] although the details are not shown here.

The magnitude of AMR is defined as the ratio of resistance change (MR ratio: $\Delta\rho/\rho$):

$$\Delta\rho/\rho = (\rho_l - \rho_t)/\rho_{av}, \tag{1.2}$$

where ρ_{av} is the average of the resistance, and is described as

$$\rho_{av} = \rho_l/3 + 2\rho_t/3 \tag{1.3}$$

by using $\langle \cos^2\theta\rangle_{av} = 1/3$ and $\langle \sin^2\theta\rangle_{av} = 2/3$.

[1] Strictly speaking, TMR had been discovered before GMR. However, the effect had been very small at room temperature, and had attracted not much attention from the practical point of view. The discovery of GMR activated the TMR study; leading to the observation of large TMR at room temperature. Based on this historical background, in Section 1.1, GMR is considered to be the origin of spintronics. More details on TMR will be given in Section 1.2.

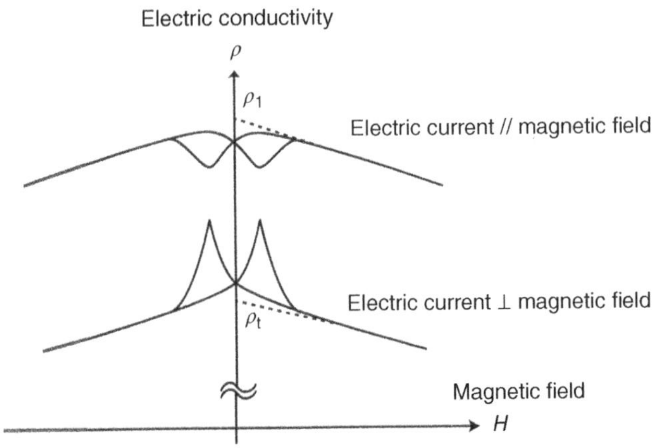

Figure 1.1 *Typical MR curves in a ferromagnetic material. The magnetic field is applied parallel or perpendicular to the direction of electric current*

Figure 1.1 shows a typical MR curve for a ferromagnetic material, where the vertical and horizontal axes represent the electric resistance and the applied magnetic field, respectively. With sweeping the magnetic field from positive to negative and then back to positive again, a hysteresis curve of the resistance shows a maximum or minimum value around the coercivity. The MR curve shown in Figure 1.1 is for the case with $\rho_l > \rho_t$. However, materials with $\rho_l < \rho_t$ also exist. In any case, $\Delta\rho/\rho$ for AMR is usually a small value from 0.1 to a few percent at room temperature [4].

1.1.2 Phenomenon of GMR Effect

The phenomenon of the GMR effect differs from AMR in ferromagnetic materials in terms of both quantitative and qualitative characteristics. First, GMR is isotropic irrespective of the angle between magnetic field and electric current. Second, the resistance decreases remarkably when the magnetization is being saturated, so the sign of MR is negative. A typical GMR curve and the corresponding magnetization curve are shown in Figure 1.2(a) and (b), respectively. The magnitude of GMR is usually defined as the MR ratio ($\Delta\rho/\rho$) and is given by

$$\Delta\rho/\rho = (\rho - \rho_s)/\rho_s, \tag{1.4}$$

where ρ is the maximum resistance around zero magnetic field and ρ_s is the resistance when the magnetization is saturated. Note that in some cases, the change in the resistance ($\Delta\rho = \rho - \rho_s$) is divided by ρ, which leads to a large difference in the MR ratio. In addition, an MR curve showing GMR sometimes contains the contribution of AMR and the difference between ρ_l and ρ_t is observed. However, compared to the total change in the resistance, the AMR effect is rather small and can often be ignored. Since GMR was discovered in Fe/Cr superlattices in 1988 [1], GMR has been reported in various superlattices consisting of ferromagnetic metals (FM) and nonmagnetic metals (NM), such as Co/Cu [5],

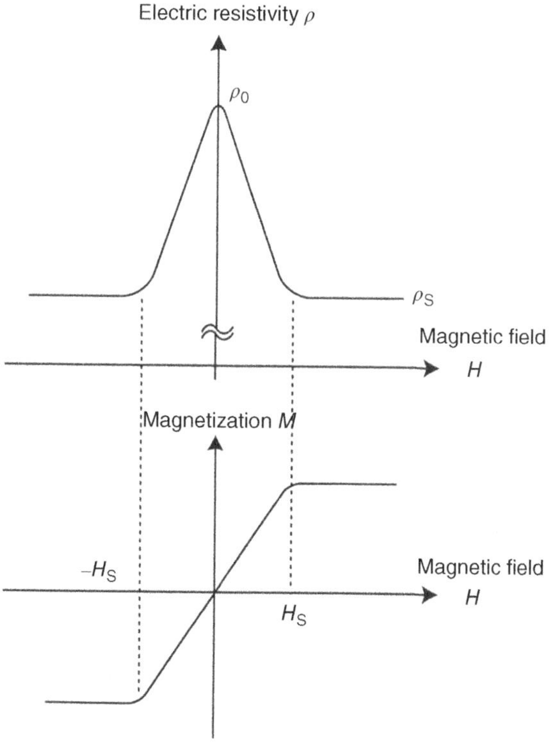

Figure 1.2 *(a) Typical MR curve and (b) magnetization curve of a superlattice showing GMR*

Co/Ag [6], Ni/Ag [7], etc. The superlattice is a thin film composed of two or more different materials, which are alternately deposited in a nanometer scale by thin film preparation methods such as molecular beam epitaxy (MBE) and sputtering.

All the superlattices with any combination of FM and NM do not show GMR. Crucial requirements should be satisfied to show GMR. One requirement is antiferromagnetic interaction between FM layers through a NM layer leading to the antiparallel alignment of magnetization vectors in adjacent FM layers at zero magnetic field. When the magnetic field is applied and increased, the alignment of magnetization turns to parallel which gives rise to the resistance decrease, i.e., the appearance of GMR. Table 1.1 summarizes the reported data in some representative superlattices showing GMR [1, 5–12]. The MR ratios are larger than 10% at room temperature, and reach a few tens to more than 100% at low temperature. GMR in metallic superlattices is caused by spin-dependent scattering of conduction electrons at the interface as mentioned in Section 1.1.3. This effect strongly depends on the condition of the interface, and the GMR is highly sensitive to the structure. Therefore, GMR is not solely determined by the combination of metal elements. Generally, however, we may say that Fe/Cr in Fe-based superlattices and Co/Cu in Co-based superlattices show the largest GMR, which is qualitatively consistent with theoretical calculations [13, 14]. In addition, GMR usually decreases with temperature.

Table 1.1 GMR effect in various kinds of superlattices

Superlattices $[A\,(x\,(\text{Å}))/B\,(y\,(\text{Å}))_{xN}$	$\Delta\rho/\rho(\%) = (\rho_0 - \rho_s)/\rho_s$	Temperature (K)	Fabrication method	Reference
$[Fe\,(30)/Cr\,(9)]_{\times 30}$	85	4.2	MBE	[1]
$[Fe\,(4.5)/Cr\,(12)]_{\times 50}$	220 42	1.5 300	MBE	[8]
$[Fe\,(20)/Cr\,(12)]_{\times 20}$	33	4.5	Sputtering	[5]
$[Co\,(15)/Cu\,(9)]_{\times 30}$	78 48	4.2 300	Sputtering	[5]
$(Co\,(8)/Cu\,(8.3)]_{\times 60}$	115 65	4.2 295	Sputtering	[10]
$[Fe\,(10.7)/Cu\,(13.7)]_{\times 15}$	26 13	4.2 R.T.	Sputtering	[11]
$[Co\,(6)/Ag\,(25)]_{\times 70}$	38 16	77 R.T.	MBE	[6]
$[Ni\,(8)/Ag\,(11)]_{\times ?}$	26	4.2	Sputtering	[7]
$[Ni_{81}Fe_{19}\,(15)/Cu\,(8)]_{\times 14}$	25 16	4.2 300	Sputtering	[12]

1.1.3 Mechanism of GMR Effect

The mechanism of GMR originates from the spin-dependent scattering of conduction electrons. The spin-dependent scattering occurs in FM layers or at FM/NM interfaces. Here we consider a two-current model where up-spin (↑) and down-spin (↓) electrons pass through two independent channels without taking into account the spin-flip scattering that causes the change of spin orientation. The scattering probability of conduction electrons depends on whether the spin orientation is parallel or antiparallel to the magnetization which is called spin-dependent scattering. The electric resistivity for spins parallel (antiparallel) to the magnetization is written as ρ_+ (ρ_-). Generally $\rho_+ \neq \rho_-$, and here we assume $\rho_+ < \rho_-$. In other words, the electrons with spins parallel to the magnetization have a low scattering possibility while the electrons with spins antiparallel to the magnetization have a high scattering possibility. (Even if we assume the opposite relationship, i.e., $\rho_+ > \rho_-$, there is no influence on the following discussion, keeping the generality.) Figure 1.3 depicts a schematic illustration of the conduction of electrons. In the case of superlattices, the thickness of each layer is of the order of nanometer. Thus the electrons do not stay in one layer during conduction, and flow across the layers even when the electric current is applied in-plane. If the magnetization vectors of adjacent FM layers are parallel, the electrons with ↑ spins parallel to the magnetizations are hardly scattered which contributes dominantly to the conduction, leading to a low electric resistance. On the other hand, if the magnetization vectors are antiparallel, both ↑ spins and ↓ spins should pass through the layers with the opposite magnetization vectors and are significantly scattered, leading to a high electric resistance. In order to discuss the above situation more quantitatively, we consider a

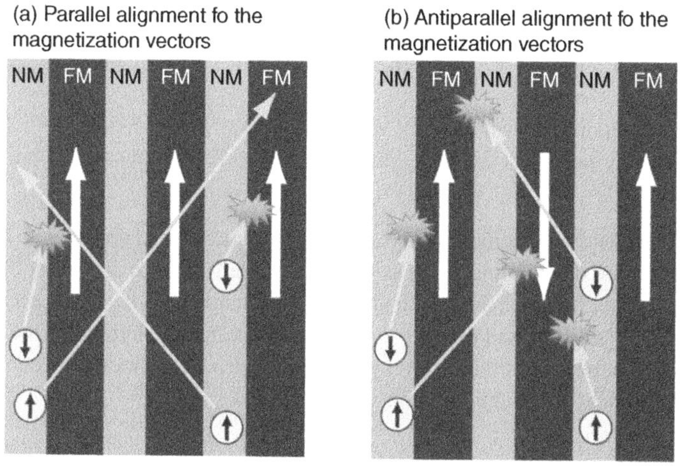

Figure 1.3 *Schematic illustration of the conduction of electrons in a superlattice with alternatively stacked a ferromagnetic metal (FM) and a nonmagnetic metal (NM). The cases for (a) parallel and (b) antiparallel alignments of the magnetization vectors are shown*

parallel circuit of the ↑ spin channel and the ↓ spin channel. Then, the electric resistivity ρ is expressed as

$$1/\rho = 1/\rho_\uparrow + 1/\rho_\downarrow, \tag{1.5}$$

where $\rho_\uparrow$ ($\rho_\downarrow$) is the resistivity for the electrons with ↑ (↓) spin. Namely, ρ is rewritten as

$$\rho = \rho_\uparrow\rho_\downarrow/(\rho_\uparrow + \rho_\downarrow). \tag{1.6}$$

When all the magnetization vectors are parallel, we consider $\rho_\uparrow \sim \rho_+$ and $\rho_\downarrow \sim \rho_-$ and the total resistance ρ_P is expressed as

$$\begin{aligned}\rho_P &= \rho_+\rho_-/(\rho_+ + \rho_-)\\ &\sim \rho_+ \quad (\text{for } \rho_+ << \rho_-).\end{aligned} \tag{1.7}$$

When the adjacent magnetization vectors are antiparallel, we consider $\rho_\uparrow \sim \rho_\downarrow \sim (\rho_+ + \rho_-)/2$, the total resistance ρ_{AP} is expressed as

$$\rho_{AP} = (\rho_+ + \rho_-)/4. \tag{1.8}$$

With Equations (1.7) and (1.8), the magnitude of GMR is given by

$$\begin{aligned}(\rho_{AP} - \rho_P)/\rho_P &= (\rho_+ - \rho_-)^2/4\rho_+\rho_-\\ &= (1-\alpha)^2/4\alpha,\end{aligned} \tag{1.9}$$

where α is a parameter representing the spin dependence of electron scattering and is defined as

$$\alpha = \rho_-/\rho_+. \tag{1.10}$$

One can see that GMR appears when $\alpha \neq 1$ and GMR is large as $\alpha >> 1$ or $\alpha << 1$.

The requirements for the appearance of GMR are summarized as follows:

(i) Antiparallel alignment of the magnetization vectors in adjacent FM layers through a NM layer.
(ii) Large spin-dependent scattering of electrons ($\alpha >> 1$ or $\alpha << 1$).
(iii) The multilayer period of the superlattice is shorter than the mean free path of conduction electrons.

The first and second requirements have already been mentioned above. The third requirement is an additional one but important as well. If the multilayer period of the superlattice is longer than the mean free path, the electron scattering inside each layer is increased for both ↑ spin and ↓ spin electrons, and the spin-*independent* scattering dominates the total electric resistance, which smears out the difference in the electric resistance between parallel and antiparallel alignments of magnetization vectors.

1.1.4 Oscillatory Behavior of Interlayer Exchange Coupling and GMR

When the antiferromagnetic interlayer exchange coupling acts on FM layers through a NM layer, the saturation field H_s in the MR curve or the magnetization curve where the electric resistance or the magnetization is saturated, is proportional to the exchange coupling energy per unit area J. The antiparallel alignment of magnetization vectors for FM layers requires $J < 0$ (antiferromagnetic coupling), and H_s corresponds to the magnetic field applied to align the magnetization vectors in parallel. The relationship between J and H_s is written as[2]

$$H_s = -4J/M_s \cdot t_{FM} \tag{1.11}$$

where M_s and t_{FM} are the saturation magnetization and the thickness of FM layer, respectively. In the case of $J > 0$ (ferromagnetic coupling), the magnetization vectors of FM layers are aligned in parallel without applied magnetic field, and the magnetization curve shows a simple ferromagnetic behavior ($H_s \sim 0$) and GMR does not appear.

The sign and the magnitude of J depend on the thickness of NM layer, t_{NM}, and have oscillatory behavior as shown in Figure 1.4. Since GMR only appears when $J < 0$, GMR also shows oscillatory behavior. The magnitude of J decreases with t_{NM}, and in general J becomes negligibly small when t_{NM} is larger than several nanometers, leading to the disappearance of GMR. The period of the oscillation in J depends on the material, and is usually a couple of nanometers [9].

As mentioned above, although the interlayer exchange coupling is closely related to GMR, it is important to notice that GMR is in nature a completely separate physical phenomenon. An important point for the appearance of GMR is that the magnetization alignment can be changed from parallel to antiparallel or from antiparallel to parallel by applying the external magnetic field, and there is no direct correlation between GMR and J. In other words, irrespective of the sign and the magnitude of J, if the magnetization alignment is controllable by a certain method, GMR may be observed which is shown in Section 1.1.5. The mechanism of the oscillation in J against t_{NM} was first discussed on the

[2] Technically, Equation (1.11) is valid only for infinite alternation of FM and NM layers. In the case of the trilayer structure FM/NM/FM, the factor is 2 instead of 4.

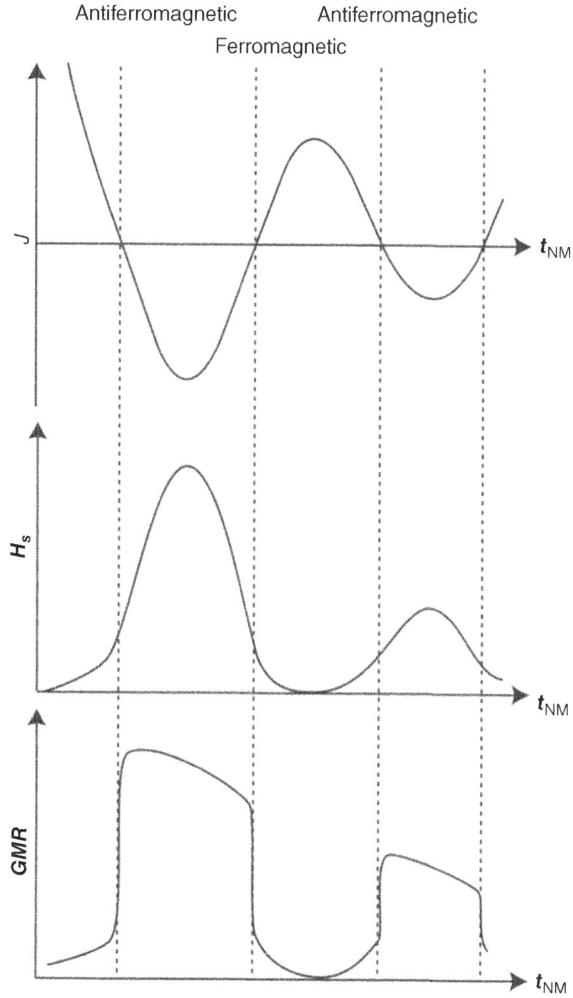

Figure 1.4 *NM layer thickness dependence of exchange coupling energy J of a FM layer through a NM layer, saturation field H_s, and GMR for a superlattice with a FM/NM structure*

basis of the RKKY interaction which successfully explained some part of the oscillatory behavior, but not completely. The phenomena that could not be interpreted only within the framework of RKKY interaction were observed, such as the oscillation in J against the thickness of FM layer t_{FM}. They are now understood by considering the quantum size effect, i.e., the formation of quantum wells caused by the multiple interferences of conduction electrons [15].

1.1.5 The Application of GMR and the Spin Valve

If the strong antiferromagnetic exchange coupling exists between adjacent FM layers, a large external magnetic field is required to align the magnetization vectors in parallel, and

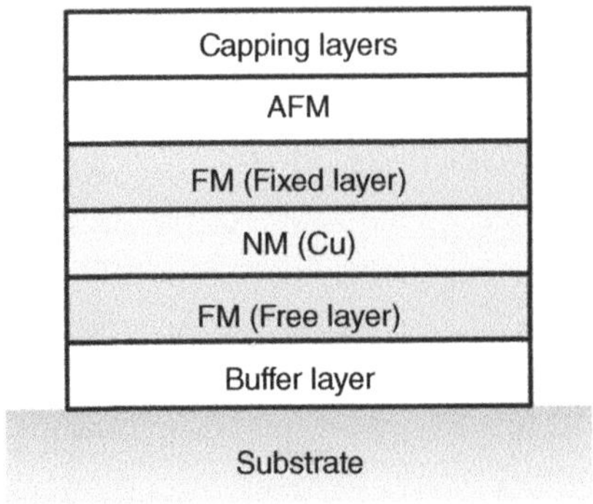

Figure 1.5 *Typical stacking structure of a spin valve*

also to obtain GMR. This was a major obstacle for practical applications of GMR. One possible way to overcome this obstacle is to eliminate the interlayer exchange coupling by increasing the thickness of NM layer and give a difference in coercivities between the adjacent FM layers to achieve the antiparallel alignment of magnetization vectors [16]. Another way is to make a spin-valve structure shown in Figure 1.5 [17]. For the spin-valve structure, one of the two FM layers (free layer) easily changes its magnetization direction with a low magnetic field whereas the other (fixed layer) has its magnetization pinned by the exchange magnetic anisotropy from the neighboring antiferromagnetic layer (AFM). Consequently, GMR appears even under the application of a low external magnetic field. In practical applications, soft magnetic materials like permalloy (Ni–Fe alloy) are usually used as FM layers, Cu is used as a NM layer, while Mn alloys like FeMn, IrMn, PtMn, or oxides like NiO and CoO are used as an AFM layer. The spin-valve structure was used for a read head of HDD in 1998, and played an important role for the remarkable enhancement of recording density as mentioned in the introductory part.

It is noted that the basic structure of a spin valve is the FM/NM/FM trilayer. The magnitude of GMR for superlattices or the MR ratio is known to depend on the number of alternated layers, which means that a structure having more layers shows a larger MR ratio. From the viewpoint of MR ratio, the trilayered structure is unfavorable. In other words, the spin valves realize the high magnetic-field sensitivity at the expense of the MR ratio. In order to improve this point, a five-layered structure of FM/NM/FM/NM/FM with AFM layers on both sides was designed which was called a dual spin valve. However, there is a limitation of the total thickness for the application to ultrahigh density recording. As a result, TMR with the same trilayered structure, showing a high MR ratio compared to GMR, has got to attract more attention, and eventually the TMR read head has replaced the GMR read head in HDD (see Section 1.2.).

1.1.6 CIP-GMR and CPP-GMR

The conventional MR measurement is carried out by measuring the change of the electric resistance under the electric current flow in the film plane. GMR depends on the magnetization configuration of the FM layers. If the electric current flows perpendicularly to the film plane and therefore all the electrons contribute to the conduction through each layer GMR is expected to become larger. The type of GMR when the electric current flows in the film plane is called current-in-plane GMR (CIP-GMR), and when the electric current flows perpendicularly to the film plane, GMR is called current-perpendicular-to-plane GMR (CPP-GMR). Because the thickness of a superlattice is at most several hundred nanometers and the resistance is small due to the metallic conductivity, the measurement of CPP-GMR is not easy. For the measurement of CPP-GMR, various techniques have been developed [18]. Nowadays superlattices are usually microfabricated into a pillar structure with a cross-sectional area of the dimension of micrometer or less, and electrodes are also attached to both the top and bottom ends of the pillar. It has experimentally been confirmed that the MR ratio of CPP-GMR is usually larger than that of CIP-GMR [19].

In the case of CIP-GMR, the mean free path is an important characteristic length as already mentioned in Section 1.1.3. In the case of CPP-GMR, on the other hand, the spin diffusion length (the distance over which the traveling electron spin keeps the initial orientation) is the characteristic length. CPP-GMR is theoretically analyzed by using the Valet–Fert model [20], in which the finite spin diffusion length is taken into account based on the two-current model. By analyzing the experimental data of the layer thickness dependence of CPP-GMR using the Valet–Fert model, the spin dependence in electron scattering is evaluated separately for the bulk and the interface [19].

CPP-GMR was studied from the fundamental point of view at the early stage. Recently, however, it has also attracted much attention in terms of practical applications, because the miniaturization of device elements leads to the increase in the device resistance. In the case of a TMR device for a read head of ultrahigh-density HDD, for example, the resistance of the device becomes so high that it will be difficult to carry out high-speed operation. On the other hand, the resistance of a CPP-GMR device is not so high thanks to all-metal structure. It is to say that the CPP-GMR device will be advantageous because of its low resistance. However, CPP-GMR devices based on the typical FM/NM/FM trilayered structure only showed a small MR ratio around 1% because of a large contribution of parasitic resistance. A variety of studies have been done to increase the MR ratio. From the viewpoint of materials development, half-metallic materials such as Heusler alloys have attracted much attention to improve the MR ratio by applying them to the FM layers, in which the conduction electrons are theoretically predicted to be 100% spin polarized. Co_2MnSi (CMS) is a representative Heusler alloy with high spin polarization, and the large GMR ratios of 36% and 67% at room temperature and 110 K, respectively, were obtained experimentally for the CMS/Ag/CMS stacks [21, 22]. Those experiments stimulated the research activity for CPP-GMR using the Heusler alloys, and recently larger GMR effects have been reported by using Co_2(Fe,Mn)Si [23, 24], and Co_2Fe(Ge,Ga) [25].

1.1.7 GMR in Granular Systems

GMR was observed not only in the layered structure of superlattice, but also in granular systems consisting of ferromagnetic nanoparticles embedded in a nonmagnetic metal matrix. The magnetization vectors of ferromagnetic particles are distributed in random directions under zero external magnetic field because of superparamagnetism caused by thermal fluctuation, or the difference in magnetic anisotropy of each particle. GMR of granular systems appears when the magnetization vectors of particles align to the same direction by applying an external magnetic field.

The granular systems are fabricated with the combination of metal elements with no solid solubility such as Cu–Co [26, 27], Ag–Co [28], Ag–Fe [29], making nonequilibrium solid solution by means of a vapor quenching method such as sputtering, and then giving heat treatment to promote the phase separation. Alternatively, granular systems can also be prepared by only sputtering at a proper substrate temperature. In addition, in the case of systems like Cu–Co which are soluble in the liquid phase, a bulk sample can be obtained by a liquid quenching method [30]. Compared to superlattices, granular systems are easily fabricated. However, low magnetic field sensitivity of MR is a major obstacle for practical applications

1.2 Tunnel Magnetoresistance (TMR) Effect

For a magnetic tunnel junction (MTJ), where two ferromagnetic electrodes are separated by an ultrathin insulating layer, its tunneling probability of electrons depends on the relative orientation of magnetization vectors of two ferromagnetic electrodes. In general, the tunneling probability is high when the magnetization vectors are aligned in parallel, while the tunneling probability is low when the magnetization vectors are aligned in antiparallel. In other words, the electric resistance of MTJ is low (high) for the parallel (antiparallel) alignment of magnetization vectors. This magnetoresistance effect is called a TMR effect.

The research for TMR has a long history. In 1975, Julliére *et al.*. reported a change in the electric resistance by 14% at 4.2 K in an MTJ consisting of Fe/Ge/Co [31]. After that, despite of pioneering work by Maekawa and Gäfvert [32] and Slonczewski [33], a room temperature TMR ratio larger than 10% had not been observed experimentally. In 1995, however, Miyazaki *et al.* [34] and Moodera *et al.* [35] independently reported TMR showing MR ratios larger than 10% at room temperature in a Fe/Al_2O_3/Fe MTJ and a FeCo/Al_2O_3/Co MTJ, respectively. After their discoveries, many experimental studies have been performed, and recently not only for basic scientific interest, but also for practical applications, such as read heads of HDD and magnetic random access memories (MRAMs). In Section 1.2, the principle and the recent progress of TMR studies are reviewed.

1.2.1 The Principle of TMR

Generally TMR arises from the spin polarization of conduction electrons at the Fermi surface for ferromagnetic electrodes. Figure 1.6 shows a schematic illustration for comprehensive explanation of the principle of TMR. With the assumption that the spin-flip scattering is negligible during electron tunneling, the electrons in an up-spin band move to the other up-spin band whereas the electrons in a down-spin band move to the other down-spin band. Therefore, when the numbers of electrons in majority and minority spin bands are different, the tunneling probability depends on the relative orientation of the magnetization vectors

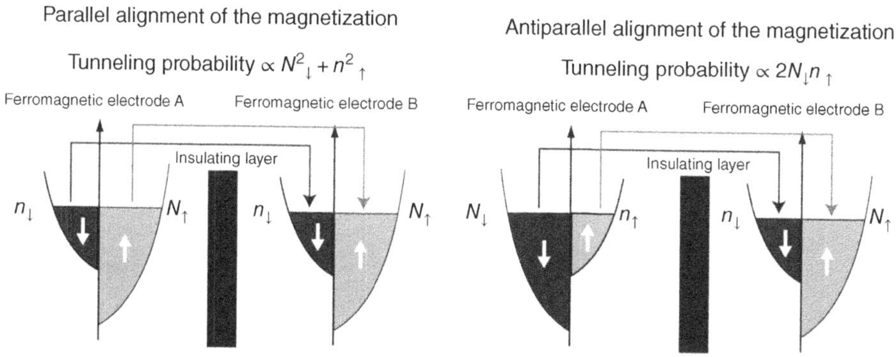

Figure 1.6 *Schematic illustration explaining the principle of TMR. $n_\downarrow$ ($N_\uparrow$) represents the density of state at the Fermi level for majority (minority) spins*

for two ferromagnetic electrodes. The relationship between the tunnel conductance G and the relative angle between the two magnetization vectors θ can be expressed as [33]

$$G = G_0(1 + P_1P_2 \cos\theta), \tag{1.12}$$

where G is the conductance at $\theta = \pi/2$. P_1 and P_2 are the spin polarization factors at the Fermi surfaces of two ferromagnetic electrodes, which are defined as

$$P_{1(2)} = (N_{1(2)} - n_{1(2)})/(N_{1(2)} + n_{1(2)}), \tag{1.13}$$

where $N_{1(2)}$ and $n_{1(2)}$ are the numbers of electrons at the Fermi level in majority and minority spin bands, respectively. Defining R_P and R_{AP} as the resistances in the cases that the two magnetization vectors are aligned in parallel ($\theta = 0$) and antiparallel ($\theta = \pi$), respectively, the magnitude of TMR can be described as the MR ratio, i.e., the ratio of the resistance difference between R_P and R_{AP} to R_P. Then, Equation (1.12) yields

$$(R_{AP} - R_p)/R_p = 2P_1P_2/(1 - P_1P_2), \tag{1.14}$$

where $0 \leq P_{1(2)} \leq 1$. Therefore, high-spin polarization factors of P_1 and P_2 give a large TMR ratio. If the two ferromagnetic electrodes are perfectly spin polarized ($P_1 = P_2 = 1$), R_{AP} reaches infinity in principle, and the infinite TMR ratio could be obtained.

1.2.2 TMR Effect in Transition Metals and Alloys with Al–O Tunnel Barrier

Many of MTJs studied previously consisted of ferromagnetic transition metals represented by Fe, Co, Ni, and their alloys, which were used as ferromagnetic electrodes, and an Al oxide that was used as an insulating material for the tunneling barrier. Most of the MTJs were prepared using conventional film preparation methods such as sputtering and vacuum evaporation. The films were patterned into a micrometer- or nanometer-sized junction employing a lithography technique, as shown in Figure 1.7. In general, an Al oxide layer for the tunneling barrier was amorphous with nonstoichiometric composition. The Al oxide

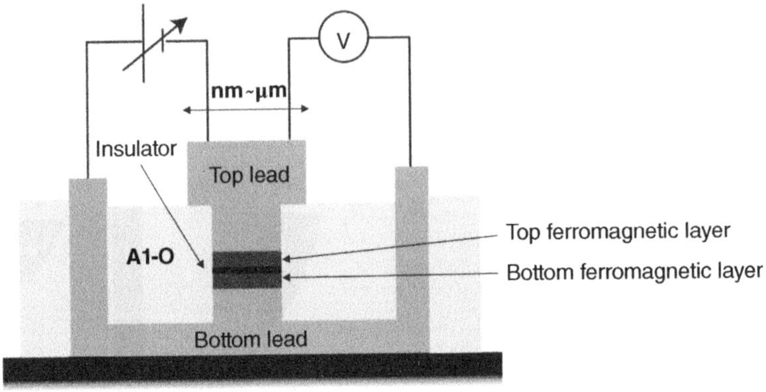

Figure 1.7 *Schematic illustration of the device structure for the measurement of TMR effect*

layer was prepared by depositing a 1–2 nm thick metallic Al film, and then oxidizing the metallic Al film. The oxidization methods are classified roughly into two kinds; one is the oxidization process carried out naturally in an O_2 atmosphere, and the other is the oxidization by using O_2 plasma. Compared to natural oxidization, an Al oxide layer is fabricated quickly by plasma oxidization. However, natural oxidization has a merit in fabricating a MTJ with a low junction resistance.

In the case of MTJ, there is no or quite small interlayer exchange coupling between two ferromagnetic electrodes, which is different from the case of metallic superlattices, and the antiparallel alignment of magnetization vectors cannot be achieved using the interlayer exchange coupling. Therefore, the following methods enable us to achieve the antiparallel orientation of magnetization vectors; making a coercivity difference between two ferromagnetic electrodes by changing the material or the thickness, or fabricating a spin-valve structure with an antiferromagnetic layer.

TMR should be measured carefully because the magnitude of TMR strongly depends on the bias voltage and the measurement temperature. These dependences are mainly attributable to the spin-flip scattering by magnons [36]. In the limit of low bias voltage and low temperature, Equation (1.14) indicates that the TMR ratio depends only on the spin polarization factors of ferromagnetic electrodes in principle. Figure 1.8 shows the plot of representative experimental values of TMR (at low temperature) for the MTJs with the ferromagnetic electrodes of transition metals and their alloys and the Al–O tunneling barrier, versus the calculated values of TMR from Equation (1.14) [37]. There are almost no differences between the experiments and the calculations. The values of spin-polarization factors are obtained independently by employing several methods: (a) preparing a tunnel junction with a target ferromagnetic metal and a superconductor, and measuring its current–voltage characteristics [38, 39], (b) fabricating a point contact with a superconductor, and measuring the Andreev reflection [40], (c) employing spin-polarized photoemission spectroscopy, and so on. The spin-polarization factors shown in the transverse axis of Figure 1.8 correspond to the values obtained with the method (a) using a tunnel junction of the ferromagnetic metal and Al superconductor with an Al oxide barrier. However, the electrons contributing to TMR are sometimes different from the

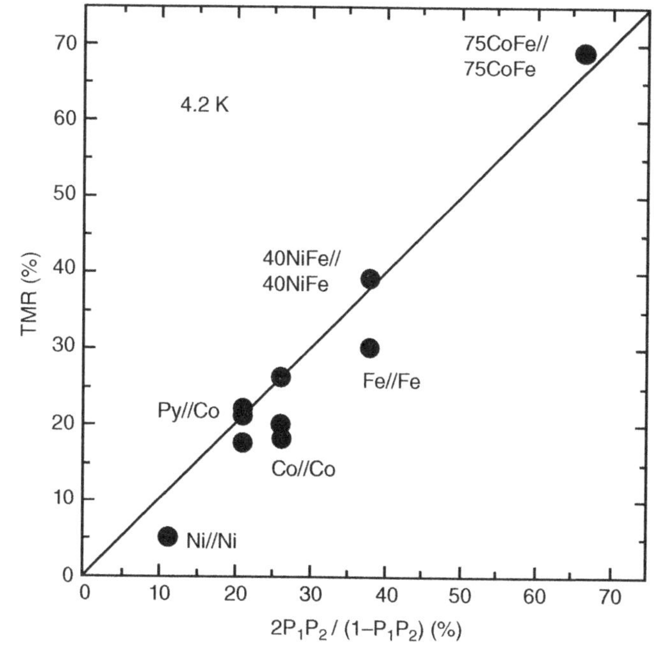

Figure 1.8 *Representative experimental values of TMR ratios obtained at low temperature for the MTJs with the ferromagnetic electrodes of transition metals and their alloys and the Al–O tunneling barrier. The values of TMR calculated from Equation (1.14) are also plotted (Ref. [37])*

electrons related to spin-polarization measurements, and it is noted that the relationship of Equation (1.14) is not always applicable.

1.2.3 TMR Effect in Half-Metallic Systems

As mentioned above, a larger TMR ratio is expected when the spin polarization becomes high, and if a 100% spin-polarized material is used as a ferromagnetic electrode, a huge (infinite in principle) TMR ratio would be obtained. A material with 100% spin polarization is called a half-metal. A half-metal, in other words, has a characteristic band structure in which either majority or minority spin band has an energy gap at the Fermi level. Half-metals are classified roughly into two groups: transition metal oxides and Heusler alloys. The oxide series include Fe_3O_4, CrO_2, and perovskite compounds represented by $LaSrMnO_3$ (LSMO). There is still no report for a large MR ratio for a MTJ with Fe_3O_4 or CrO_2. For the perovskite compounds, a large MR ratio of 1800% was reported at 4.2 K in an MTJ with LSMO/ $SrTiO_3$/LSMO where $SrTiO_3$ (STO) acted as a tunneling barrier [41]. However, LSMO is not suitable for practical application because the Curie temperature is low, and almost no TMR appears at room temperature. In the case of Heusler alloys, there are two types of Heusler alloys: half-Heusler alloys having $C1_b$ structures, and full-Heusler alloys having $L2_1$ structure. NiMnSb is a typical half-Heusler alloy, but a large TMR ratio has not been reported yet. For the full-Heusler alloys, on the other hand, large

TMR effects have been reported. At the early stage of the study on the MTJs with the full-Heusler alloys, the TMR ratios of 20%–40% at room temperature were reported for Co_2MnAl [42], Co_2MnSi [43] and $Co_2(Cr, Fe)Al$ [44] In 2006, the MTJ consisting of Co_2MnSi/Al–O/ Co_2MnSi showed a TMR ratio of 570% at low temperature [45], suggesting the half-metallic feature of Co_2MnSi. Following the experimental demonstration of large TMR effect in the full-Heusler alloy huge TMR ratios were achieved at low temperature for several full-Heusler alloys, e.g., 832% at 9 K for $Co_2Fe(Al,Si)$/MgO/$Co_2Fe(Al,Si)$ [46], and 1995% at 4.2 K for Co_2MnSi/MgO/Co_2MnSi [47], where the crystalline MgO tunnel barriers were used. The role of MgO barrier is described in Section 1.2. Although the huge TMR effects are obtained at low temperature, the MTJs with full-Heusler alloys show remarkable temperature dependence of the TMR ratio, and it is largely reduced at room temperature [45]. This is a remaining important problem for the full-Heusler alloys.

1.2.4 TMR Effect with Coherent Tunneling

Although Equation (1.14) simply describes the relationship between the TMR and the spin polarization of ferromagnetic electrodes, in realistic cases the relationship between the TMR and the electronic structure is not so simple. If tunneling electrons are scattered strongly at the interface between the electrode and the tunneling barrier, so-called *diffusive limit*, Equation (1.14) well explains the TMR effect by directly using the spin-polarization factor at the interface for each electrode as $P_{1(2)}$. If scattering rates are very small and tunneling electrons travel ballistically, however, the symmetry of wave functions and quantum-mechanical interference affects the TMR significantly.

Yuasa *et al.*, fabricated a MTJ with a single-crystal Fe electrode and an atomically flat Al–O tunneling barrier, and found that the MR ratio was varied for three crystal orientations of (100), (110), and (211). This means that the TMR effect depends on the crystal orientations of ferromagnetic electrodes [48]. It was also revealed that when the thickness of Fe electrodes was smaller than ~ 10 monolayers, the electronic states were modulated by the quantum size effect, and the TMR showed oscillatory behavior due to the formation of quantum well states [49]. Furthermore, for a Co(001)/Cu(001)/Al–O/NiFe MTJ where the nonmagnetic Cu(001) layer was inserted between the Co(001) single crystal electrode and the Al–O tunneling barrier, the TMR ratio showed oscillatory behavior with the sign reversal as a function of the Cu layer thickness [50]. This behavior is attributable to the spin-polarized resonant tunneling via the quantum-well states formed in the Cu interlayer. This cannot be explained only by the spin polarization at the interfaces, but is explained by the quantum-mechanical interference effects, which is called coherent tunneling.

Although an amorphous Al–O layer was used as a tunneling barrier in most cases previously, the TMR effect with coherent tunneling has been found to give rise to a large MR ratio. Since the lattice mismatch between Fe(001) and MgO(001) is very small and Fe(001) grows on MgO(001) epitaxially, this combination is suitable for the fabrication of the epitaxial MTJ. Theoretical calculations also predict a huge TMR ratio of 1000% in a Fe(001)/MgO(001)/Fe(001) MTJ due to a perfectly spin-polarized Δ_1 band of Fe [51, 52]. Yuasa *et al.* fabricated a fully epitaxial Fe(001)/MgO(001)/Fe(001) MTJ, and observed a large MR ratio of ~ 180% at room temperature and over 200% at low temperature [53].

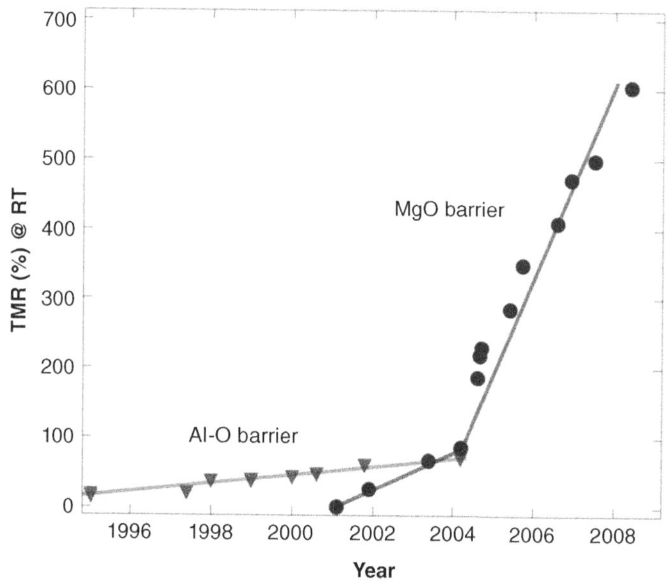

Figure 1.9 Yearly change of TMR at room temperature with Al–O and MgO tunnel barriers

They also reported an oscillation of the TMR as a function of the MgO layer thickness due to a coherent effect. Large TMR ratios reaching 200% at room temperature were also observed in a polycrystalline CoFe [54] and an amorphous CoFeB [55], in which the amorphous CoFeB was crystallized into bcc-CoFe by annealing. As in the case of MTJs with Fe/MgO/Fe, the large TMR observed for the CoFe electrode is interpreted within the framework of the coherent tunneling effect. Figure 1.9 shows yearly change in TMR at room temperature with Al–O and MgO tunnel barriers. Huge TMR ratios of 604% and 1144% at room temperature and 5 K, respectively, were obtained for CoFeB/MgO/CoFeB [56].

1.2.5 TMR Effect in Granular Systems

A TMR effect is also observed in granular systems, where magnetic nanoparticles are dispersed in an insulating matrix. If the insulator separating two nanoparticles is thinner than 1–2 nm, electrons move from one particle to another via tunneling. The magnetization vectors for nanoparticles are randomly distributed without magnetic field. When the magnetization vectors are aligned to the same direction by applying an external magnetic field, the tunneling probability of electrons increases, resulting in the decrease of resistance. The phenomenon that the resistance is decreased by the application of magnetic field is the same as that in granular systems with metallic matrices (see Section 1.1.6). However, a characteristic feature for insulating matrices is that the resistance decrease is not caused by spin-dependent scattering but by spin-dependent tunneling. In addition, the magnitudes of electric resistances are quite different. The electric resistances for metallic matrices are of the order of 10–10^2 μΩ cm; on the other hand, those for insulating matrices are about 10^5–10^6 μΩ cm (at room temperature). The TMR effects in granular systems are observed in Co, Fe, or CoFe nanoparticles embedded in Al oxide, Si oxide, Mg oxide, rare-earth oxide, or Mg fluoride insulating matrix. For details, see Ref. [57].

References

1. M. N. Baibich, J. M. Broto, A. Fert, *et al.*, *Phys. Rev. Lett.*, **61** (1988) 2472.
2. P. Grünberg, R. Schreiber, Y. Pang, *et al.*, *Phys. Rev. Lett.*, **57** (1986) 2442.
3. G. Binasch, P. Grünberg, F. Saurenbach, and W. Zinn, *Phys. Rev. B*, **39** (1989) 4828.
4. T. R. McGuire and R. I. Potter, *IEEE Trans. Magn.*, **MAG-11** (1975) 1018.
5. D. H. Mosca, F. Petroff, A. Fert, *et al.*, *J. Magn. Magn. Mater.*, **94** (1991) L1.
6. S. Araki, K. Yasui, and Y. Narumiya, *J. Phys. Soc. Jpn.*, **60** (1991) 2827.
7. C. A. dos Santos, B. Rodmacq, M. Vaezzadeh, and B. George, *Appl. Phys. Lett.*, **59** (1991) 126.
8. R. Shad, C. D. Potter, P. Beliën, *et al.*, *Appl. Phys. Lett.*, **64** (1994) 3500.
9. S. S. P. Parkin, N. More, and K. P. Roche, *Phys. Rev. Lett.*, **64** (1990) 2304.
10. S. S. P. Parkin, Z. G. Li, and D. J. Smith, *Appl. Phys. Lett.*, **58** (1991) 2710.
11. K. Saito, K. Takanashi, S. Mitani, and H. Fujimori, Technical reports of Institute for Materials Research, Tohoku University, **17** (1997) 55.
12. S. S. P. Parkin, *Appl. Phys. Lett.*, **60** (1992) 2152.
13. J. Inoue, H. Itoh, and S. Maekawa, *J. Phys. Soc. Jpn*, **61** (1992) 1149.
14. H. Itoh, J. Inoue, and S. Maekawa, *Phys. Rev. B*, **47** (1993) 5809.
15. S. N. Okuno and K. Inomata, *Phys. Rev. Lett.*, **72** (1994) 1553
16. T. Shinjo and H. Yamamoto, *J. Phys. Soc. Jpn.*, **59** (1990) 3061.
17. B. Dieny, V. S. Speriosu, S. S. P. Parkin, *et al.*, *Phys. Rev. B*, **43** (1991) 1297.
18. *Spin Dependent Transport in Magnetic Nanostructures*, (eds. S. Maekawa and T. Shinjo), p. 29, Taylor & Francis, 2002.
19. For review, J. Bass and W. P. Pratt Jr., *J. Magn. Magn. Mater.*, **200** (1999) 274.
20. T. Valet and A. Fert, *Phys. Rev., B*, **48** (1993) 7099.
21. T. Iwase, Y. Sakuraba, S. Bosu, *et al.*, *Appl. Phys. Exp.*, **2** (2009) 063003.
22. Y. Sakuraba, K. Izumi, T. Iwase, *et al.*, *Phys. Rev. B*, **82** (2010) 094444.
23. J. Sato, M. Oogane, H. Naganuma, and Y. Ando, *Appl. Phys. Exp.*, **4** (2011) 113005.
24. Y. Sakuraba, M. Ueda, Y. Miura, *et al.*, *Appl. Phys. Lett.*, **101** (2012) 252408.
25. Y. K. Takahashi, A. Srinivasan, B. Varaprasad, *et al.*, *Appl. Phys. Lett.*, **98** (2011) 152501.
26. A. E. Berkowitz, J. R. Mitchell, M. J. Carey, *et al.*, *Phys. Rev. Lett.*, **68** (1992) 3745.
27. J. Q. Xiao, J. S. Jiang, and C. L. Chien, *Phys. Rev. Lett.*, **68** (1992) 3749.
28. J. Q. Xiao, J. S. Jiang, and C. L. Chien, *Phys. Rev., B*, **46** (1992) 9266.
29. A. Tsoukatos, H. Wan, G. C. Hadjipanayis, and Z. G. Li, *Appl. Phys. Lett.*, **61** (1992) 3059.
30. N. Kataoka, H. Endo, K. Fukamichi, and Y. Shimada, *Jpn. J. Appl. Phys.*, **32** (1993) 1969.
31. M. Julliere, *Phys. Lett.*, **54A** (1975) 225.
32. S. Maekawa and U. Gäfvert, *IEEE Trans. Magn.*, **18** (1982) 707.
33. J. C. Slonczewski, *Phys. Rev. B*, **39** (1989) 6995.
34. T. Miyazaki and N. Tezuka, *J. Magn. Magn. Mater.*, **139** (1995) L231.

35. J. S. Moodera, L. R. Kinder, T. M. Wong, and R. Meservey, *Phys. Rev. Lett.*, **74** (1995) 3273.
36. X. F. Han, A. C. C. Yu, M. Oogane, J. Murai, *et al.*, *Phys. Rev. B*, **63** (2001) 224404.
37. T. Miyazaki, *J. Mag. Soc. Jpn.*, **25** (2001) 471.
38. R. Meservey and P. M. Tedrow, *Phys. Rep.*, **238** (1994) 173.
39. D. J. Monsma and S. S. P. Parkin, *Appl. Phys. Lett.*, **77** (2000) 720.
40. R. J. Soulen Jr., J. M. Byers, M. S. Osofsky, *et al.*, *Science*, **282** (1998) 85.
41. M. Bowen, M. Bibes, A. Barthélémy, *et al.*, *Appl. Phys. Lett.*, **82** (2003) 233.
42. H. Kubota, J. Nakata, M. Oogane, *et al.*, *Jpn. J. Appl. Phys.*, **43** (2004) L984.
43. A. Hütten, S. Kämmerer, J. Schmalhorst, *et al.*, *Phys. Stat. Sol.*, **A201** (2004) 3271.
44. K. Inomata, S. Okamura, R. Goto, and N. Tezuka, *Jpn. J. Appl. Phys.*, **42** (2003) L419.
45. Y. Sakuraba, M. Hattori, M. Oogane, *et al.*, *Appl. Phys. Lett.*, **88** (2006) 192508.
46. N. Tezuka, N, Ikeda, F. Mitsuhashi, and S. Sugimoto, *Appl. Phys. Lett.*, **94** (2009) 162504.
47. H. Liu, Y. Honda, T. Taira, *et al.*, *Appl. Phys. Lett.*, **101** (2012) 132418.
48. S. Yuasa, T. Sato, E. Tamura, *et al.*, *Europhys. Lett.*, **52** (2000) 344.
49. T. Nagahama, S. Yuasa, Y. Suzuki, and E. Tamura, *Appl. Phys. Lett.*, **79** (2001) 4381.
50. S. Yuasa, T. Nagahama, and T. Suzuki, *Science*, **297** (2002) 234.
51. J. Mathon and A. Umerski, *Phys. Rev. B*, **63** (2001) 220403R.
52. W. H. Butler, X.-G. Zhang, T. C. Schulthess, and J. M. Maclaren, *Phys. Rev. B*, **63** (2001) 054416.
53. S. Yuasa, T. Nagahama, A. Fukushima, *et al.*, *Nature Mater.*, **3** (2004) 868.
54. S. S. P. Parkin, C. Kaiser, A. Panchula, *et al.*, *Nature Mater.*, **3** (2004) 862.
55. D. D. Djayaprawira, K. Tsunekawa, M. Nagai, *et al.*, *Appl. Phys. Lett.*, **86** (2005) 092502.
56. S. Ikeda, J. Hayakawa Y. Ashizawa, *et al.*, *Appl. Phys. Lett.*, **79** (2001) 4381.
57. H. Fujimori, S. Mitani, and K. Takanashi, *Mater. Sci. Eng. A*, **267** (1999) 184.

2

Spintronics Materials with High-Spin Polarization

Y.K. Takahashi and K. Hono
Magnetic Materials Unit, National Institute for Materials Science, Tsukuba, Japan

Here, we review our work on the exploration of quaternary ferromagnetic (FM) Heusler alloys with high-spin polarization and their applications to magnetoresistive devices. Although current spin polarization of ternary Co-based Heusler alloys deduced from point contact Andreev reflection measurements were below 60% at 4.2 K, some quaternary Heusler alloys, i.e., $Co_2Y(Z_{1-x}Z'_x)$, demonstrated high-spin polarization. We found that $Co_2Mn(Ge_{0.75}Ga_{0.25})$ and $Co_2Fe(Ga_{0.5}Ge_{0.5})$ were rather promising materials of the various alloys we explored. Current-perpendicular-to-plane giant magnetoresistance (CPP-GMR) and lateral spin valve (LSV) devices using these FM alloys exhibited superior properties compared to similar devices with conventional FM materials.

2.1 Introduction

Spintronics is an area using two freedoms of electrons, i.e., charge and spin. One of the most successful spintronics devices are tunneling magnetoresistance (TMR) devices for read sensor applications in hard disk drives (HDDs). Due to the invention of giant magnetoresistance (GMR) in ferromagnetic/nonmagnetic multilayers by Baibich [1] and Grünberg [2] and the subsequent invention of spin valves [3], the areal density of HDDs has increased dramatically due to improvements to the sensitivity of GMR sensors. After the breakthrough with the spin-filtering effect on MgO-based magnetic tunnel junctions (MTJs) by Yuasa [4] and Parkin [5] in 2004, GMR sensors were replaced by TMR sensors, which are still extending the increase in the areal density of HDDs. MTJ-based devices are attracting

Spintronics for Next Generation Innovative Devices, First Edition. Edited by Katsuaki Sato and Eiji Saitoh.
© 2015 John Wiley & Sons, Ltd. Published 2015 by John Wiley & Sons, Ltd.

many applications in addition to read sensor applications, such as those in magnetic random access memories (MRAMs), spin metal-oxide-semiconductor field-effect-transistors (spin MOSFETs), and biosensors. Highly spin polarized materials that can generate spin polarized current are crucial to achieve the high levels of performance in these applications.

Although a great deal of effort has been devoted to developing half-metals, there have been no reports on 100% spin polarization at room temperature (RT). For example, MTJs using $La_{0.7}Sr_{0.3}MnO_3$ (LSMO) have demonstrated large TMRs of 1800% at 4.2 K [6], which indicated that LSMO was half-metal at low temperature. However, the TMR output is completely lost at room temperature (RT) due to the low Curie temperature (T_c). High-spin polarization at RT is required for practical applications. One candidate for RT half-metallic ferromagnetic (FM) materials is Co-based Heusler alloy because its T_c is far above RT so that the half-metallicity predicted by *ab initio* calculations for 0 K is retained at RT.

Sakuraba *et al.* [7] demonstrated large TMR of 570% using MTJs with Co_2MnSi FM electrodes and Al–O barriers at 2 K, which indicates that Co_2MnSi is half-metal at low temperature. However, TMR is substantially degraded at RT, which questions the half-metallicity at RT. The large decrease in spin polarization against temperature is the most critical issue in applying FM Heusler alloys to spintronics devices.

The equilibrium structure of Co-based Heusler alloys with the chemical formula of X_2YZ is $L2_1$, as shown in Figure 2.1. However, disordering to B2 (disordering between Y and Z), DO_3 (disordering between X and Y) and A2 (disordering between X, Y,

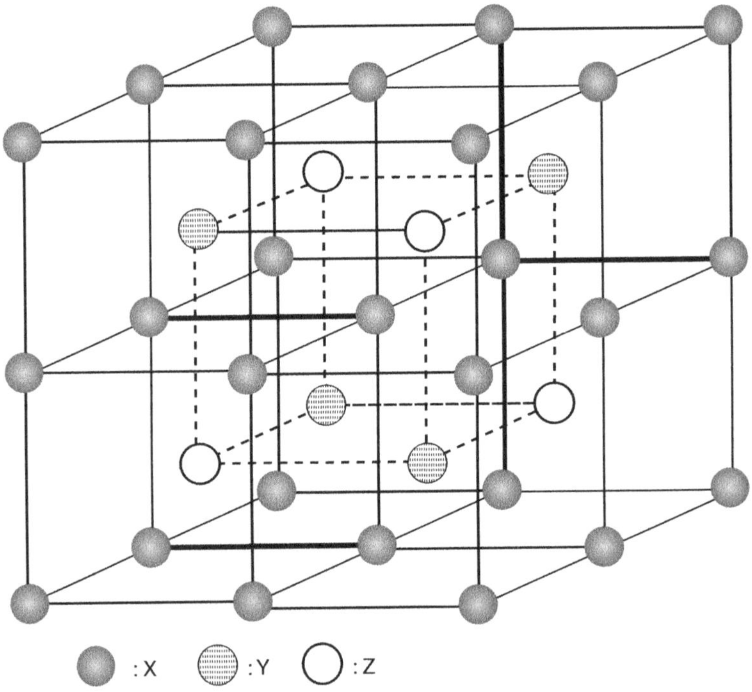

Figure 2.1 *Crystal structure of $L2_1$ ordered Co-based Heusler alloy*

and Z) structures occurs as the thermodynamic driving force for the $L2_1$ ordering is low, which causes reduced spin polarization [8]. Since it is difficult to obtain a perfect $L2_1$ ordered structure in this system even in bulk alloys, it is not easy to achieve a high degree of $L2_1$ ordering in thin film devices. Therefore, we need to accomplish high-spin polarization under the presence of a certain degree of a disordered state. According to Fecher *et al.* [9], if Fermi energy, E_F, is located close to one of the minority gap edges, spin polarization decreases due to the smearing of states close to E_F by disordering, defects and lattice parameters change. However, half-metallicity may become robust against disordering and temperature by tuning E_F to the center of the band-gap. The theoretical calculations for $Co_2Fe(Al_xSi_{1-x})$ have revealed that E_F shifts from the top edge to the center of the band gap by Al being substituted for Si in Co_2FeSi, while the half-metallic band gap is maintained because the number of valence electrons of Al is smaller than that of Si by one [9]. Experimentally measured spin polarization has demonstrated maximum at $Co_2Fe(Al_{0.5}Si_{0.5})$ due to E_F tuning [10]. Galanakis [11] also suggested that E_F engineering by alloying the fourth element to the ternary alloys effectively enables the new half-metallic Heusler alloys to be explored. Based on this concept, we explored quaternary Heusler alloys with high-spin polarization. In addition, we demonstrated the effectiveness of the newly developed materials using current-perpendicular-to-plane giant magnetoresistive (CPP-GMR) devices.

2.2 Development of Highly Spin Polarized Materials

The spin polarization of materials has to be promptly measured to explore highly spin polarized materials. We chose the point contact Andreev reflection (PCAR) method for this purpose [12]. PCAR is used to measure the bias dependent conductance of point contact between FM samples and superconducting material. As seen in Figure 2.2(a), when electrons enter a superconductor from a normal conductor, they should form Cooper pairs when the energy of electrons is smaller than the quasi-particle state. Since the normal conductor has the same number of spin-up and spin-down electrons at E_F, all the electrons form Cooper pairs and enter the superconducting condensate state as a super current. Holes are left behind at that time to retain the momentum of the system, which is called Andreev reflection. As a result, conductance doubles in the super conducting gap. Andreev reflection is suppressed for the conduction between FM and the superconductor, as shown in Figure 2.2(b), because the number of spin-up and spin-down electrons is not equal at E_F. As a result, conductance is suppressed with the increase in spin polarization and becomes zero in a half-metal, as can be seen from Figure 2.2(c). The conductance curves were fitted with the modified Blonder, Tinkham, and Klapwijk (BTK) model [13] to estimate spin polarization. Since the depth of the dip in the conductance curve with the zero interfacial scattering factor (Z) corresponds to the spin polarization of the material, we can directly discuss relative spin polarization by using the conductance curves. Point contact is created by mechanical contact between the Nb needle and the ferromagnetic sample, and the electron conductance through such contact mainly occurs due to the diffusive mode. Therefore, spin polarization determined by PCAR is

$$P = \frac{v_F^2 D_\uparrow - v_F^2 D_\downarrow}{v_F^2 D_\uparrow + v_F^2 D_\downarrow}, \tag{2.1}$$

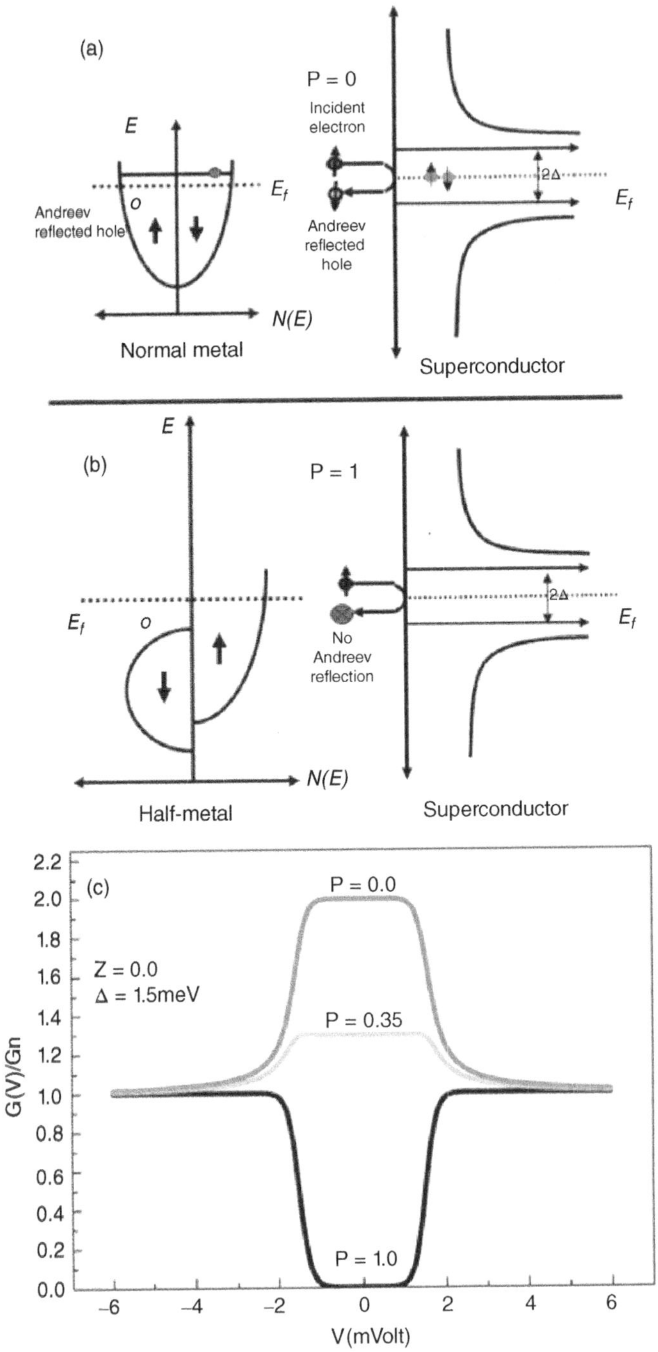

Figure 2.2 *Schematic images of Andreev reflections with point contact of (a) normal conductor and superconductor and (b) half-metal and superconductor. (c) Calculated conductance curves with different spin polarization*

Table 2.1 *Spin polarization deduced by PCAR for various ternary and quaternary Heusler alloys and their Curie temperatures. (Reprinted with permission from B. S. D. Ch. S Varaprasad, A. Srinivasan, Y. K. Takahashi, M. Hayashi, A. Rajanikanth, and K. Hono,* Acta Mater. ***60****, 6257 (2012). ©2012 Elsevier B.V.)*

Quaternary alloys	*P*	T_c (K)	Ternary alloys	*P*	T_c (K)
$Co_2Mn(Ge_{0.75}Ga_{0.25})$	0.74	895	Co_2CrAl	0.62	600
$Co_2Mn(Ga_{0.5}Sn_{0.5})$	0.72	770	Co_2CrGa	0.61	–
$Co_2Fe_{0.75}Ge_{0.25})$	0.70	990	Co_2MnSn	0.60	800
$Co_2Fe(Ga_{0.5}Ge_{0.5})$	0.69	1080	Co_2MnAl	0.60	–
$Co_2(Cr_{0.02}Fe_{0.98})Ga$	0.67	–	Co_2MnGa	0.60	700
$Co_2Mn(Ge_{0.25}Sn_{0.75})$	0.67	–	Co_2FeSi	0.60	1100
$Co_2(Mn_{0.95}Fe_{0.05})Sn$	0.65	–	Co_2FeAl	0.59	–
$(Co_{1.93}Fe_{0.062})MnGe$	0.68	–	Co_2MnGe	0.58	900
$Co_2(Mn_{0.5}Fe_{0.5})Ga$	0.70	990	Co_2FeGe	0.58	1000
$Co_2(Cr_{0.02}Fe_{0.98})Si$	0.65	990	Co_2FeGa	0.58	1100
$Co_2Mn(Ti_{0.25}Sn_{0.75})$	0.64	480	Co_2TiSn	0.57	364
$Co_2Mn(Al_{0.5}Sn_{0.5})$	0.63	–	Co_2MnSi	0.56	900
$Co_2Mn(Ga_{0.25}Si_{0.75})$	0.63	–	Fe_2VAl	0.56	–
$Co_2Mn(Si_{0.25}Ge_{0.75})$	0.63	–	Co_2VAl	0.48	–
$Co_2(Mn_{0.5}Fe_{0.5})Si$	0.61	–			
$Co_2Mn(Al_{0.5}Si_{0.5})$	0.60	–			
$Co_2Fe(Ga_{0.5}Si_{0.5})$	0.60	–			
$Co_2Fe(Al_{0.5}Si_{0.5})$	0.60	–			

where v_F is the Fermi velocity and D is the density of states (DOS) with up and down spins. Hence, spin polarization deduced by PCAR in the diffusive resume is equivalent to spin polarization in electron current, $P = (\sigma_\uparrow - \sigma_\downarrow)/(\sigma_\uparrow + \sigma_\downarrow)$, where σ is the conductance of electrons. Therefore, the spin polarization determined by PCAR does not necessarily equal the spin polarization of the density of states at 0 K deduced from *ab initio* calculations.

Table 2.1 summarizes the spin polarization deduced by the PCAR and T_c of ternary and newly developed quaternary Heusler alloys. [14]. The spin polarization of ternary alloys is around 60% at 4.2 K. The spin polarizations of quaternary alloys, on the other hand, developed by substituting Z elements with other similar elements with different valences ranges from 60% to 70% with the highest spin polarization of 74% for the $Co_2Mn(Ge_{0.75}Ga_{0.25})$ alloy. Here, we show the experimental results for the two typical alloys, $Co_2Mn(Ge_{1-x}Ga_x)$ [15] and $Co_2Fe(Ge_{1-x}Ga_x)$ alloys [14, 16].

Although $L2_1$-ordered Co_2MnGe was predicted to be half-metal [17], there was no experimental evidence for high-spin polarization in Co_2MnGe [18, 19]. Our PCAR measurements demonstrated 59% spin polarization. The reason for the low-spin polarization was the DO_3 disordering. A Co antisite, where a Mn site is replaced by a Co atom, induces a spin-down in-gap state, according to theoretical calculations [20]. In addition, since the formation energy of the Co antisite is relatively low, DO_3 disordering occurs easily and causes reduced spin polarization. However, Co_2MnGa had a large number of spin-up DOSs around E_F, although the alloy was not predicted to be half-metal [17]. High-spin polarization was expected by substituting Ge with Ga in Co_2MnGe due to the increased spin-up state

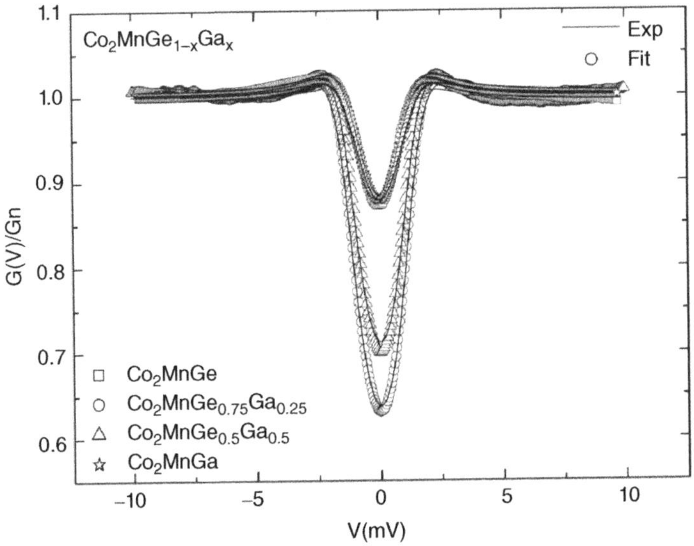

Figure 2.3 *Normalized conductance curves of $Co_2Mn(Ge_{1-x}Ga_x)$ ($x = 0$, 0.25, 0.5 and 1) alloys*

around E_F. In addition, since the number of valence electrons of Ga was smaller than that of Ge by 1, substituting Ga with Ge in Co_2MnGa was expected to change E_F. The theoretical calculations revealed that high-spin polarization larger than 90% was expected for $x \leq 0.5$ in $Co_2Mn(Ge_{1-x}Ga_x)$ [15]. An $L2_1$ ordered structure was observed in all the alloy compositions in the bulk sample. The DO_3 disordering estimated with Rietveld analysis was 12.6% for Co_2MnGe and 5.2% for $Co_2Mn(Ge_{0.75}Ga_{0.25})$. The DO_3 disordering decreased with increased Ga. Differential thermal analysis (DTA) revealed that $Co_2Mn(Ge_{1-x}Ga_x)$ alloy with $x \leq 0.75$ is an intermetallic compound, which means the $L2_1$ phase is stable up to melting temperature. Figure 2.3 shows the normalized conductance curves of these alloys. Since Z is zero in these curves, we can directly compare spin polarization from the dip in the conductance curve. The minimum in the conductance curves is 0.85 for Co_2MnGe and Co_2MnGa, whereas that for $Co_2Mn(Ge_{0.75}Ga_{0.25})$ alloy is 0.63. This indicates that the spin polarization of $Co_2Mn(Ge_{0.75}Ga_{0.25})$ is much higher than that of terminal ternary alloys. Fitting with the modified BTK model revealed that the spin polarization of $Co_2Mn(Ge_{0.75}Ga_{0.25})$ was 74%, which is the highest value in the Heusler alloys reported thus far. Since $Co_2Mn(Ge_{0.75}Ga_{0.25})$ alloy has the highest spin polarization and relatively high T_c (895 K), it is expected to be a promising candidate as a spin polarized current source.

Co_2Fe-based Heusler alloys have lower kinetics of $L2_1$ ordering, unlike Co_2Mn-based Heusler alloys. However, Co_2Fe-based Heusler alloys in CPP-GMR devices have better tolerance against interdiffusion with an Ag spacer layer at elevated temperatures. Therefore, highly spin polarized Co_2Fe-based Heusler alloys have the potential to act as the ferromagnetic electrodes of CPP-GMR devices. Co_2FeGe was predicted to be half-metal [14]. However, the $L2_1$ single phase was not experimentally obtained. Co_2FeGa, on the other hand, formed an $L2_1$ single phase although the spin polarization measured

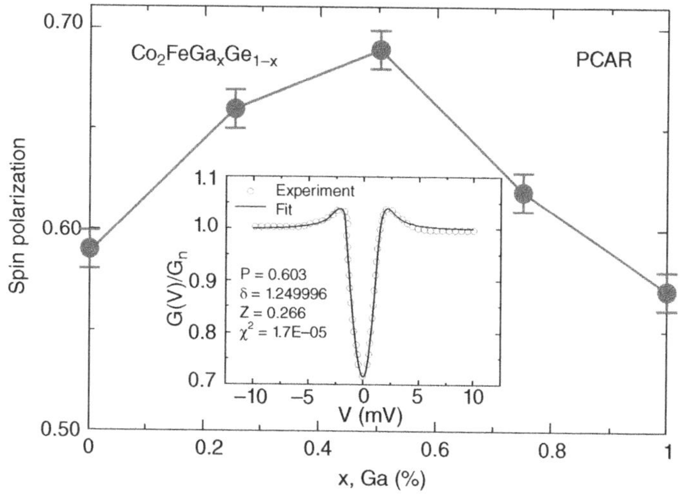

Figure 2.4 *Change in spin polarization of $Co_2Fe(Ge_{1-x}Ga_x)$ alloy as a function of x (Ga) content. Inset shows typical conductance and fitted curves of $Co_2Fe(Ge_{0.5}Ga_{0.5})$ alloy. (Reprinted with permission from Ref. [16]. ©2011 American Institute of Physics)*

with PCAR was only 59% [21]. Hence, the substitution of Ge with Ga in Co_2FeGe was expected to stabilize the $L2_1$ phase. In addition, the theoretical calculations indicated that the $Co_2Fe(Ge_{0.5}Ga_{0.5})$ alloy was half-metal even in the B2 ordered structure. That the $L2_1$ phase had formed was confirmed in $Co_2Fe(Ge_{1-x}Ga_x)$ by the partial substitution of Ge with Ga. The degrees of B2 and $L2_1$ ordering estimated from XRD were ~0.7 and 0.3 for all the alloys. Figure 2.4 plots the change in the spin polarization of $Co_2Fe(Ge_{1-x}Ga_x)$ alloys as a function of Ga concentration. The inset plots a typical conductance curve for the $Co_2Fe(Ge_{0.5}Ga_{0.5})$ alloy. The terminal compositions of Co_2FeGa and Co_2FeGe indicate 59% spin polarization. The spin polarization increases with increasing Ga composition having a maximum value of 69% at $x = 0.5$, which is the second highest spin polarization in Co_2Fe-based Heusler alloys after $Co_2Fe(Si_{0.75}Ge_{0.25})$. In addition to the high-spin polarization, $Co_2Fe(Ge_{0.5}Ga_{0.5})$ alloy has a high T_c of 1080 K. Therefore, $Co_2Fe(Ge_{0.5}Ga_{0.5})$ alloy is considered to be a promising material for spintronics devices.

2.3 Device Applications

TMR devices are the most successful spintronics device applications that have thus far been industrially implemented. CoFeB/MgO-based TMR devices are currently used as read sensors in HDDs. After the Fe/MgO spin-filtering effect [4, 5] had been discovered, Djayaprawira *et al.* applied CoFeB to MgO-based MTJs as an FM electrode instead of Fe for practical applications [22]. They achieved a coherent interface of CoFe(001)/MgO(001) through the heteroepitaxial growth of CoFe when amorphous CoFeB was crystallized. Recent work by Ikeda *et al.* obtained a large TMR of 604% at RT in pseudo-spin valve type MTJs [23]. The TMR ratio is often expressed as $TMR = 2P_1P_2/(1 - P_1P_2)$, where P_1 and P_2 are the spin polarizations of electrodes. However, it should be noted that both

P_1 and P_2 are in fact the spin polarization of tunneling currents, which varies depending on the barrier/electrode interface structure rather than the intrinsic spin polarization of materials. In fact, the tunneling spin polarization of Fe combined with an MgO barrier is higher than 0.8, although the intrinsic spin polarization of Fe is considered to be around 0.4. Because of this confusion, Heusler alloys were considered to be promising as FM electrodes for MTJs because of their intrinsic half-metallicity. In fact, Sakuraba *et al.* demonstrated a large TMR using a Co_2MnSi/Al–O MTJ [7]. However, the TMR was substantially degraded at RT. Recent TMR devices using Heusler alloys mostly utilize MgO barriers. Kawami *et al.* reported a large TMR ratio of 462% at RT [24] by utilizing these barriers; however, it was not clear whether this was due to the intrinsic spin polarization of materials or the spin filtering effect by MgO. Considering the industrial viability of CoFeB/MgO barriers that can be grown on Si substrates, there appears to be little chance that Heusler alloy-based MTJs would replace CoFeB/MgO-based MTJs as read sensors. However, recent investigations have reported that the damping constant of Heusler alloys is one order of magnitude smaller than that for standard FM alloys. There may be some possibility of developing MTJs that can have spin-transfer torque switched at lower currents by taking advantage of the low damping constant of Heusler alloys.

Although CoFeB/MgO-MTJs are still extending the areal density of HDDs, they face huge problems with further increases. According to Takagishi's calculations [25] in Figure 2.5, a read sensor with low device resistance and high output is required as a reader for >2 Tbit/in^2 because of impedance matching with preamplifiers to achieve a good high-frequency response. The gray line in Figure 2.5 plots the ΔRA values (resistance-area product, output of MR devices) of currently used MTJs with MgO barriers. The ΔRA reduces substantially with the decrease in RA (device resistance) due to the unavoidable formation of pin holes in thin MgO barriers of less than 1 nm. However, current-perpendicular-to-plane (CPP)-GMR devices that consist of all metallic layers can easily achieve low RA. However, only small ΔRA values less than 2 mΩ μm^2 have been reported in CPP-GMR devices using conventional FM materials such as CoFe. [26–28]. According to the Valet–Fert model [29], ΔRA is approximated by the following equation under the assumption of $t_F \ll l_{sf}$ (spin diffusion length):

$$\Delta RA \approx 2\beta^2 \rho_F^* t_F + 4\gamma^2 AR_{F/N}^*, \tag{2.2}$$

where β and γ are the bulk and interface spin asymmetries, ρ_F and ρ_N are the resistivities of FM and spacer materials ($\rho_F^* = \rho_F/(1-\beta^2)$), t_N and t_F are the thickness of the spacer and FM layers, respectively, and $AR_{F/N}$ is the interfacial resistivity between FM and spacer layers ($AR_{F/N}^* = AR_{F/N}/(1-\gamma^2)$). Therefore, high output can be expected in CPP-GMR using highly spin polarized materials, i.e., materials with $\beta \sim 1$ and $\gamma \sim 1$.

2.3.1 CPP-GMR Devices using Highly Spin Polarized Heusler Alloys

The first report on CPP-GMR using Heusler alloys was the NiMnSb/Cu system obtained by Caballero *et al.* [30]. Although NiMnSb was predicted to be half-metal, they obtained a GMR ratio of 8% at 4.2 K. Later, Hoshiya and Hoshino made CPP-GMR devices using a full Heusler alloy of CoMnGe [31]. By using CoFe/CoMnGe/CoFe as a free layer, they increased ΔRA/ to 1.6 mΩ μm^2. Yakushiji *et al.* fabricated epitaxial $Co_2MnSi/Cr/Co_2MnSi$ CPP-GMR devices and obtained a large ΔRA of 19 mΩ μm^2

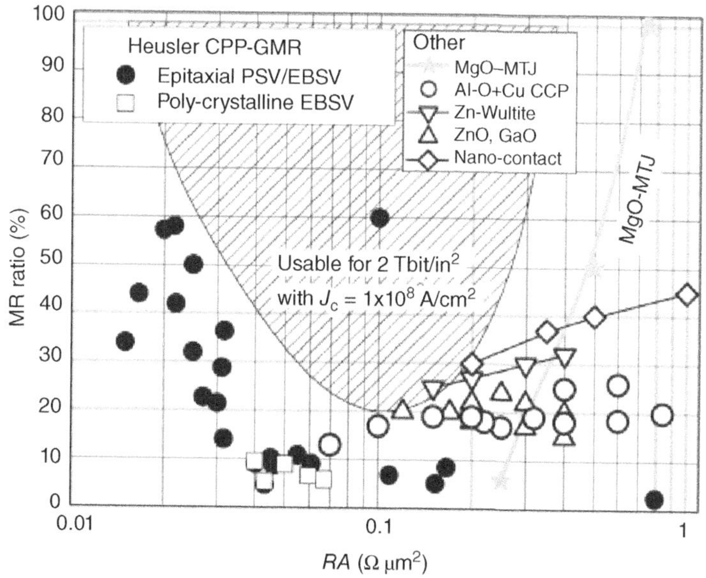

Figure 2.5 *Usable range of MR ratio and RA in 2 Tb/in² [25]. Gray line plots data for MgO-based MTJs [50]. Closed circles and rectangles correspond to MR ratios for CPP-GMR of epitaxial [16, 35, 41, 51–53] and polycrystalline [43, 54–56] systems. Open circles, triangles, inverted triangles and rhombuses correspond to MR ratios for Al–O+Cu-based current-confined-path (CCP) type [57–59], ZnO and GaO spacer [60], Zn wüstite spacer [61] and nano-contact [62] CPP-GMR devices*

[32], which is considered to have been overestimated due to the current crowding effect. To enhance ΔRA further, Furubayashi *et al.* demonstrated CPP-GMR devices using $Co_2FeAlSi$ quaternary Heusler alloys as FM layers and Ag as a spacer layer [33]. They obtained an MR ratio of 6.9% and 14% at RT for the former and 6 K for the latter. Although the MR ratio was relatively large at that time, it was much smaller than that expected from the half-metallicity of the FM layer. Note that Ag was used as a spacer layer for the first time in their work and it has become a standard choice for Heusler-based CPP-GMR due to exceptional band matching. We applied newly developed $Co_2Mn(Ge_{0.75}Ga_{0.25})$ (CMGG) and $Co_2Fe(Ge_{0.5}Ga_{0.5})$ (CFGG) alloys to the FM layer of CPP-GMR devices to further enhance the MR ratio.

Figure 2.6 plots the MR curves of CPP-PSV using CFGG after annealing at 500° C [16]. The film stack was MgO(001) sub./Cr(10)/Ag(100)/Heusler/Ag(5)/Heusler/Ag(5)/Ru(8), where the numbers in parentheses are the thickness of each layer in nanometers. All the layers were epitaxially grown with a [001] orientation perpendicular to the film plane and the stack was annealed to chemically order the Heusler layers after the film had been fabricated. The multilayers were microfabricated into ellipsoidal pillars with dimensions ranging from 0.70 × 0.14 to 0.20 × 0.40 μm² by electron beam lithography and Ar ion etching. The triangles in Figure 2.6 represent the device resistance (RA) at 300 K and the circles represent those at 10 K. The low and high resistances correspond to the parallel and antiparallel configurations of the two CFGG magnetizations. We confirmed that both CFGG layers

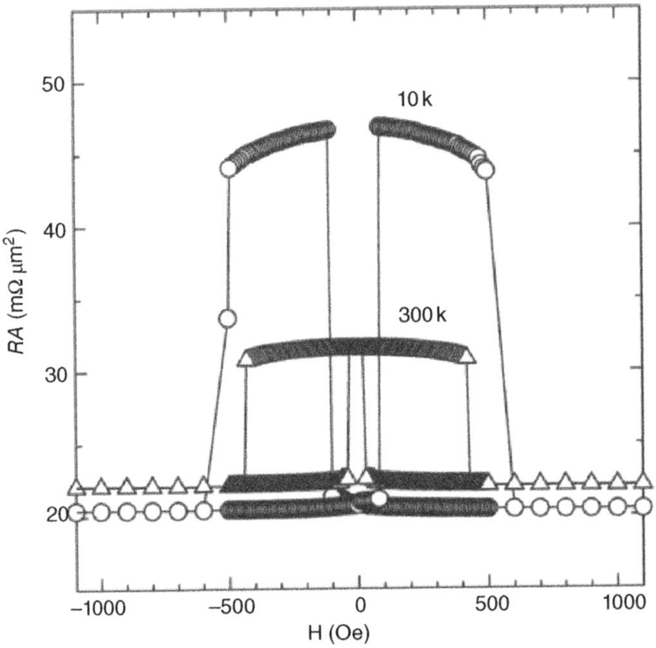

Figure 2.6 *MR curves for $Co_2Fe(Ge_{0.5}Ga_{0.5})$(12 nm)/Ag(5 nm)/$Co_2Fe(Ge_{0.5}Ga_{0.5})$ (12 nm) PSV at 300 K (triangles) and 10 K (circles)*

had a B2 structure by analyses with X-ray diffraction (XRD) and transmission electron microscopy (TEM). The ΔRA and MR ratios corresponded to 9.5 mΩμm^2 and 42% at 300 K and 26.4 mΩμm^2 and 129% at 10 K [16]. These values were the highest values at that time. To understand the origin of the large MR output, β and γ were estimated with the Valet–Fert model. [34]. Figure 2.7 plots the change in ΔRA as a function of the thickness of the CFGG layers. The ΔRA increases with the thickness of the CFGG layers and saturates around 5 nm due to the short spin diffusion length of CFGG. The dashed line represents the fitting result using the Valet–Fert equation. The β and γ were estimated to be 0.73 and 0.60 at RT and 0.9 and 0.84 at 10 K. The 0.73 of β at RT was one of the highest values reported at that time. Therefore, the large ΔRA at RT was attributed to the large β of CFGG. Note that β at 10 K is 0.9, which indicates that CFGG is a half-metal at low temperature even in the B2 ordered structure.

Detailed microstructure analyses using scanning transmission electron microscopy/energy-dispersive-spectroscopy (STEM/EDS) revealed the segregation of Ga and Ge at the Ru/Ag interface in the cap layers [34]. This has been interpreted to be the result of the large interfacial energy of the Ru/Ag capping layers. Annealing was carried out under this assumption just after the top CFGG layer had been deposited. Then, ΔRA was enhanced to 12.0 mΩμm^2 in the device annealed at 600° C [35]. Since the CFGG layer exhibited an $L2_1$ structure in the nanobeam diffraction pattern, the enhanced ΔRA was attributed to the increased spin polarization of CFGG. Further annealing above 600° C led to a decrease in ΔRA because of the degraded surface roughness. The β and γ were estimated to be 0.83 and 0.63 at RT and 0.93 and 0.88 at 10 K by using the Valet–Fert model.

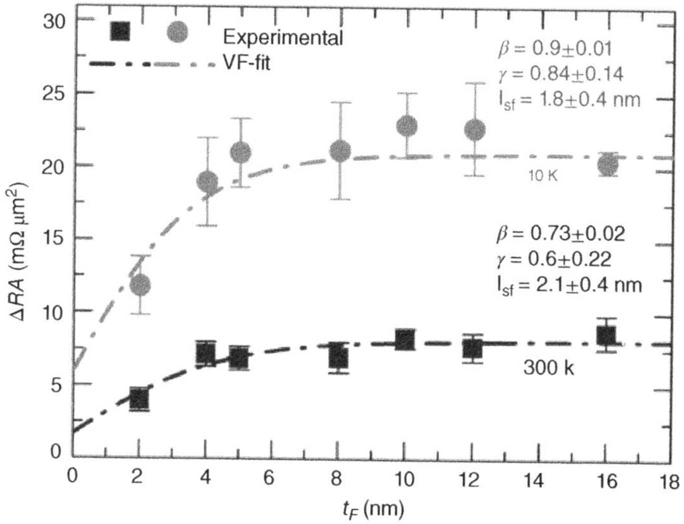

Figure 2.7 *ΔRA change as a function of CFGG thickness (t_F) plot of $Co_2Fe(Ga_{0.5}Ge_{0.5})(t_F)/Ag(5\ nm)/Co_2Fe(Ga_{0.5}Ge_{0.5})(t_F)$ PSVs measured at 10 K (circles) and 300 K (squares) and dashed line represents curves fitted with Valet–Fert's model. (Reprinted with permission from H. S. Goripati, T. Furubayashi, Y. K. Takahashi, and K. Hono,* J. Appl. Phys. ***113****, 0439001 (2013). ©2013 American Institute of Physics)*

This suggests that the $L2_1$ ordered CFGG was virtually half-metal. Figure 2.8 plots the dependence of ΔRA on temperature in the CPP-GMR using CFGG that was annealed at 600° C. The ΔRA of 33 mΩ μm² at 10 K is substantially reduced to 12 mΩ μm² at RT, which is attributed to the decrease in the spin polarization of not only the bulk but also the interface at finite temperature. It is important to understand the reason for the large temperature degradation of β and γ by considering the intrinsically large spin polarization of CFGG at low temperature. Note that the Curie temperature of CFGG is 1080 K and a very small decrease in magnetization was detected at RT. The strong dependence on temperature is also a huge problem in Heusler-based MTJs. Weak exchange [36] and magnon excitation [37] at the FM/Al–O interface have been proposed to be the reason for the large temperature degradation of TMR in Heusler-based MTJs. Further investigations into the origin of the large dependence on temperature are necessary in CFGG-based CPP-GMR.

Figure 2.9 plots the change in ΔRA as a function of annealing temperature in CMGG (5 nm)/Ag (5 nm)/CMGG (5 nm) CPP-PSVs. The ΔRA values of other Heusler CPP-GMR devices are also plotted in the figure. The ΔRA increases with the annealing temperature, which is due to the enhanced spin polarization of CMGG resulting from improvements to the degree of $L2_1$ ordering. However, ΔRA has a maximum of 6.1 mΩ μm² at 400° C and then decreases even though the degree of $L2_1$ ordering is improved by annealing at higher temperatures. The detailed microstructures of films annealed at 400° C and 550° C were investigated to better understand the degradation of ΔRA by high-temperature annealing [38]. Both samples exhibited atomically sharp interfaces without any evidence of interdiffusion or the destruction of layer structures based on analyses of the microstructure

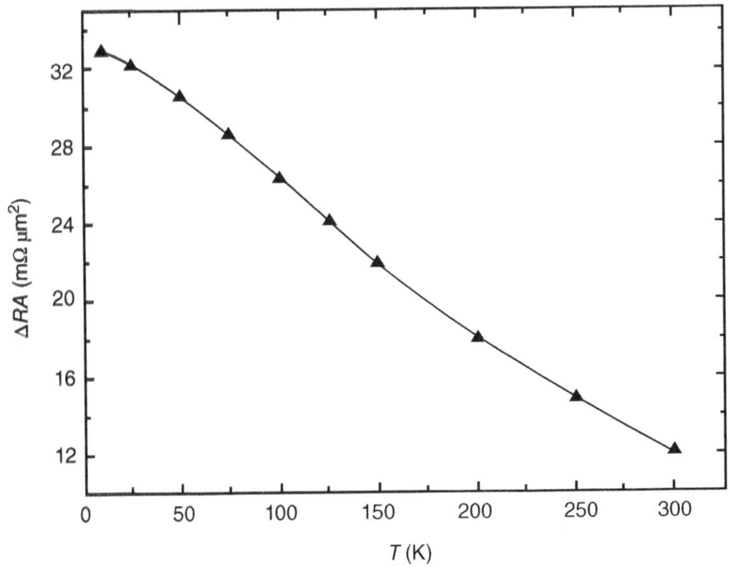

Figure 2.8 *Dependence of ΔRA of $Co_2Fe(Ga_{0.5}Ge_{0.5})$(10 nm)/Ag(5 nm)/$Co_2Fe(Ga_{0.5}Ge_{0.5})$(10 nm) CPP-PSVs on temperature in-situ annealed at 600°C. (Reprinted with permission from S. Li, Y. K. Takahashi, T. Furubayashi, and K. Hono,* Appl. Phys. Lett. ***103****, 042405 (2013). ©2013 American Institute of Physics)*

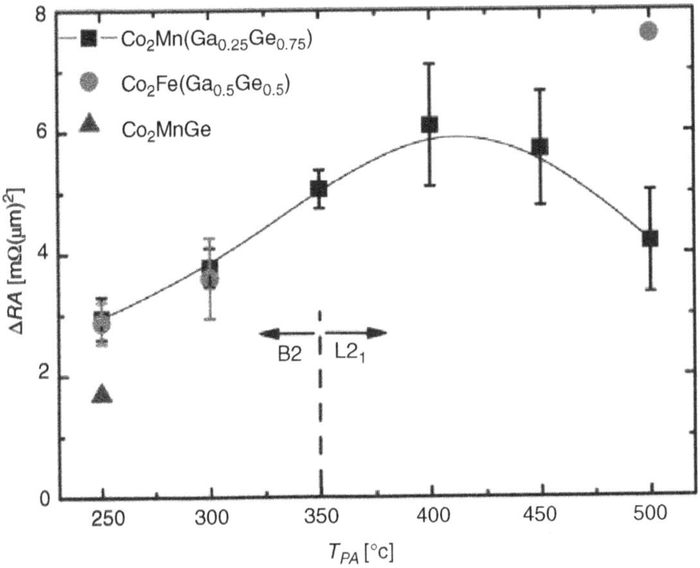

Figure 2.9 *Dependence of ΔRA of $Co_2Mn(Ga_{0.25}Ge_{0.75})$ (squares), several established Heusler alloys ($Co_2Fe(Ga_{0.5}Ge_{0.5})$) (circles), and polycrystalline Co_2MnGe (triangles) on annealing temperature. (Reprinted with permission from Y. K. Takahashi, N. Hase, M. Kodzuka, A. Itoh, T. Koganezawa, T. Furubasyashi, S. Li, B. S. D. Ch. S. Varaprasad, T. Ohkubo, and K. Hono,* J. Appl. Phys. ***113****, 223901 (2013). ©2013 American Institute of Physics)*

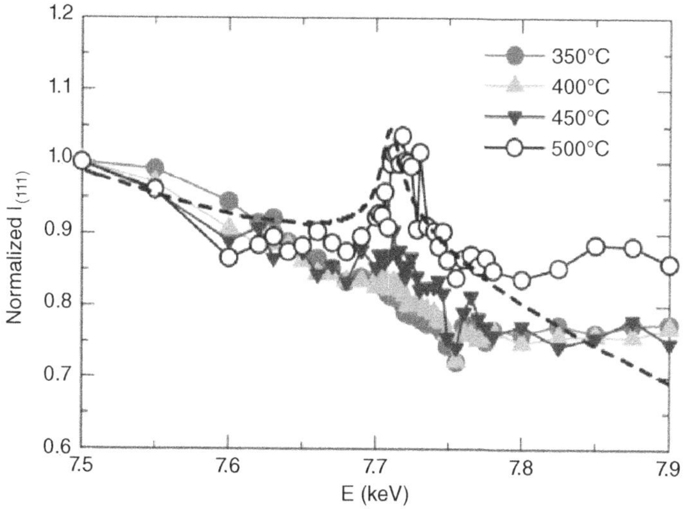

Figure 2.10 Dependence of {111} peak intensity on energy with samples annealed at various annealing temperatures. Dashed line represents calculated curve under assumption of 18% DO_3 disordering with subtraction of empirical offset line. (Reprinted with permission from Y. K. Takahashi, N. Hase, M. Kodzuka, A. Itoh, T. Koganezawa, T. Furubasyashi, S. Li, B. S. D. Ch. S. Varaprasad, T. Ohkubo, and K. Hono, J. Appl. Phys. **113**, 223901 (2013). ©2013 American Institute of Physics)

with TEM and a three-dimensional atom probe (3DAP). The DO_3 disordering creates additional DOS in the half-metallic band gap according to Picozzi *et al.* [20] resulting in decreased spin polarization. Therefore, we investigated the possibility of DO_3 disordering in CMGG film to understand the decrease in ΔRA by annealing at temperatures higher than 400° C.

It is difficult to estimate DO_3 disordering with standard XRD using Cu Kα due to similarities in the atomic scattering factors of Co and Mn. However, we can detect it quantitatively with the anomalous scattering dispersion effect. Figure 2.10 plots the dependence of the {111} peak intensity on energy in 20 nm-thick CMGG films at various annealing temperatures. The peak intensity changes linearly against the energy at 400° C, which indicates that the CMGG film does not contain DO_3 disordering at this annealing temperature. The {111} peak intensity at the energy of the Co edge gradually increases for the samples annealed at temperatures higher than 450° C and a large increase in peak intensity can be observed in the sample annealed at 500° C. Compared to the calculated results plotted by the dashed line, DO_3 disordering is estimated to be about 18%. Therefore, we can conclude that the degraded ΔRA is due to increased DO_3 disordering. Since the film composition was $Co_{52.7}Mn_{25.6}Ga_{6.1}Ge_{15.6}$ in atomic percent and was slightly Co rich, it may have created a Co anti-site, which induced a spin-down in-gap state and caused spin polarization to reduce. The Mn-rich off-stoichiometric composition CMGG film may have led to high ΔRA because the Mn antisite, where the Co site was replaced by an Mn atom, did not induce additional DOS in the spin-down band gap. A high TMR ratio was achieved in the Mn-rich composition in the Co_2MnSi-based MTJ [39].

In addition to CFGG and CMGG, the $Co_2(Fe_{0.4}Mn_{0.6})Si$ (called CFMS after this) reported by Kubota *et al.* is a promising quaternary Heusler alloy for spintronics devices [40]. Recently, Sato *et al.* reported a very large MR of 74.8% at RT in CPP-GMR devices using CFMS [41]. However, the RA value for their CPP-GMR pillar was more than three times larger than the typical value for Heusler/Ag pseudo-spin valves. Hence, there is a possibility that the interface was partially oxidized forming a narrow current path, causing enhanced MR of the current-confined-pass effect. Later, Sakuraba *et al.* carried out systematic investigations into CPP-GMR devices using CFMS and reported ~13 mΩ μm^2 and 58% at RT [42], which are nearly the same values as those reported for CFGG/Ag PSV.

The closed circles in Figure 2.5 plot the MR ratio of Heusler-based CPP-GMR. Although the MR ratio has increased due to the highly spin polarized Heusler alloys that have been newly developed, it is still beyond the usable range in view of the low RA. Further improvements need to be made to ΔRA and it is essential to optimize RA so that the MR ratio is within the usable range. There are several ways of increasing the RA values. The first is by inserting a very thin oxide-based granular layer to confine the electron path. The second is by utilizing all-Heusler CPP-GMR, which is thought to achieve high ΔRA and RA. All Heusler CPP-GMR using $Co_2MnGe/Rh_2CuSn/Co_2MnGe$ was first reported by Nikolaev *et al.* [43]. They obtained a relatively high ΔRA of 4 mΩ μm^2 and an RA of 60 mΩ μm^2. The main advantage of a Heusler spacer is thought to be high γ due to the same crystal symmetry that introduces excellent band matching. Although one concern is the short spin diffusion length of the Heusler alloy spacer due to the large resistivity of non-magnetic (NM) Heusler alloy, large ΔRA and RA would be possible by using a thin NM Heusler spacer. The MR ratio of all Heusler CPP-GMR would increase by using quaternary Heusler alloy with high-spin polarization.

2.3.2 Narrow Read Sensor for High Density Recording

Since the resolution of read sensors depends on the gap in the magnetic shield, the total device thickness should be reduced to 20 nm for 2 Tbit/in^2, which is difficult to achieve in currently used exchange-biased spin valves due to the requirement for a relatively thick antiferromagnetic layer [25]. There have been some proposals to overcome this problem, such as trilayer sensors [44] and lateral spin valve (LSV) devices [45]. The trilayer sensors are composed of two FM layers sandwiching a thin NM spacer whose thickness is controlled so that the two FM layers are antiferromagnetically interlayer exchange coupled. Since no anitiferromagnetic layer for exchange bias is required, the total thickness of the device can be narrowed to ~10 nm. Nakatani *et al.* demonstrated a relatively large ΔRA of 5.6 mΩ μm using antiferromagnetically interlayer exchange coupled $Co_2Fe(Al,Si)$ [46]. However, the LSV device was composed of two FM wires bridged by NM wire, as shown in Figure 2.11. When spin-polarized current was injected from the FM1 wire to the NM wire, spin accumulation occurred near the contact region in the NM wire. This spin accumulation decayed exponentially with the distance from the contact. When another FM wire (FM2) was placed within the spin diffusion length of the NM wire, spin polarization occurred in the electrochemical potential of the FM2 wire. Therefore, a spin accumulation signal in the NM wire could be detected by measuring the voltage between NM and FM2 wires. Since the spin signal depended on the configuration of the two FM wires, the spin detection part worked as a sensor. Figure 2.12 illustrates the proposed read sensor using LSV.

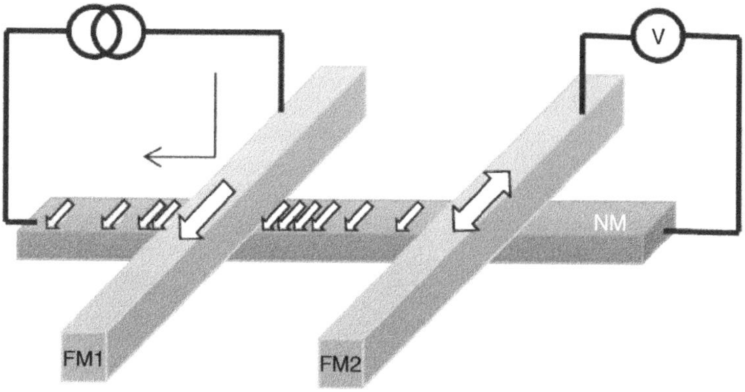

Figure 2.11 *Schematic of lateral spin valve (LSV) device*

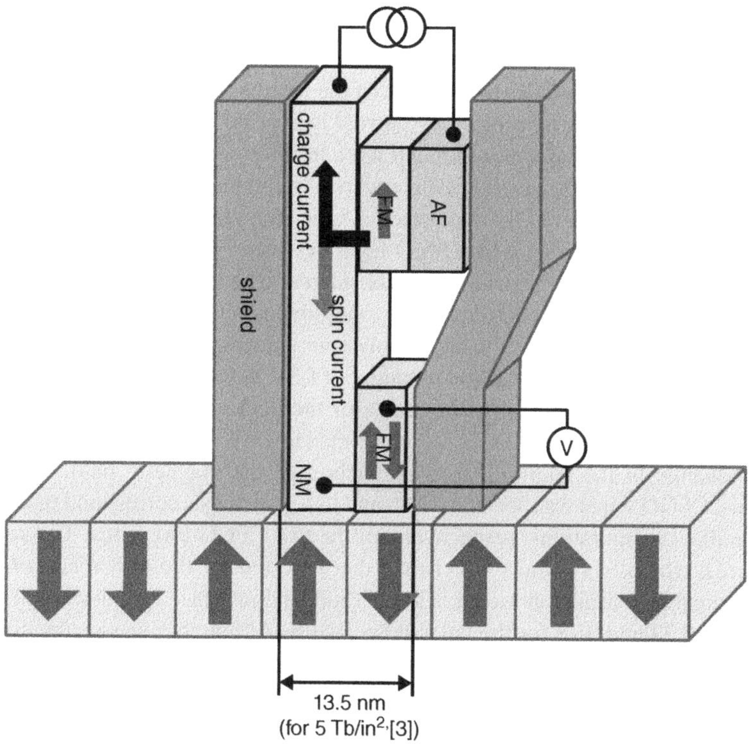

Figure 2.12 *Schematic of read sensor using LSV device. Since spin injection and detection parts are laterally separated, shield-to-shield gap that corresponds to head resolution can be dramatically reduced. It should be less than 13 nm for 5 Tb/in^2 [25]. (Reprinted with permission from Y. K. Takahashi, S. Kasai, S. Hirayama, S. Mitani, and K. Hono,* Appl. Phys. Lett. ***100****, 052405 (2012). ©2012 American Institute of Physics)*

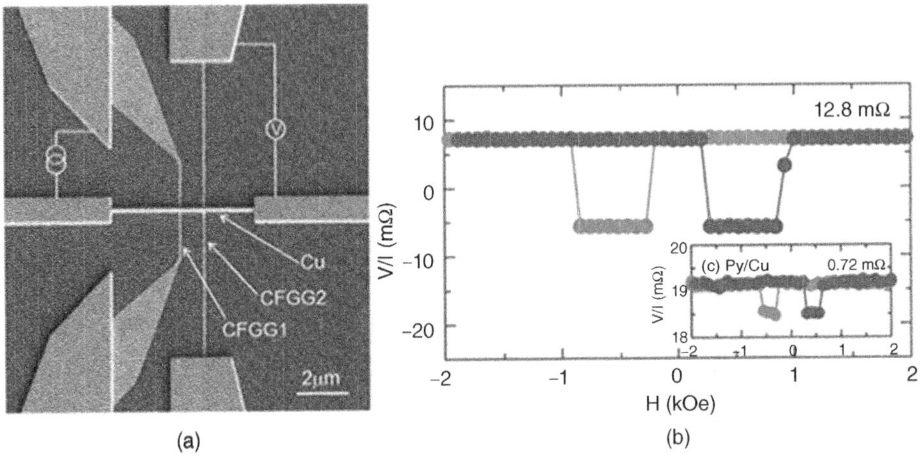

Figure 2.13 *(a) SEM image of typical LSV device using CFGG and Cu. (b) Spin signal in CFGG/Cu LSV devices*

It has an advantage as a narrow reader because the sensor part only has two metallic layers. In addition, one can increase the volume of the antiferromagnetic material in the spin injection part. Also, the noise induced by current is expected to be reduced as no current flows in the sensor part. However, not much work has been done on LSV devices for read sensor applications, since the spin signal is orders of magnitude smaller than that required for read sensors. Fukuma *et al.* [47] reported a large spin signal of 50 mΩ at RT in 2011 that they achieved by inserting a thin MgO layer between FM and NM layers. However, LSV devices with low device resistance are considered to be necessary for the read sensors for high-density recording. According to the one-dimensional spin diffusion model [48], a large spin signal is expected by using highly spin polarized material even in an ohmic system. Therefore, we investigated the potential of LSV devices using CFGG.

Figure 2.13(a) shows a typical SEM image of the device [49]. There are two CFGG wires connected by a Cu wire. CFGG was confirmed to have a B2 structure. Figure 2.13(b) shows the change in the spin signal as a function of the magnetic field. The distance between two CFGG wires was 350 nm. High and low resistances correspond to the parallel and antiparallel configurations, respectively, of the two CFGG FM wires. The spin signal was 12.8 mΩ, which is 17 times larger than that reported for the Py/Cu case. Using the one dimensional spin diffusion model, β and γ (bulk and interface spin asymmetries) were deduced to be 0.73 and 0.28 for the latter, respectively. Although β was in agreement with that estimated by CPP-GMR, γ was too low. The output (ΔV) of an LSV device for 2 Tb/in^2 should be more than 1 mV. However, the current device only had 0.07 mV due to the low RA. For example, the spin signal would be enhanced further by improving γ by using Ag. Since CPP-GMR and LSV are basically the same, the combination of FM and NM materials that give rise to high ΔRA should result in a high-spin signal. However, as the spin signals of LSVs heavily depend on microfabrication processes, we expect further progress toward attaining LSVs with higher spin signals will be made in the near future.

2.4 Summary

We reviewed recent developments in highly spin polarized ferromagnetic Heusler alloys and their device applications. Some of the newly developed quaternary Heusler alloys, e.g., $Co_2Mn(Ge_{0.75}Ga_{0.25})$ and $Co_2Fe(Ge_{0.5}Ga_{0.5})$, were found to be promising FM materials in spintronics devices such as CPP-GMR and LSV. Our search for highly spin polarized materials revealed that the PCAR technique was effective because of its availability and there was no need for device fabrication. The spin polarization that was estimated with PCAR demonstrated excellent correlation with the spin polarization of FM layers that was estimated with CPP-GMR and LSV devices.

Acknowledgements

The authors thank Dr. A. Rajanikanth and Dr. B.S.D.Ch.S. Varaprasad for PCAR measurements, Dr. N. Hase, Dr. T.M. Nakatani, Dr. S. Li, Dr. H.S. Goripati and Dr. T. Furubayashi for the CPP-GMR experiments, Dr. S. Kasai and Dr. S. Mitani for the LSV experiment and Dr. T. Koganezawa for the synchrotron radiation experiment. This work was partially supported by PREST-JST, the Grant-in-Aid for Scientific Research (A) (#22246091), (B)(#20360322) and IDEMA-ASTC.

References

1. M.N. Baibich, J.M. Broto, A. Fert, *et al.*, Giant magnetoresistance of (001)Fe/(001)Cr magnetic superlattices, *Phys. Rev. Lett.* **61**, 2472 (1988).
2. Magnetfeldsensor MIT ferromagnetischer, dünner Schicht, Patent EP0442407A1.
3. B. Dieny, V.S. Speriosu, S.S.P. Parkin, *et al.*, Giant magnetoresistive in soft ferromagnetic multilayers, *Phys. Rev. B* **43**, 1297 (1991).
4. S. Yuasa, T. Nagahama, A. Fukushima, *et al.*, Giant room-temperature magnetoresistance in single-crystal Fe/MgO/Fe magnetic tunnel junctions, *Nature Mater.* **3**, 868 (2004).
5. S.S.P. Parkin, C. Kaiser, A. Panchula, *et al.*, Giant tunneling magnetoresistance at room temperature with MgO (100) tunnel barriers, *Nature Mater.* **3**, 682 (2004).
6. M. Bowen, M. Bibes, A. Barthelemy, *et al.*, Nearly total spin polarization in $La_{2/3}Sr_{1/3}MnO_3$ from tunneling experiments, *Appl. Phys. Lett.* **82**, 233 (2003).
7. Y. Sakuraba, M. Hattori, M. Oogane, *et al.*, Giant tunneling magnetoresistance in $Co_2MnSi/Al{-}O/Co_2MnSi$ magnetic tunnel junctions, *Appl. Phys. Lett.* **88**, 192508 (2006).
8. Y. Miura, K. Nagao, and M. Shirai, Atomic disorder effects on half-metallicity of the full Heusler alloys $Co_2(Cr_{1-x}Fe_x)Al$: A first-principles study, *Phys. Rev. B* **69**, 144413 (2004).
9. G.H. Fecher, B. Balke, S. Ouardi, *et al.*, High energy, high resolution photoelectron spectroscopy of $Co_2Mn_{1-x}Fe_xSi$, *J. Phys. D: Appl. Phys.* **40**, 1576 (2007).

10. T.M. Nakatani, A. Rajanikanth, Z. Gercsi, *et al.*, Structure, amgnetic property and spin polarization of $Co_2FeAl_xSi_{1-x}$ Heusler alloys, *J. Appl. Phys.* **102**, 033916 (2007).
11. I. Galanakis, Apperance of half-metallicity in the quaternary Heusler alloys, *J. Phys.: Condens. Matter* **16**, 3089 (2004).
12. R.J. Soulen Jr., J.M. Byers, M.S. Osofsky, *et al.*, Measuring the spin polarization of a metal with a superconducting point contact, *Science* **282**, 85 (1998).
13. G.J. Strijkers, Y. Ji, F.Y. Yang, *et al.*, Andreev reflections at metal/superconductor point contacts: Measurement and analysis, *Phys. Rev. B* **63** 104510 (2001).
14. B.S.D.C.S. Varaprasad, A. Srinivasan, Y.K. Takahashi, *et al.*, Spin polarization and Gilbert damping of $Co_2Fe(Ga_xGe_{1-x})$ Heusler alloys, *Acta Mater.* **60**, 6257 (2012).
15. B.S.D.C.S. Varaprasad, A. Rajanikanth, Y.K. Takahashi, and K. Hono, Enhanced spin polarization of Co_2MnGe Heusler alloy by substitution of Ga for Ge, *Appl. Phys. Express* **3**, 23002 (2010).
16. Y.K. Takahashi, A. Srinivasan, B. Varaprasad, *et al.*, Large magnetoresistance in current-perpendicular-to-plane pseudo spin valve using $Co_2Fe(Ge_{0.5}Ga_{0.5})$ Heusler alloy, *Appl. Phys. Lett.* **98**, 152501 (2011).
17. I. Galanakis, P.H. Dederichs, and N. Papanikolaou, Slater–Pauling behavior and origin of the half-metallicity of the full-Heusler alloys, *Phys. Rev. B* **66**, 174429 (2002).
18. T. Taira, T. Ishikawa, N. Itabashi, *et al.*, Influence of annealing on spin-dependent tunneling characteristics of fully epitaxial $Co_2MnGe/MgO/Co_{50}Fe_{50}$ magnetic tunnel junctions, *Appl. Phys. Lett.* **94**, 072510 (2009).
19. A. Rajanikanth, Y.K. Takahashi, and K. Hono, Spin polarization of Co_2MnGe and Co_2MnSi thin films with A2 and $L2_1$ structures, *J. Appl. Phys.* **101**, 023901 (2007).
20. S. Picozzi, A. Continenza, and A.J. Freeman, Role of structural defects on the half-metallic character of Co_2MnGe and Co_2MnSi Heusler alloys, *Phys. Rev. B* **69**, 94423 (2004).
21. M. Zhang, E. Bruck, F.R. de Boer, *et al.*, The magnetic and transport properties of the Co_2FeGa Heusler alloy, *J. Phys. D Appl. Phys.* **37**, 2049 (2004).
22. D.D. Djayaprawira, K. Tsunekawa, M. Nagai, *et al.*, 230% room-temperature magnetoresistance in CoFeB/MgO/CoFeB magnetic tunnel junctions, *Appl. Phys. Lett.* **86**, 092502 (2005).
23. S. Ikeda, J. Hayakawa, Y. Ashizawa, *et al.*, Tunnel magnetoresistance of 604% at 300 K by suppression of Ta diffusion in CoFeB/MgO/CoFeB pseudo-spin-valves annealed at high temperature, *Appl. Phys. Lett.* **93**, 082508 (2008).
24. T. Kawami, H.X. Liu, Y. Honda, *et al.*, 37th Annual Meeting of Magntic Society of Japan, 3pC-8 (2013).
25. M. Takagishi, K. Yamada, H. Iwasaki, *et al.*, Magnetoresistance ratio and resistance area design of CPP-MR film for 2–5 Tb/in^2 read sensors, *IEEE Trans. Magn.* **46**, 2086 (2010).
26. A.C. Reilly, W. Park, R. Slater, *et al.*, Perpendicular giant magnetoresistance of $Co_{91}Fe_9/Cu$ exchange-biased spin-valves: Further evidence for a unified picture, *J. Magn. Magn. Mater.* **195**, L269 (1999).
27. M. Covington, M. AlHajDarwich, Y. Ding, *et al.*, Current-induced magnetization dynamics in current perpendicular to the plane spin valves, *Phys. Rev. B* **69**, 184406 (2004).

28. J.R. Childress, M.J. Carey, S.I. Kiselev, *et al.*, Dual current-perpendicular-to-plane giant magnetoresistive sensors for magnetic recording heads with reduced sensitivity to spin-torque-induced noise, *J. Appl. Phys.* **99**, 08S305 (2006).
29. T. Valet and A. Fert, Theory of the perpendicular magnetoresistance in magnetic multilayers, *Phys. Rev. B* **48**, 7099 (1993).
30. J.A. Caballero, A.C. Reilly, Y. Hao, *et al.*, Effect of deposition parameters on the CPP-GMR of NiMnSb-based spin-valve structures, *J. Magn. Magn. Mater.* **198–199**, 55 (1999).
31. H. Hoshiya and K. Hoshino, Current-perpendicular-to-the-plane giant magnetoresistance in structures with half-metal materials laminated between CoFe layers, *J. Appl. Phys.* **95**, 6774 (2004).
32. K. Yakushiji, K. Saito, S. Mitani, *et al.*, Current-perpendicular-to-plane magnetoresistance in epitaxial $Co_2MnSi/Cr/Co_2MnSi$ trilayers, *Appl. Phys. Lett.* **88**, 222504 (2006).
33. T. Furubayashi, K. Kodama, H. Sukegawa, *et al.*, Current-perpendicular-to-plane giant magnetoresistance in spin-valve structures using epitaxial $Co_2FeAl_{0.5}Si_{0.5}/Ag/Co_2FeAl_{0.5}Si_{0.5}$ trilayers, *Appl. Phys. Lett.* **93**, 122507 (2008).
34. H.S. Goripati, T. Furubayashi, Y.K. Takahashi, and K. Hono, Current-perpendicular-to-plane giant magnetoresistance using $Co_2Fe(Ga_{1-x}Ge_x)$ Heusler alloy, *J. Appl. Phys.* **113**, 0439001 (2013).
35. S. Li, Y.K. Takahashi, T. Furubayashi, and K. Hono, Enhancement of giant magnetoresistance by $L2_1$ ordering in $Co_2Fe(Ge_{0.5}Ga_{0.5})$ Heusler alloy current-perpendicular-to-plane pseudo spin valves, *Appl. Phys. Lett.* **103**, 042405 (2013).
36. A. Sakuma, Y. Toga, and H. Tsuchiura, Theoretical study on the stability of magnetic structures in the surface and interfaces of Heusler alloys, Co_2MnAl and Co_2MnSi, *J. Appl. Phys.* **105**, 07C910 (2009).
37. T. Ishikawa, N. Itabashi, T. Taira, *et al.*, Critical role of interface states for spin-dependent tunneling in half-metallic Co_2MnSi-based magnetic tunnel junctions investigated by tunneling spectroscopy, *Appl. Phys. Lett.* **94**, 092503 (2009).
38. Y.K. Takahashi, N. Hase, M. Kodzuka, *et al.*, Structure and magnetoresistance of surrent-perpendicular-to-plane pseudo spin valves using $Co_2Mn(Ga_{0.25}Ge_{0.75})$ Heusler alloy, *J. Appl. Phys.* **113**, 223901 (2013).
39. M. Yamamoto, T. Ishikawa, T. Taira, *et al.*, Effect of defects in Heusler alloy thin films on spin-dependent tunneling characteristics of $Co_2MnSi/MgO/Co_2MnSi$ and $Co_2MnGe/MgO/Co_2MnGe$ magnetic tunnel junctions, *J. Phys: Condens Mater.* **22**, 164212 (2010).
40. T. Kubota, S. Tsunegi, M. Oogane, *et al.*, Half-metallicity and Gilbert damping constant in $Co_2Fe_xMn_{1-x}Si$ Heusler alloys depending on the film composition, *Appl. Phys. Lett.* **94**, 122504 (2009).
41. J. Sato, M. Oogane, H. Naganuma, and Y. Ando, Large magnetoresistance effect in epitaxial $Co_2Fe_{0.4}Mn_{0.6}Si/Ag/Co_2Fe_{0.4}Mn_{0.6}Si$ devices, *Appl. Phys. Express* **4**, 113005 (2011).
42. Y. Sakuraba, M. Ueda, Y. Miura, *et al.*, Extensive study of giant magnetoresistance properties in half-metallic Co_2(Fe,Mn)Si-based devices, *Appl. Phys. Lett.* **101**, 252408 (2012).

43. K. Nikolaev, P. Kolbo, T. Pokhil, *et al.*, "All-Heusler alloy" current-perpendicular-to-plane giant magnetoresistance, *Appl. Phys. Lett.* **94**, 222501 (2009).
44. R. Lamberton, M. Seigler, K. Pelhos, *et al.*, Current-in-plane GMR trilayer head design for hard-disk-drives: Characterization and extendibility, *IEEE Trans. Magn.* **43**, 645 (2007).
45. T.D. Boone, Jr., B.A. Gurney, and N. Smith, U.S. patent US2010/0296202, Nov.25, 2010. K. Ito and H. Takahashi, Japanese Patent Publication No. 2004-342241. Dec.2, 2004.
46. T.M. Nakatani, S. Mitani, T. Furubayashi, and K. Hono, Oscillatory antiferromagnetic interlayer exchange coupling in $Co_2Fe(Al_{0.5}Si_{0.5})/Ag/Co_2Fe(Al_{0.5}Si_{0.5})$ films and its application to trilayer magnetoresistive sensor, *Appl. Phys. Lett.* **99**, 182505 (2011).
47. Y. Fukuma, L. Wang, H. Idzuchi, *et al.*, Giant enhancement of spin accumulation and long-distance spin precession in metallic lateral spin valves, *Nature Mater.* **10**, 527 (2011).
48. S. Takahashi and S. Maekawa, Spin injection and detection in magnetic nanostructures, *Phys. Rev. B* **67**, 052409 (2003).
49. Y.K. Takahashi, S. Kasai, S. Hirayama, *et al.*, All-metallic laterial spin valves using $Co_2Fe(Ge_{0.5}Ga_{0.5})$ Heusler alloy with a large spin signal, *Appl. Phys. Lett.* **100**, 052405 (2012).
50. H. Maehara, K. Nishimura, Y. Nagamine, *et al.*, Tunnel magnetoresistance above 170% and resistance-area product of $1\Omega(\mu m)^2$ attained by in situ annealing of ultra-thin MgO tunnel barrier, *App. Phys. Express* **4**, 033002 (2011).
51. T. Iwase, Y. Sakuraba, S. Bosu, *et al.*, Large interface spin-asymmetry and magnetoresistance in fully epitaxial $Co_2MnSi/Ag/Co_2MnSi$ current-perpendicular-to-plane magnetoresistive devices, *Appl. Phys. Express* **2**, 063003 (2009).
52. T. M. Nakatani, N. Hase, H. S. Goripati, *et al.*, Co-based Heusler alloys for APP-GMR spin-valves with large magnetoresistive outputs, *IEEE Trans. Magn.* **48**, 1751 (2012).
53. T. M. Nakatani, T. Furubayashi, S. Kasai, *et al.*, Bulk and interfacial scatterings in current-perpendicular-to-plane giant magnetoresistance with $Co_2Fe(Al_{0.5}Si_{0.5})$ Heusler alloy layers and Ag spacer, *Appl. Phys. Lett.* **96**, 212501 (2010).
54. A. Jogo, K. Nagasaka, T. Ibusuki, *et al.*, Current-perpendicular spin valves with high-resistivity ferromagnetic metals for ultrahigh-density magnetic recording, *J. Magn. Magn Mater.* **309**, 80 (2007).
55. S. Maat, M. J. Carey, and J. R. Childress, Magnetotransport properties and spin-torque effects in current perpendicular to the plane spin valves with Co–Fe–Al magnetic layers, *J. Appl. Phys.* **101**, 093905 (2007).
56. S. Maat, M. J. Carey, and J. R. Childress, Current perpendicular to the plane spin-valves with CoFeGe magnetic layers, *Appl. Phys. Lett.* **93**, 143505 (2008).
57. H. Yuasa, M. Hara, S. Murakami, *et al.*, Enhancement of magnetoresistance by hydrogen ion treatment for current-perpendicular-to-plane giant magnetoresistive films with a current-confined-path nano-oxide layer, *Appl. Phys. Lett.* **97**, 112501 (2010).
58. K. Nakamoto, H. Hoshiya, H. Katada, *et al.*, CPP-GMR heads with a current screen layer for 300 Gb/in^2 recording, *IEEE Trans. Magn.* **44**, 95 (2008).
59. H. Fukuzawa, H. Yuasa, and H. Iwasaki, CPP-GMR films with a current-confined-path nano-oxide layer (CCP-NOL), *J. Phys. D Appl. Phys.* **40**, 1213 (2007).

60. K. Shimazawa, Y. Tsuchiya, T. Mizuno, *et al.*, CPP-GMR film with Zn-O-based novel spacer for future high-density magnetic recording, *IEEE Trans. Magn.* **46**, 1487 (2010).
61. Y. Fuji, M. Hara, H. Yuasa, *et al.*, Enhancement of magnetoresistance by ultra-thin Zn wüstite layer, *Appl. Phys. Lett.* **99**, 132103 (2011).
62. M. Takagishi, H. Iwasaki, S. Hashimoto, and H. NiuFuke, High Sensitivity and High Resolution Reader Design for over 2 Tbpsi, TMRC11 (August 29–31, 2011, Minneapolis).

3
Spin Current

Ken-ichi Uchida[1] *and Eiji Saitoh*[1,2]
[1]*Institute for Materials Research, Tohoku University, Sendai, Japan*
[2]*WPI Advanced Institute for Materials Research, Tohoku University, Sendai, Japan*

In this chapter, we introduce the basic concept of spin current. We begin with an introduction to the general concept of spin and spin current, which is followed by a discussion of particular spin currents, such as incoherent, exchange, topological, and thermal spin currents.

3.1 Introduction

Beside charge, an electron has internal angular momentum. This internal angular momentum, similar to the rotation of a classical particle, is named spin. Spin is the dominant origin of magnetism, thus, when spins of electrons in a solid are aligned to some extent in the same direction, the solid becomes a magnet.

A flow of electron charge is a charge current, or an electric current. The physics of charge current has been developed in the previous century and is now an essential contributor to our understanding of electronics. Since an electron carries both charge and spin, the existence of a charge current naturally implies the existence of a flow of spin. This flow is called a spin current.

Experiments carried out in the previous century did not focus on spin currents because of its relatively short decay timescale τ. However, the rapid progress in nanofabrication technology in this century has allowed researchers to access spin currents. From the theoretical point of view, the detailed formulation of spin currents is not simple and is still a challenging undertaking. Nevertheless, a spin current is a very useful and versatile concept; it has given birth to many phenomena in condensed matter science and spintronics.

Spintronics for Next Generation Innovative Devices, First Edition. Edited by Katsuaki Sato and Eiji Saitoh.
© 2015 John Wiley & Sons, Ltd. Published 2015 by John Wiley & Sons, Ltd.

3.2 Concept of Spin Current

In this section, we introduce a concept of spin cursrent. Before dealing with spin currents, we do a quick review of the definition of charge currents for comparison.

A charge current is defined in terms of the charge conservation law. Consider a region enclosed by a closed surface Ω. When the total charge Q within this region is increasing, the increase is due to the charge flowing into this region across the surface Ω, owing to the charge conservation law. This flow of charge is described by the equation

$$\iiint_V \dot{\rho} d\boldsymbol{r} = -\iint_\Omega \boldsymbol{j}_c \cdot \mathrm{d}\Omega, \tag{3.1}$$

where ρ is charge density and $\boldsymbol{j}_c$ is charge current density. The left-hand side of this equation is the increase in charge in the volume surrounded by the surface Ω. By applying Gauss's theorem to this equation, we obtain

$$\dot{\rho} = -\,\mathrm{div}\boldsymbol{j}_c. \tag{3.2}$$

This equation called the continuity equation of charge, which is a representation of the charge conservation law, defines a charge current density [2].

We are now in a position to consider a spin current (see Figure 3.1). Spin current density $\mathbf{j}_s$ is introduced similarly in terms of spin angular momentum conservation. If spin angular momentum is fully conserved, the continuity equation for spin angular momentum can be written as

$$\frac{\mathrm{d}\boldsymbol{M}}{\mathrm{d}t} = -\,\mathrm{div}\boldsymbol{j}_s, \tag{3.3}$$

and the spin current density is defined via this equation. $\boldsymbol{M}$ denotes local magnetization (magnetic-moment density). Since a spin current has two orientations, the spatial flowing orientation and the spin orientation, the expectation value of spin current density is not a vector but a second-rank tensor.

In practice, in nanoscale in solids, this angular momentum conservation is often a good approximation. However, in general, spin is not conserved completely due to the spin

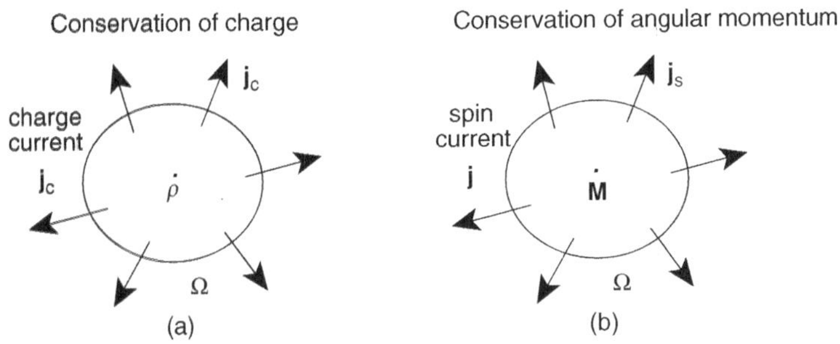

Figure 3.1 The sum of the charge variations over the volume surrounded by a surface is equal to the sum of the currents penetrating the surface

relaxation, and thus it obeys the following modified equation of continuity:

$$\frac{d\boldsymbol{M}}{dt} = -\operatorname{div}\boldsymbol{j}_s + \boldsymbol{T}. \tag{3.4}$$

The last term $\boldsymbol{T}$ represents the nonconservation of spin angular momentum, namely, the relaxation and generation of spin angular momentum. As shown later, in fact, $\boldsymbol{T}$ can be calculated if system Lagrangian is fully given. However, if not, this term should be treated phenomenologically. The simplest phenomenological model for the relaxation is

$$\boldsymbol{T} = -(\boldsymbol{M} - \boldsymbol{M}_0)0\tau, \tag{3.5}$$

which is called the single pole relaxation. τ is a decay time constant and $(\boldsymbol{M} - \boldsymbol{M}_0)$ is nonequilibrium magnetization measured from its equilibrium value $\boldsymbol{M}_0$.

3.3 An Exact Definition of Spin Current

In some case, when a proper Lagrangian is given concretely, a spin current can be exactly defined in terms of the conservation law. In this section, we show an example of such a definition. We start with reviewing a quantum mechanical definition of charge current.[1]

3.3.1 Microscopic Description of Conduction Electrons

For free electrons, the Hamiltonian described using field operators becomes

$$\mathcal{H} = \int \sum_{\sigma} \left[\frac{\hbar^2}{2m} |\nabla c_\sigma(\boldsymbol{r})|^2 - \mu c_\sigma^\dagger(\boldsymbol{r}) c_\sigma(\boldsymbol{r}) \right] d\boldsymbol{r}, \tag{3.6}$$

using the electron mass m. Electrons are fermions, therefore c and $c^\dagger$ anticommute. σ is an index to show the two spin states. μ is the chemical potential. Fourier transformation of this Hamiltonian yields

$$\mathcal{H} = \sum_{\boldsymbol{k},\sigma} \left(\frac{\hbar^2 k^2}{2m} - \mu \right) c^\dagger_{\boldsymbol{k},\sigma}(\boldsymbol{r}) c_{\boldsymbol{k},\sigma}(\boldsymbol{r}). \tag{3.7}$$

For the following discussion, the Lagrangian formalism is more convenient. The Lagrangian of the system of electrons defined with the operators $c^\dagger$ and c can be written using the Hamiltonian $\mathcal{H}$ as

$$L = i\hbar \int \sum_{\sigma} c_\sigma^\dagger \frac{\partial c_\sigma}{\partial t} - \mathcal{H}\ d\boldsymbol{r}. \tag{3.8}$$

[1] This section draws heavily from "Basics of spintronics" by. G. Tatara (Baifukan, Co., Ltd., Japan, 2009).

3.3.2 Conservation of Charge

First, we show that the charge conservation law is related to the rotational symmetry in phase factor of wave function: the U(1) gauge symmetry. Important physical quantities in electron systems are the charge current density $\boldsymbol{j}_c$ and the charge density. As is shown, these quantities satisfy the charge conservation rule, or the continuity equation of charge. Let us confirm this using the U(1) symmetry. For the Lagrangian of free electrons

$$L = \int i\hbar c^{\dagger}\frac{\partial c}{\partial t} - \frac{\hbar^2}{2m}\nabla c^{\dagger}\nabla c \ d\boldsymbol{r}, \tag{3.9}$$

we consider the phase transformation of the electron field

$$\begin{aligned} c(\boldsymbol{r},\ t) &\rightarrow e^{i\varphi(\boldsymbol{r},\ t)}c(\boldsymbol{r},\ t), \\ c^{\dagger}(\boldsymbol{r},\ t) &\rightarrow e^{-i\varphi(\boldsymbol{r},\ t)}c^{\dagger}(\boldsymbol{r},\ t). \end{aligned} \tag{3.10}$$

The scalar quantity $\varphi(\boldsymbol{r}, t)$ is the phase which is dependent on spacetime. The derivative of the electron field is converted into

$$\frac{\partial c}{\partial x_\mu} \rightarrow e^{i\varphi}\left(\frac{\partial}{\partial x_\mu} + i\frac{\partial \varphi}{\partial x_\mu}\right)c \tag{3.11}$$

by the phase transformation. In turn, the Lagrangian is converted into

$$L = \int i\hbar c^{\dagger}\left(\frac{\partial}{\partial t} + i\frac{\partial}{\partial t}\varphi\right)c - \frac{\hbar^2}{2m}(\nabla - i\nabla\varphi)c^{\dagger}(\nabla + i\nabla\varphi)c \ d\boldsymbol{r}. \tag{3.12}$$

If the phase φ is a single-valued function of spacetime and is differentiable, physical phenomena must be invariant with phase transformation. Therefore, the Lagrangian must be invariant, thus the first-order term of φ in Equation (3.12) must be zero. Using the variation of the action, or time integral of the Lagrangian $I \equiv \int_{-\infty}^{\infty} L dt$, integration by parts results in

$$\delta I = \int_{-\infty}^{\infty}\int \hbar\left[\frac{\partial\left(c^{\dagger}c\right)}{\partial t} - \frac{i\hbar}{2m}\,\mathrm{div}(c^{\dagger}\overleftrightarrow{\nabla}c)\right]\varphi \ d\boldsymbol{r}dt = 0. \tag{3.13}$$

In other words, the charge conservation law

$$\frac{\partial \rho}{\partial t} + \mathrm{div}\boldsymbol{j}^{(0)} = 0 \tag{3.14}$$

is obtained. Here,

$$\rho \equiv e\langle c^{\dagger}c\rangle, \ j_i^{(0)} \equiv e\langle -i\frac{\hbar}{2m}c^{\dagger}\overleftrightarrow{\nabla}c\rangle \tag{3.15}$$

are the charge density and current density, respectively. Equation (3.11) is called a covariant differential. A U(1) gauge field is defined as $A_\mu \equiv \partial_\mu\varphi$ and the Equation (3.12) is the Lagrangian of electrons interacting with the gauge field. However, the quantity

$F_{\mu\nu} \equiv \partial_\mu A_\nu - \partial_\nu A_\mu$ corresponding to the physical field satisfies $F_{\mu\nu} = 0$ if φ is single-valued and is differentiable. Clearly this reflects the fact that continuous transform of phase does not change the phenomena. In contrast, if there is an electromagnetic field, φ is multivalued or is not differentiable. In this case $F_{\mu\nu}$ becomes a finite nonzero value because the partial differential of φ depends upon the order of differential. The following discussion will consider situations where magnetic or electric fields exist, and the U(1) gauge field A_{em} corresponding to the electromagnetic field is separated from the differentiable part of the phase degree of freedom φ. The Lagrangian of free electrons including the electromagnetic field is

$$L_{\mathrm{em}} = \int i\hbar c^\dagger \frac{\partial c}{\partial t} - e\phi c^\dagger c - \frac{\hbar^2}{2m}\left(\nabla + i\frac{e}{\hbar}\boldsymbol{A}_{\mathrm{em}}\right) c^\dagger \left(\nabla - i\frac{e}{\hbar}\boldsymbol{A}_{\mathrm{em}}\right) c \; d\boldsymbol{r}, \tag{3.16}$$

where $\phi \equiv \hbar A_{em\,t}$ and $A_{em\,i}$ are the scalar potential and vector potential, respectively.

The continuity equation of a charge current is also derived in the case when there are spin-orbit interaction and an external electromagnetic field. The Lagrangian becomes

$$L_{\mathrm{em},so} = \int i\hbar c^\dagger \frac{\partial c}{\partial t} - e\phi c^\dagger c - \frac{\hbar^2}{2m}\left(\nabla + i\frac{e}{\hbar}\boldsymbol{A}_{\mathrm{em}}\right) c^\dagger \left(\nabla - i\frac{e}{\hbar}\boldsymbol{A}_{\mathrm{em}}\right) c \; d\boldsymbol{r} - \mathcal{H}_{\mathrm{so}}. \tag{3.17}$$

$$\begin{aligned} \mathcal{H}_{so} &= - i\frac{e\hbar^2}{4m^2c^2} \int c^\dagger \left\{ \nabla\phi_{\mathrm{so}} \cdot \left[\left(\nabla - \frac{e}{\hbar}\boldsymbol{A}_{em}\right) \times \sigma \right] \right\} c \; d\boldsymbol{r} \\ &= - i\lambda_{\mathrm{so}} \int c^\dagger \left\{ \nabla\phi_{\mathrm{so}} \cdot \left[\left(\nabla - \frac{e}{\hbar}\boldsymbol{A}_{em}\right) \times \sigma \right] \right\} c \; d\boldsymbol{r}. \end{aligned} \tag{3.18}$$

Requirement of invariance under phase transformation of the electron field 3.10 analogous to the derivation of Equation (3.14) for free electrons yields

$$\frac{\partial \rho}{\partial t} + \mathrm{div}\boldsymbol{j}_c = 0. \tag{3.19}$$

Here, ρ is given by Equation (3.15) as in the case of free electrons. The current density is

$$j_i \equiv e\left\langle -i\frac{\hbar}{2m} c^\dagger \overleftrightarrow{\nabla}_i c - \frac{e}{m} A_{em\;i} c^\dagger c - \frac{\lambda_{so}}{\hbar} \sum_{jk} \varepsilon_{ijk} \nabla_j \phi_{so} c^\dagger \sigma_k c \right\rangle. \tag{3.20}$$

3.3.3 Conservation of Spin and Spin Current

The law of the spin conservation is derived by looking at the change in $L_{em,so}$ under the rotation in the spin space (SU(2) rotation)

$$\begin{aligned} c(\boldsymbol{r}, t) &\to e^{i\varphi(\boldsymbol{r},\, t)\cdot\sigma} c(\boldsymbol{r}, t), \\ c^\dagger(\boldsymbol{r}, t) &\to c^\dagger(\boldsymbol{r}, t) e^{-i\varphi(\boldsymbol{r},\, t)\cdot\sigma}, \end{aligned} \tag{3.21}$$

where φ is a three-component vector. In the absence of spin-orbit interaction, the free electron part L_{em} has a similar form as Equation (3.19) as

$$\frac{\partial \rho_s^{(0)\alpha}}{\partial t} + div\boldsymbol{j}_s^{(0)\alpha} = 0, \tag{3.22}$$

$$\hat{j}_{s_i}^{(0)\alpha} \equiv e\left(-i\frac{\hbar}{2m}c^\dagger \overleftrightarrow{\nabla} i\, \sigma_\alpha c - \frac{e}{m}A_{\mathrm{em}}c^\dagger \sigma_\alpha c\right), \tag{3.23}$$

$$\rho_s^{(0)\alpha} \equiv e\langle c^\dagger \sigma_\alpha c\rangle. \tag{3.24}$$

This is a continuity equation of spin, where the spin current is defined as $\boldsymbol{j}_s^{\alpha(0)}$. This equation represents the spin angular momentum conservation under the Lagrangian 3.16, which is due to the action, the integral of the Lagrangian, $I_{em} = \int_{-\infty}^{\infty} dtL_{em}$, is unchanged by the rotational transformation in the spin sector 3: the spin rotational symmetry.

However, in the presence of the spin-orbit interaction term, total spin is not conserved because the interaction breaks the spin rotational symmetry. In fact, the variation of the spin-orbit interaction is

$$\delta \mathcal{H}_{\mathrm{so}} = \frac{\hbar}{e}\int \sum_\alpha \varphi_\alpha (div\delta \boldsymbol{j}_{\mathrm{s}}^\alpha - T^\alpha)d\boldsymbol{r}. \tag{3.25}$$

There is a term $\boldsymbol{T}$ that cannot be written as a divergence. Here,

$$\delta \hat{j}_{s_i}^\alpha \equiv -\frac{e}{\hbar}\lambda_{\mathrm{so}}\sum_j \varepsilon_{ij\alpha}(\nabla_j \phi_{\mathrm{so}})(c^\dagger c) \tag{3.26}$$

and

$$T^\alpha \equiv \frac{2m}{\hbar^2}\lambda_{\mathrm{so}}\sum_{ijk\beta}\varepsilon_{ijk}\varepsilon_{\alpha\beta k}\langle(\nabla_i\phi_{\mathrm{so}})\hat{j}_{s_j}^{(0)\beta}\rangle \tag{3.27}$$

Then, requirement of invariance under spin rotation of the electron filed analogous to the derivation of Equation (3.19) yields

$$\frac{\partial \rho_s^\alpha}{\partial t} + div\boldsymbol{j}_{\mathrm{s}}^\alpha = T^\alpha. \tag{3.28}$$

Here, the density of the total spin current is

$$j_{\mathrm{s}i}^\alpha \equiv e\langle \hat{j}_{s_i}^{(0)\alpha} + \delta \hat{j}_{s_i}^\alpha\rangle. \tag{3.29}$$

The spin-orbit interaction adds a correction δj_{si}^α to the spin current density, resulting in a term $\boldsymbol{T}$ corresponding to nonconservation of spin. This term corresponds to a source or sink of spin. In the presence of the spin-orbit interaction, the spin conservation is only an approximation because of this effect. This nonconservation of spin is due to the angular momentum transfer of spin into the orbit and eventually into macroscopic degrees of freedom such as the lattice system.

3.4 Incoherent Spin Current

3.4.1 Fermi-Dirac Distribution [3, 4, 5]

The thermal equilibrium properties of a free N-electron system at temperature T is calculated using the distribution function or occupation probability $f(E,T)$. We consider an N-electron system with one-electron energy levels E_i. The degeneracy of the levels g_i and their occupation number n_i must satisfy $n_i \leq g_i$ because of the Pauli exclusion principle. In the equilibrium condition, the free energy F of the total system must be minimized with respect to a variation in the relative occupation numbers of the levels as

$$\delta F = \sum_i \frac{\partial F}{\partial n_i} \delta n_i = 0, \tag{3.30}$$

with the conservation of the electron number:

$$\sum_i \delta n_i = 0. \tag{3.31}$$

For a simple case of exchange of electrons between two levels, k and l, Equations (3.30) and (3.31) become $(\partial F/\partial n_k)\delta n_k + (\partial F/\partial n_l)\delta n_l = 0$ and $\delta n_k + \delta n_l = 0$, respectively. These conditions are satisfied only when $\partial F/\partial n_k = \partial F/\partial n_l$. As shown in this simple example, in the equilibrium condition, all $\partial F/\partial n_i$ must be equal. Thus, we define the chemical potential μ^c as

$$\frac{\partial F}{\partial n_i} = \mu^c. \tag{3.32}$$

The free energy discussed above is expressed as $F = U - TS$ with the internal energy $U = \sum_i n_i E_i$ and the entropy $S = k_B \ln\ W$. W is the number of ways of distributing the N electrons among the states. Since electrons cannot be distinguished from each other, ways of distributing n_i electrons in the energy level E_i is $g_i!/[n_i!(g_i - n_i)!]$. Thus, the number of ways W of the total system is

$$W = \prod_i \frac{g_i!}{n_i!(g_i - n_i)!}, \tag{3.33}$$

which yields the entropy as

$$S = k_B \sum_i [\ln\ g_i! - \ln\ n_i! - \ln\ (g_i - n_i)!]. \tag{3.34}$$

For large n, the factorials can be replaced by using Stirling's approximate formula, $\ln\ n! \approx n \ln\ n - n$, and one finds the chemical potential as

$$\mu^c = \frac{\partial F}{\partial n_i} = E_i + k_B T \ln\ \frac{n_i}{g_i - n_i}, \tag{3.35}$$

or the occupation number n_i:

$$n_i = g_i \frac{1}{e^{(E_i-\mu^c)/k_BT} + 1}. \tag{3.36}$$

Here, the probability that a quantum mechanical state is occupied is given by

$$f(E,T) = \frac{1}{e^{(E-\mu^c)/k_BT} + 1}, \tag{3.37}$$

which is known as the Fermi-Dirac distribution function.

The behavior of the Fermi-Dirac function $f(E,T)$ is shown in Figure 3.2. At $T = 0, f(E, T = 0)$ becomes the step function; at $T = 0$, all the states with energy lower than μ^c are occupied and all the states with energy higher than μ^c are empty. Thus at $T = 0$, the chemical potential is equal to the Fermi energy E_F:

$$\mu^c(T = 0) = E_F, \tag{3.38}$$

where E_F is defined as the energy of the highest occupied quantum state in a system at $T = 0$. At finite temperature T, some of the electrons are excited to the states above E_F and $f(E,T)$ deviates from the step function only in the thermal energy range of the order k_BT around $\mu^c(T)$.

The chemical potential μ^c is temperature dependent; μ^c is determined by the constraint that the total number of electrons must remain constant. Using the Fermi-Dirac function in Equation (3.37), one can calculate the particle number density as

$$n = \int_{-\infty}^{\infty} N(E)f(E)dE, \tag{3.39}$$

where $N(E)$ is the density of states and $N(E)dE$ is the number of electrons per unit volume of r-space with energies between E and $E + dE$. At almost all temperatures of interest in

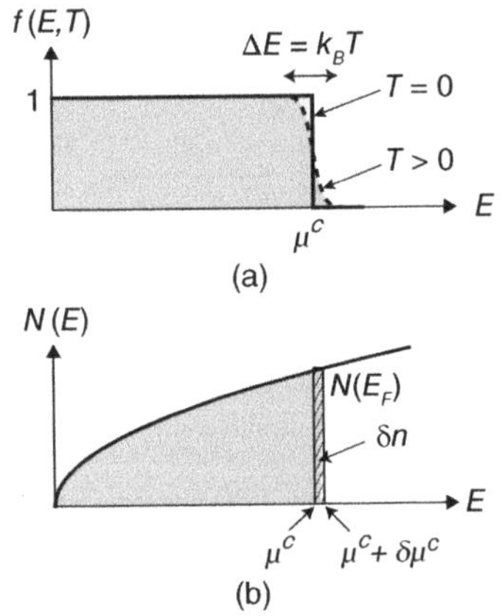

Figure 3.2 (a) The Fermi-Dirac function $f(E,T)$ at $T = 0$ (solid line) and $T > 0$ (dotted curve). The two curves differ only in a region of order k_BT around μ^c. At $T > 0$, some electrons just below E_F have been excited to levels just above E_F. (b) The density of state $N(E)$ for free electrons

metals, T is much less than the Fermi temperature $T_F = E_F/k_B : T \ll T_F$. Since $f(E,T)$ at temperature T differs from $f(E, T = 0)$ only in a region of the order k_BT around μ^c, the difference between $\int_{-\infty}^{\infty} N(E)f(E)dE$ and their zero temperature values, $\int_{-\infty}^{E_F} N(E)dE$, is entirely determined by the form of $N(E)$ near $E = \mu^c$. If $N(E)$ does not vary rapidly in the energy range of the order of k_BT around $E = \mu^c$, Equation (3.39) can be approximated using the Taylor expansion of $N(E)$ around $E = \mu^c$ and then we can use the Sommerfeld expansion:

$$\int_{-\infty}^{\infty} H(E)f(E)dE = \int_{-\infty}^{\mu_c} H(E)dE + \frac{\pi^2}{6}(k_BT)^2 \left.\frac{dH(E)}{dE}\right|_{E=\mu_c} + O(T^4), \tag{3.40}$$

By expanding the upper limit of the integral about E_F, and keeping terms up to second order in T, we have

$$n = \int_0^{E_F} N(E)dE + (\mu^c - E_F)N(E_F) + \frac{\pi^2}{6}(k_BT)^2 \left.\frac{dN(E)}{dE}\right|_{E=E_F}. \tag{3.41}$$

Since the first term is equal to n, we obtain

$$\mu^c = E_F - \frac{\pi^2}{6}(k_BT)^2 \frac{d}{dE} \ln N(E)\Big|_{E=E_F}. \tag{3.42}$$

For free electrons, $N(E) \propto E^{1/2}$, this becomes

$$\mu^c = E_F\left[1 - \frac{1}{3}\left(\frac{\pi k_BT}{2E_F}\right)^2\right] = E_F\left[1 - \frac{\pi^2}{12}\left(\frac{T}{T_F}\right)^2\right]. \tag{3.43}$$

This relation gives μ^c as a function of T and shows that the temperature dependence of μ^c is small for the free-electron gas because of $T \ll T_F$.

As described in Equation (3.32), the chemical potential is the energy necessary to add a particle to the system. When the particle density n becomes $n + \delta n$, the chemical potential changes from μ^c to $\mu^c + \delta\mu^c$. Assuming $N(E) \approx N(E_F)$ around $E = \mu^c$ and using Equation (3.41), we have

$$\delta\mu^c = \frac{\delta n}{N(E_F)}. \tag{3.44}$$

This relation can be seen in Figure 3.2(b).

3.4.2 Diffusion Equation

When electron density is nonuniform, the gradient of the electron density drives a current called a diffusive current. For simplicity, we consider one-dimensional electron diffusion as shown in Figure 3.3(a). Let the electron density at position x at time t be $n(t, x)$. Here, we assume that, at each step Δt, each electron at position x (1) moves to the right by a step a with probability p, (2) moves to the left by a step a with probability p, and (3) stay on x with probability $1 - 2p$. At time t, there are $a \cdot n(t, x)$ electrons in the region from $x - a/2$ to $x + a/2$. After the time evolution of $\Delta t, pa \cdot n(t, x)$ electrons move to the right and $pa \cdot n(t, x)$ electrons move to the left, respectively. Similarly, $pa \cdot n(t, x + a)$ electrons

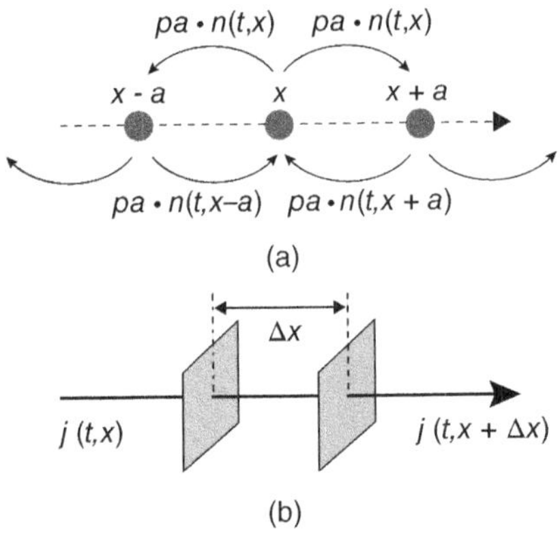

Figure 3.3 *(a) Electron diffusion in one-dimensional system. Each electron at position x moves to neighboring sites by a step a with probability p or stay on with probability 1 – 2p. (b) The flow of flux j of n*

move to x from the right and $pa \cdot n(t, x - a)$ electrons move to x from the left. Thus we have

$$a \cdot n(t + \Delta t, x) = a \cdot n(t, x) - 2pa \cdot n(x, t) + pa \cdot n(t, x - a) + pa \cdot n(t, x + a). \qquad (3.45)$$

Now, we assume that the step a is very small compared to the scales on which n varies. We then have a Taylor expansion

$$n(t, x \pm a) \simeq n(t, x) \pm a\frac{\partial n(t, x)}{\partial x} + \frac{a^2}{2}\frac{\partial^2 n(t, x)}{\partial x^2}. \qquad (3.46)$$

If n changes only slightly during the time step Δt, we can approximate

$$n(t + \Delta t, x) \simeq n(t, x) + \Delta t\frac{\partial n(t, x)}{\partial t}. \qquad (3.47)$$

Substituting Equations (3.46) and (3.47) into Equation (3.45), in the limit of $a \to 0$ and $\Delta t \to 0$ with keeping $a^2/\Delta t$ constant, we obtain the diffusion equation

$$\frac{\partial}{\partial t}n(t, x) = D\frac{\partial^2}{\partial x^2}n(t, x), \qquad (3.48)$$

where $D = pa^2/\Delta t$ is the diffusion constant.

Let $j(t,x)$ denote the flux of n, the net rate at which n is passing from the left of x to the right of x at time t (see Figure 3.3(b)). Since the electron density n is a conserved quantity, for arbitrary Δx,

$$\frac{d}{dt}\int_x^{x+\Delta x} n(t, x)dx = j(t, x) - j(t, x + \Delta x). \qquad (3.49)$$

In the limit of $\Delta x \to 0$, Equation (3.49) yields

$$\frac{\partial n}{\partial t} = -\frac{\partial j}{\partial x}, \qquad (3.50)$$

or in a three-dimensional system

$$\frac{\partial}{\partial t}n(t, \mathbf{x}) + div\boldsymbol{j}(t, \boldsymbol{x}) = 0, \tag{3.51}$$

which is known as the continuity equation. The diffusion equation (Equation (3.48)) and the continuity equation (Equation (3.50)) give a current that is proportional to the local gradient in the density:

$$j_{\text{diffusion}} = -D\frac{\partial n}{\partial x}. \tag{3.52}$$

This is known as Fick's law. Equation (3.52) says that electrons diffuse on average from regions of high density toward regions of low density.

3.4.3 Spin Diffusion Equation [6]

Here, we will discuss a diffusion spin current due to spatial inhomogeneous spin density and a drift spin current in the absence of coherent dynamics of spin. Conduction electrons in a semiconductor or in a metal can be regarded as an electron gas. First, we consider spinless electrons. In the presence of an electric field **E**, the drift current density is given by $\boldsymbol{j}_{\text{drift}} = \sigma\mathbf{E}$. The sum of the drift and diffusion current density $\boldsymbol{j} = \boldsymbol{j}_{\text{drift}} + \boldsymbol{j}_{\text{diffusion}}$ is

$$\boldsymbol{j} = \sigma\boldsymbol{E} + eD\nabla n, \tag{3.53}$$

where $\boldsymbol{j}_{\text{diffusion}} = eD\nabla n$ is the diffusion current density obtained from Equation (3.52). σ is electrical conductivity and $e = 1.602 \times 10^{-19}$ C is the elementary charge. Since $N(E_F)\nabla\mu^c = \nabla n$, a gradient in the electrochemical potential $\mu = \mu^c - e\phi$ is

$$\nabla\mu = e\boldsymbol{E} + \frac{\nabla n}{N(E_F)}. \tag{3.54}$$

Thus, for $\nabla\mu = 0$, the total current density

$$\boldsymbol{j} = (\sigma - e^2N(E_F)D)\boldsymbol{E}, \tag{3.55}$$

must be zero and thus one obtains the Einstein relation:

$$\sigma = e^2N(E_F)D. \tag{3.56}$$

Because of Equations (3.54) and (3.56), we can write

$$\boldsymbol{j} = \frac{\sigma}{e}\nabla\mu. \tag{3.57}$$

This relation expresses the fact that the driving force for a current in this system is a gradient of the electrochemical potential $\nabla\mu$.

Next, we consider the spin degree of freedom. The driving force for a diffusion or drift spin current is a gradient of the difference in the spin-dependent electrochemical potential μ_σ for spin up ($\sigma =\uparrow$) and spin down ($\sigma =\downarrow$). The current density $\boldsymbol{j}_\sigma$ for spin channel $\sigma(\sigma =\uparrow, \downarrow)$ is expressed as

$$\boldsymbol{j}_\sigma = \frac{\sigma_\sigma}{e}\nabla\mu_\sigma, \tag{3.58}$$

where $\mu_\sigma = \mu_\sigma^c - e\phi$ is the spin-dependent electrochemical potential. Here, we introduce a charge current $\boldsymbol{j}_c = \boldsymbol{j}_\uparrow + \boldsymbol{j}_\downarrow$ and a spin current $\boldsymbol{j}_s = \boldsymbol{j}_\uparrow - \boldsymbol{j}_\downarrow$, which are rewritten as

$$\boldsymbol{j}_c = \frac{1}{e}\nabla(\sigma_\uparrow\mu_\uparrow + \sigma_\downarrow\mu_\downarrow), \tag{3.59}$$

$$\boldsymbol{j}_s = \frac{1}{e}\nabla(\sigma_\uparrow\mu_\uparrow - \sigma_\downarrow\mu_\downarrow). \tag{3.60}$$

In nonmagnetic metals or semiconductors, the electrical conductivity is spin-independent, $\sigma_\uparrow = \sigma_\downarrow = (1/2)\sigma_N$, and thus $\boldsymbol{j}_s = (\sigma_N/2e)\nabla(\mu_\uparrow - \mu_\downarrow)$.

The continuity equation for charge is

$$\frac{d}{dt}\rho = -div\boldsymbol{j}_c. \tag{3.61}$$

The continuity equation for spins can be written as

$$\frac{d}{dt}M_z = -div\boldsymbol{j}_s + T_z, \tag{3.62}$$

where M_z is the z component of magnetization. z is defined as the quantization axis. T_z represents spin relaxation, which can be written as $T_z = e(n_\uparrow - \bar{n}_\uparrow)/\tau_{\uparrow\downarrow} - e(n_\downarrow - \bar{n}_\downarrow)/\tau_{\downarrow\uparrow}$ using the single-pole-relaxation approximation. $\bar{n}_\sigma$ is the equilibrium carrier density with spin σ and $\tau_{\sigma\sigma'}$ is the scattering time of an electron from the spin state σ to σ'. Note that the detailed balance principle imposes that $N_\uparrow/\tau_{\uparrow\downarrow} = N_\downarrow/\tau_{\downarrow\uparrow}$, so that in equilibrium no net spin scattering takes place, where N_σ denotes the spin-dependent density of states at the Fermi energy. This implies that, in general, in a ferromagnet, $\tau_{\uparrow\downarrow}$ and $\tau_{\downarrow\uparrow}$ are not the same. In the equilibrium condition, $d\rho/dt = dM_z/dt = 0$, substituting Equations (3.59), (3.60), and $N_\uparrow/\tau_{\uparrow\downarrow} = N_\downarrow/\tau_{\downarrow\uparrow}$ into Equations (3.61) and (3.62), we have

$$\nabla^2(\sigma_\uparrow\mu_\uparrow + \sigma_\downarrow\mu_\downarrow) = 0, \tag{3.63}$$

$$\nabla^2(\mu_\uparrow - \mu_\downarrow) = \frac{1}{\lambda^2}(\mu_\uparrow - \mu_\downarrow). \tag{3.64}$$

Equation (3.64) is known as the spin-diffusion equation. Here, $\lambda = \sqrt{D\tau_{sf}}$ is the spin diffusion length and $D = D_\uparrow D_\downarrow(N_\uparrow + N_\downarrow)/(N_\uparrow D_\uparrow + N_\downarrow D_\downarrow)$ is the spin-averaged diffusion constant with D_σ being the spin-dependent diffusion constant. The spin relaxation time τ_{sf} is given by $1/\tau_{sf} = 1/\tau_{\uparrow\downarrow} + 1/\tau_{\downarrow\uparrow}$.

Now, we consider a simple example of a spin current in a ferromagnetic/nonmagnetic (F/N) junction with a charge current passing through the interface as shown in Figure 3.4(a). The general solution of Equations (3.63) and (3.64) is

$$\mu_\uparrow^F = A_F + B_F x + \frac{C_F}{\sigma_\uparrow^F}\exp\left(\frac{x}{\lambda_F}\right) + \frac{D_F}{\sigma_\uparrow^F}\exp\left(-\frac{x}{\lambda_F}\right), \tag{3.65}$$

$$\mu_\downarrow^F = A_F + B_F x - \frac{C_F}{\sigma_\downarrow^F}\exp\left(\frac{x}{\lambda_F}\right) - \frac{D_F}{\sigma_\downarrow^F}\exp\left(-\frac{x}{\lambda_F}\right), \tag{3.66}$$

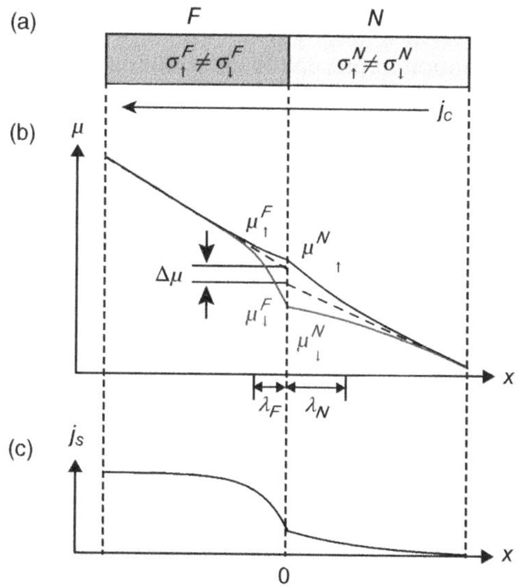

Figure 3.4 *(a) A ferromagnetic/nonmagnetic (F/N) junction with a charge current j_c passing through the interface. (b) The spatial variation of the electrochemical potential $\mu_\sigma^{F(N)}$ for spin-up and spin-down electrons. (c) The spatial variation of a spin current j_s*

$$\mu_\uparrow^N = A_N + B_N x + \frac{C_N}{\sigma_\uparrow^N}\exp\left(\frac{x}{\lambda_N}\right) + \frac{D_N}{\sigma_\uparrow^N}\exp\left(-\frac{x}{\lambda_N}\right), \tag{3.67}$$

$$\mu_\downarrow^N = A_N + B_N x - \frac{C_N}{\sigma_\downarrow^N}\exp\left(\frac{x}{\lambda_N}\right) - \frac{D_N}{\sigma_\downarrow^N}\exp\left(-\frac{x}{\lambda_N}\right), \tag{3.68}$$

where $\mu_\sigma^{F(N)}$, $\lambda_{F(N)}$, and $\sigma_\sigma^{F(N)}$ are the electrochemical potential, the spin diffusion length, and the electrical conductivity for the $F(N)$ layer, respectively. From Equation (3.58), the current density $j_\sigma^{F(N)}$ is

$$j_\sigma^{F(N)} = \frac{\sigma_\sigma^{F(N)}}{e}\frac{\partial}{\partial x}\mu_\sigma^{F(N)}. \tag{3.69}$$

In the F layer, the electrical conductivity is spin dependent and thus $\sigma_\uparrow^F + \sigma_\downarrow^F = \sigma_F$. In contrast, in the N layer, the electrical conductivity is spin independent: $\sigma_\uparrow^N = \sigma_\downarrow^N = \sigma_N/2$. The coefficients $A_{F(N)}, B_{F(N)}, C_{F(N)}$, and $D_{F(N)}$ are determined by boundary conditions. Without loss of generality, we can define first boundary conditions as

$$\mu_\uparrow^F(x = -\infty) = \mu_\downarrow^F(x = -\infty), \tag{3.70}$$

$$\mu_\uparrow^N(x = \infty) = \mu_\downarrow^N(x = \infty). \tag{3.71}$$

These conditions yield $D_F = 0$ and $C_N = 0$. An applied charge current density j_c is

$$j_\uparrow^F + j_\downarrow^F = j_\uparrow^N + j_\downarrow^N = j_c, \tag{3.72}$$

which gives $B_F = ej_c/(\sigma_\uparrow^F + \sigma_\downarrow^F) = ej_c/\sigma_F$ and $B_N = ej_c/(\sigma_\uparrow^N + \sigma_\downarrow^N) = ej_c/\sigma_N$. At the *F*/*N* interface, the boundary conditions representing the continuity of $\mu_\sigma^{F(N)}$ and the conservation of $j_\sigma^{F(N)}$ are

$$\mu_\sigma^F(x=0) = \mu_\sigma^N(x=0), \tag{3.73}$$

$$j_\sigma^F(x=0) = j_\sigma^N(x=0). \tag{3.74}$$

Setting $A_F = 0$ and using these boundary conditions, one finds

$$\mu_\uparrow^F = \frac{ej_c}{\sigma_F}x - \frac{ej_c P\lambda_N(1-P^2)\sigma_F}{2\sigma_\uparrow^F\sigma_N\left(1+(1-P^2)\frac{\sigma_F\lambda_N}{\sigma_N\lambda_F}\right)}\exp\left(\frac{x}{\lambda_F}\right), \tag{3.75}$$

$$\mu_\downarrow^F = \frac{ej_c}{\sigma_F}x + \frac{ej_c P\lambda_N(1-P^2)\sigma_F}{2\sigma_\downarrow^F\sigma_N\left(1+(1-P^2)\frac{\sigma_F\lambda_N}{\sigma_N\lambda_F}\right)}\exp\left(\frac{x}{\lambda_F}\right), \tag{3.76}$$

$$\mu_\uparrow^N = \frac{ej_c P^2\lambda_N}{\sigma_N\left(1+(1-P^2)\frac{\sigma_F\lambda_N}{\sigma_N\lambda_F}\right)} + \frac{ej_c}{\sigma_N}x - \frac{ej_c P\lambda_N}{\sigma_N\left(1+(1-P^2)\frac{\sigma_F\lambda_N}{\sigma_N\lambda_F}\right)}\exp\left(-\frac{x}{\lambda_N}\right), \tag{3.77}$$

$$\mu_\downarrow^N = \frac{ej_c P^2\lambda_N}{\sigma_N\left(1+(1-P^2)\frac{\sigma_F\lambda_N}{\sigma_N\lambda_F}\right)} + \frac{ej_c}{\sigma_N}x + \frac{ej_c P\lambda_N}{\sigma_N\left(1+(1-P^2)\frac{\sigma_F\lambda_N}{\sigma_N\lambda_F}\right)}\exp\left(-\frac{x}{\lambda_N}\right), \tag{3.78}$$

where $P = (\sigma_\uparrow^F - \sigma_\downarrow^F)/(\sigma_\uparrow^F + \sigma_\downarrow^F)$ is the spin polarization of the *F* layer. Figure 3.4(b) shows the spatial variation of the electrochemical potential for spin-up and spin-down electrons with a current through a *F*/*N* interface. In the *N* layer, a spin current j_s driven by $\nabla(\mu_\uparrow^N - \mu_\downarrow^N)$ flows from the interface toward the inside of the *N* layer as shown in Figure 3.4(c). The decay length of j_s is characterized by the spin diffusion length λ_N. In the *F* layer, a spin-polarized current is suppressed near the interface ($\sim \lambda_F$) due to the back flow of spin-polarized electrons induced by the spin accumulation at the interface.

The spin polarization of the current at the interface $\alpha = (j_\uparrow^N - j_\downarrow^N)/(j_\uparrow^N + j_\downarrow^N) = (j_\uparrow^F - j_\downarrow^F)/(j_\uparrow^F + j_\downarrow^F)$ is obtained as

$$\alpha = P\frac{1}{1+(1-P^2)\dfrac{\sigma_F\lambda_N}{\sigma_N\lambda_F}}. \tag{3.79}$$

Note that the spin polarization α of a current injected into the *N* layer is different from the bulk polarization *P* of the *F* layer.

Although $\mu_\uparrow$ and $\mu_\downarrow$ are continuous at the interface, the slope of the electrochemical potentials can be discontinuous at the interface (see the dotted lines in Figure 3.4(b)). This voltage drop at the interface $\Delta\mu$ gives spin-coupled interface resistance $R_s = \Delta\mu/(ej_c)$:

$$R_s = P^2 \frac{\lambda_N \sigma_N^{-1}}{1 + (1 - P^2)\dfrac{\sigma_F \lambda_N}{\sigma_N \lambda_F}}. \tag{3.80}$$

The above equations show that the magnitude of the spin polarization and the spin-coupled resistance contain the same factor $(\sigma_F \lambda_N)/(\sigma_N \lambda_F)$. In many cases, the spin-diffusion length of F is much shorter than that of N, $\lambda_F \ll \lambda_N$, and, in this case, λ_F is a limiting factor to obtain a large spin polarization. This problem becomes serious when a ferromagnetic metal is used to inject spin-polarized currents into semiconductors. In this case, the electrical conductivity, $\sigma_N \ll \sigma_F$, drastically limits the polarization. This problem is known as the conductivity mismatch problem. A way to overcome the conductivity mismatch problem of spin injection into a semiconductor is to use a ferromagnetic semiconductor as a spin source. Another way is to insert a spin-dependent interface resistance at a metal/semiconductor interface.

There are some methods for experimentally detecting pure spin currents, spin currents without accompanying charge currents. One direct method is the utilization of the inverse spin-Hall effect, a method which was demonstrated first by spin pumping [7] and nonlocal technique [8, 9]. In semiconductors, optical method was also demonstrated (see Ref. 10). As an alternative way, one can infer spin-current generation indirectly by measuring spin accumulation.

3.5 Exchange Spin Current

A spin current is carried also by a spin wave, a collective excitation of magnetization in magnets. In this section, we first rewrite exchange interaction in magnets by introducing a concept of an exchange spin current and then formulate a spin-wave spin current.

3.5.1 Magnetic Order and Exchange Interaction

States of matters can be classified into several types in terms of magnetic properties. In paramagnetic and diamagnetic states, matters have no magnetic order and exhibit zero magnetization in the absence of external magnetic fields. By applying a magnetic field, matters in paramagnetic states exhibit magnetization parallel to the external field while those in diamagnetic states exhibit magnetization antiparallel to the field. The other types of material states contain magnetic orders. In ferromagnetic states, the permanent magnetic moments of atoms or ions align parallel to a certain direction and the matters exhibit finite magnetization even in the absence of external magnetic fields. Antiferromagnetic states refer to states in which the permanent magnetic moments align antiparallel and cancel out each other and the net magnetization is zero in the absence of magnetic fields. In ferrimagnets, the moments align antiparallel but the cancellation is not perfect and net magnetization appears.

The interaction that aligns spins is called the exchange interaction. One typical model for the exchange interaction is Heisenberg's Hamiltonian: $H = -J \sum \mathbf{s}_i \cdot \mathbf{s}_j$, where $\mathbf{s}_i$ represents the spin operator of an atom or an ion at the position labeled by i. J is the interaction coefficient; for $J > 0$ case, parallel alignment of s_i and s_j reduces the energy. When this energy reduction is greater than the thermal fluctuation energy, ferromagnetic states can appear. The summation runs over all the combination of nearest neighboring i and j.

3.5.2 Exchange Spin Current

3.5.2.1 *Landau-Lifsitz-Gilbert equation*

In this section, we derive an equation which describes spin or magnetization dynamics. Consider a system described by the Hamiltonian

$$\mathcal{H} = -\boldsymbol{M} \cdot \boldsymbol{H}_{\text{eff}}, \tag{3.81}$$

which describes the fact that a spin $\boldsymbol{s}$ tends to align parallel to the external magnetic field $\boldsymbol{H}$ due to Zeemen's interaction. Magnetization $\boldsymbol{M}$ satisfies the following commutation relation of angular momentum:

$$[M_i, M_j] = i\gamma\hbar\varepsilon_{ijk}M_k. \tag{3.82}$$

The dynamics of $\boldsymbol{M}$ is described by a Heisenberg equation of motion [11]

$$\frac{d\boldsymbol{M}}{dt} = -\frac{i}{\hbar}[\boldsymbol{M}, \mathcal{H}]. \tag{3.83}$$

Substituting Equation (3.82) into this equation, the following result is obtained.

$$\frac{d\boldsymbol{M}}{dt} = -\gamma\,\boldsymbol{M} \times \boldsymbol{H}. \tag{3.84}$$

This equation describes the dynamics of an isolated spin magnetic moment; a spin keeps undergoing a precession motion around the magnetic field $\boldsymbol{H}$, as shown in Figure 3.5. However, as shown in Equation (3.81), the energy is minimized when the magnetic moment is aligned parallel to the external magnetic field. Therefore the precession motion should be relaxed to this energy minimized state before too long. This relaxation is due to interaction of spins with environment degrees of freedom, such as conduction electrons and/or lattice

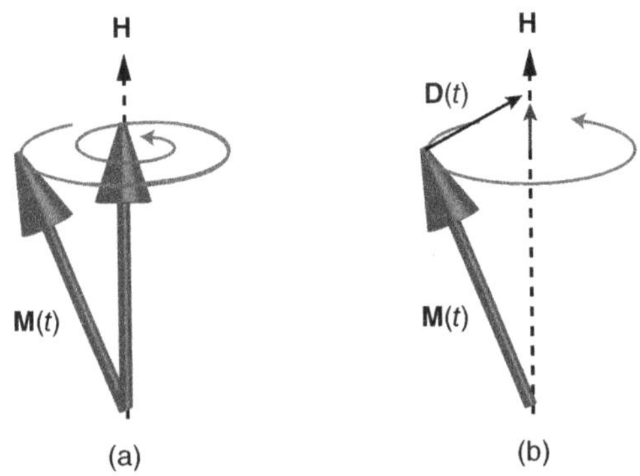

Figure 3.5 *(a) Concept of spin precession and decay. (b) Directions of Gilbert term* **D**(*t*)

vibrations in the matter. This relaxation is often taken into consideration by adding a term called the Gilbert term

$$\boldsymbol{D} = -\frac{\alpha}{M}\boldsymbol{M} \times \frac{d\boldsymbol{M}}{dt} \tag{3.85}$$

to the equation of motion. Note that the Gilbert term is always directed toward the magnetic field direction, or the precession axis. The equation motion then becomes

$$\frac{d\boldsymbol{M}}{dt} = -\gamma \boldsymbol{M} \times \boldsymbol{H} + \frac{\alpha}{M}\boldsymbol{M} \times \frac{d\boldsymbol{M}}{dt}, \tag{3.86}$$

which is called the Landau-Lifsitz-Gilbert (LLG) equation.

Next, we consider general interactions acting on a single spin magnetic moment. For small-angle spin dynamics, the interaction can be introduced into the LLG equation simply by replacing $\boldsymbol{H}$ with an effective magnetic field $\boldsymbol{H}_{\text{eff}}$ as follows

$$\boldsymbol{H}_{\text{eff}} = -\frac{\delta E_i(\boldsymbol{S})}{\delta \boldsymbol{s}_i}. \tag{3.87}$$

$E_i(\boldsymbol{S})$ is the energy on the i site electron as a function of spin direction of the system. $\boldsymbol{H}_{\text{eff}}$ describes total interactions acting on the spin including external magnetic fields, magnetic anisotropy, and exchange interaction.

3.5.2.2 *Rewriting Landau-Lifsitz-Gilbert Equation*

We are now in a position to consider ferromagnetic interaction described by Heisenberg's Hamiltonian for a ferromagnet

$$E_i = -2J\sum_j \boldsymbol{s}_i \cdot \boldsymbol{s}_j \quad (J > 0). \tag{3.88}$$

Let us apply the continuum approximation to $\boldsymbol{s}_i$ to rewrite $\boldsymbol{s}$ into a field value $\boldsymbol{s}(\boldsymbol{r})$ where $\boldsymbol{r}$ represents a position vector. When $\boldsymbol{s}_i = \boldsymbol{s}(\boldsymbol{r})$, a neighboring $\boldsymbol{s}_j$ is written as $\boldsymbol{s}(\boldsymbol{r}+\boldsymbol{a})$ where $\boldsymbol{a}$ is the displacement vector of the j-site measured from the i-site. $\boldsymbol{s}(\boldsymbol{r}+\boldsymbol{a})$ is expanded as

$$\boldsymbol{s}(\boldsymbol{r}+\boldsymbol{a}) = \boldsymbol{s}(\boldsymbol{r}) + \frac{\partial \boldsymbol{s}(\boldsymbol{r})}{\partial \boldsymbol{r}} \cdot \boldsymbol{a} + \frac{1}{2}\frac{\partial^2 \boldsymbol{s}(\boldsymbol{r})}{\partial \boldsymbol{r}^2}\boldsymbol{a}^2 + \cdots. \tag{3.89}$$

The second term of the expansion is vanished in Equation 3.88 due to the summation of i and j since there are same atoms or ions at $r = a$ and $r = -a$ and, via the summation, the second terms of the expansion for these atoms are canceled out. Therefore, the third term is dominant term and we will neglect the higher order terms. We then obtain

$$\boldsymbol{H}_{\text{eff}} = -2Ja^2\frac{\partial \boldsymbol{s}(\boldsymbol{r})}{\partial \boldsymbol{r}^2} \equiv A\nabla^2 \boldsymbol{M}(\boldsymbol{r}). \tag{3.90}$$

A is called the spin-stiffness constant. Then the LLG equation in which the exchange interaction is taken into consideration becomes

$$\frac{\partial}{\partial t}\boldsymbol{M}(\boldsymbol{r}) = -A\gamma \boldsymbol{M}(\boldsymbol{r}) \times \nabla^2 \boldsymbol{M}(\boldsymbol{r}) + \frac{\alpha}{m}\boldsymbol{M}(\boldsymbol{r}) \times \frac{\partial}{\partial t}\boldsymbol{M}(\boldsymbol{r}). \tag{3.91}$$

Next, let us rewrite the LLG equation to a form of a continuity equation. By using a mathematical formula of vector analysis: $\boldsymbol{A} \times \nabla^2 \boldsymbol{A} = div(\boldsymbol{A} \times \nabla \boldsymbol{A})$, Equation (3.91) for exchange-interacting spins is rewritten as

$$\frac{\partial}{\partial t}\boldsymbol{M}(\boldsymbol{r}) = -\mathrm{div}[A\gamma \boldsymbol{M}(\boldsymbol{r}) \times \nabla \boldsymbol{M}(\boldsymbol{r})] + \frac{\alpha}{m}\boldsymbol{M}(\boldsymbol{r}) \times \frac{\partial}{\partial t}\boldsymbol{M}(\boldsymbol{r}). \tag{3.92}$$

For now, we neglect once the Gilbert relaxation term for simplicity, say,

$$\frac{\partial}{\partial t}\boldsymbol{M}(\boldsymbol{r}) = -\mathrm{div}[A\gamma \boldsymbol{M}(\boldsymbol{r}) \times \nabla \boldsymbol{M}(\boldsymbol{r})]. \tag{3.93}$$

This equation has the same form as a continuity equation. In fact, by defining the current $\boldsymbol{j}_s$ as $\boldsymbol{j}_s = A\gamma \boldsymbol{M} \times \nabla \boldsymbol{M}$, the equation becomes

$$\frac{\partial \boldsymbol{M}}{\partial t} = -\mathrm{div}\boldsymbol{j}_s, \tag{3.94}$$

which represents the conservation rule of spin angular momentum. $\boldsymbol{j}_s$ is interpreted as a flow of spin, say, a spin current; called an exchange spin current or a magnetization current. Since the above equation is a simple rewriting of the LLG equation and exchange interaction, it means that the exchange interaction between spins is equivalent to a flow of an exchange spin currents $\boldsymbol{j}_s$. Then, we restore the Gilbert damping: $\dot{\boldsymbol{M}} = -\mathrm{div}\boldsymbol{j}_s + \frac{\alpha}{m}\boldsymbol{M} \times \frac{\partial}{\partial t}\boldsymbol{M}$. Steady states described by this equation is obtained by $\partial \boldsymbol{M}/\partial t = 0$ as

$$\mathrm{div}\boldsymbol{j}_s = \mathrm{div}(A\gamma \boldsymbol{M} \times \nabla \boldsymbol{M}) = 0, \tag{3.95}$$

which implies a uniform alignment of magnetization.

In this way, the exchange interaction can be rewritten in terms of the exchange spin current, which derives a ferromagnetic order in cooperation with the damping. In a uniform magnetization state, there are no exchange spin currents. In steady nontrivial magnetic structure, for instance in a magnetic domain wall, the torque due to the exchange spin current is balanced with magnetic-anisotropy torque.

3.5.3 Spin-Wave Spin Current

The exchange spin current can be driven in an inequilibrium manner by exciting spin waves, or magnons. Spin wave is an elementary excitation from magnetically ordered states, which can be generated by, for instance, applying a microwave. At finite temperature, a spin current is generated also as a thermal fluctuation.

3.5.3.1 Spin Wave Formulation

Now we consider low energy excitations from a ferromagnetic ground state. Let us assume that spins are coupled with nearest neighbor spins via the exchange interaction. If one of spins is tilted against the ground-state direction, the neighboring spins tend to follow this tilt and the whole system will start to perform a collective motion just as the linear chain of masses connected by springs as shown in Figure 3.6; the masses become the magnetic moments of the spins and the role of the springs is taken by the exchange interaction. This collective excitation of spins are the spin waves.

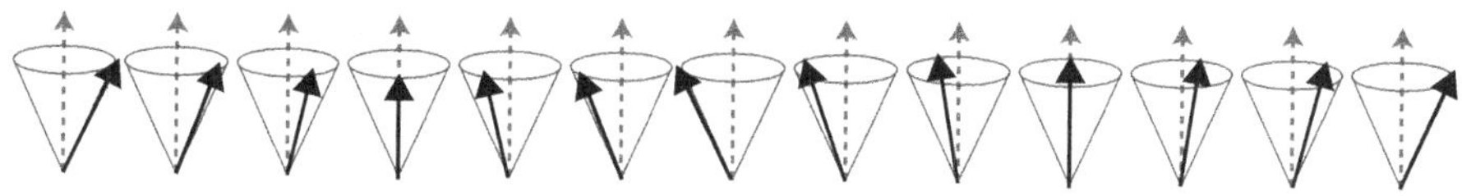

Figure 3.6 Spin wave of a one-atomic chain

In a simple situation where only nearest neighbor interactions are important and all nearest exchange interactions are equal, the Hamiltonian is

$$\mathcal{H} = -2J \sum_{\langle i,j \rangle} \boldsymbol{s}_i \cdot \boldsymbol{s}_j - g\mu_B H \sum_i s_{iz}. \tag{3.96}$$

The second term represents the Zeeman energy. We will further assume $s = 1/2$ and $J > 0$. Equation (3.96) can be written in a convenient form using the spin raising and lowering operators for the ith spin site:

$$s_i^+ = s_{ix} + is_{iy}, \tag{3.97}$$

$$s_i^- = s_{ix} - is_{iy}. \tag{3.98}$$

Now consider a state $|s, M\rangle$, which is an eigenstate of the spin operators $\boldsymbol{s}_i^2$ and s_{iz}: $\boldsymbol{s}_i^2|s, M\rangle = s(s+1)|s, M\rangle$ and $s_{iz}|s, M\rangle = M|s, M\rangle$. This yields

$$s_i^+|s, M\rangle = [s(s+1) - M(M+1)]^{1/2}|s, M+1\rangle, \tag{3.99}$$

$$s_i^-|s, M\rangle = [s(s+1) - M(M-1)]^{1/2}|s, M-1\rangle. \tag{3.100}$$

In terms of these operators, the Hamiltonian becomes

$$\mathcal{H} = -2J \sum_{\langle i,j \rangle} \left[\frac{1}{2}\left(s_i^+ s_j^- + s_i^- s_j^+\right) + s_{iz}s_{jz}\right] - g\mu_B H \sum_i s_{iz}. \tag{3.101}$$

The dynamics of $\boldsymbol{s}_j$ is obtained using the Heisenberg equation of motion

$$\frac{d\boldsymbol{s}_j}{dt} = \frac{i}{\hbar}[\mathcal{H}, \boldsymbol{s}_j] = -\frac{1}{\hbar}(\epsilon_j \times \boldsymbol{s}_j), \tag{3.102}$$

where

$$\varepsilon_j = 2J \sum_{i \neq j} \boldsymbol{s}_i + g\mu_B \boldsymbol{H} \tag{3.103}$$

is an effective magnetic field acting on $\boldsymbol{s}_j$. The second relation of Equation (3.102) can be obtained from the commutation relations for spin operators.

When the external magnetic field is $\boldsymbol{H} = (0, 0, H)$, each spin is aligned along the z-axis. In the case of low-energy excited states, the deviations of the spin from the z direction is small and the change of the z component of $\boldsymbol{s}_i$ is a small quantity of second order. Then one can approximate $s_{jz} \simeq s$. Substituting Equation (3.103) with $\boldsymbol{H} = (0, 0, H)$ into Equation (3.102), we obtain

$$\hbar \frac{ds_{jx}}{dt} = -2Js \sum_i (s_{iy} - s_{jy}) + g\mu_B H s_{jy}, \tag{3.104}$$

$$\hbar\frac{ds_{jy}}{dt} = -2Js\sum_i (s_{jx} - s_{ix}) - g\mu_B H s_{jx}. \tag{3.105}$$

Using $s_j^{\pm} = s_{jx} \pm i s_{jy}$, we have

$$\hbar\frac{ds_j^{\pm}}{dt} = \pm i\left[-2Js\sum_i \left(s_j^{\pm} - s_i^{\pm}\right) - g\mu_B H s_j^{\pm}\right]. \tag{3.106}$$

The local change of spins propagates through the whole spin system through the first term on the right-hand side. The coupled motion of neighboring spins can be decoupled by exploiting the periodicity with the Bloch representation:

$$s_{\boldsymbol{k}}^{\pm} = \frac{1}{\sqrt{N}}\sum_j e^{-i\boldsymbol{k}\cdot\boldsymbol{R}_j} s_j^{\pm}. \tag{3.107}$$

With these normal coordinates, we arrive at

$$\frac{\hbar}{i}\frac{ds_{\boldsymbol{k}}^{\pm}}{dt} = \left[2Js\sum_i \left(1 - e^{-i\boldsymbol{k}\cdot(\boldsymbol{R}_i - \boldsymbol{R}_j)}\right) + g\mu_B H\right] s_{\boldsymbol{k}}^{\pm}. \tag{3.108}$$

Assuming $s_{\boldsymbol{k}}^{-} \propto \delta s_{\boldsymbol{k}} e^{i\omega_{\boldsymbol{k}} t + i\alpha}$, we find the eigenfrequency of Equation (3.107) as

$$\hbar\omega_{\boldsymbol{k}} = 2JsZ(1 - \gamma_{\boldsymbol{k}}) + g\mu_B H, \tag{3.109}$$

where Z is the number of nearest neighbors and $\gamma_{\boldsymbol{k}} = (1/Z)\sum_{\boldsymbol{R}_i - \boldsymbol{R}_j} e^{i\boldsymbol{k}\cdot(\boldsymbol{R}_i - \boldsymbol{R}_j)}$. The above discussion shows that the whole spin configuration of a crystal behaves as an oscillatory motion with the frequency $\omega_{\boldsymbol{k}}$ and the wave vector $\boldsymbol{k}$. This collective mode is the spin wave that is mediated by the exchange interaction. This collective mode corresponds to a coherent precession of the individual spins around the direction of the ferromagnetic orientation. It is completely analogous to the lattice modes in a solid.

For a cubic lattice where the nearest neighbors are along the $\pm x$, $\pm y$, and $\pm z$ axes at a distance a, $\gamma_{\boldsymbol{k}} = (1/3)(\cos k_x a + \cos k_y a + \cos k_z a)$. Therefore, for small k, we have a quadratic dispersion relation for spin waves:

$$\hbar\omega_{\boldsymbol{k}} = g\mu_B H + 2Jsa^2k^2. \tag{3.110}$$

This shows a quadratic dependence on the wave vector around the minimum at $\boldsymbol{k} = 0$.

3.5.3.2 Spin Current Carried by Spin Wave

We here show that spin-wave propagation carries spin angular momentum. In the following, the Gilbert damping term is neglected for simplicity. We can rewrite the LLG equation as

$$\frac{\partial}{\partial t}\boldsymbol{M}(\boldsymbol{r},t) = \gamma\boldsymbol{H}_{\text{eff}} \times \boldsymbol{M}(\boldsymbol{r},t) - \text{div}\boldsymbol{j}^{M_\alpha}(\boldsymbol{r},t), \tag{3.111}$$

where $\boldsymbol{j}^{M_\alpha}$ is an exchange spin current, defined above, whose components are

$$j_\beta^{M_\alpha} = \frac{D}{M_s}[\boldsymbol{M} \times \nabla_\beta \boldsymbol{M}]_\alpha. \tag{3.112}$$

The z-component of the LLG equation (3.111) gives a continuity equation for the exchange spin current: $\partial M_z/\partial t + \mathrm{div}\boldsymbol{j}^{M_z} = 0$, which represents spin-angular momentum conservation, when $\alpha = 0$.

Here we consider an exchange spin current carried when a spin wave is excited. We introduce a spin-wave wave function $\psi(\boldsymbol{r},t) = M_+(\boldsymbol{r},t) = M_x(\boldsymbol{r},t) + iM_y(\boldsymbol{r},t)$ and its conjugate complex $\psi^*(\boldsymbol{r},t)$. The z-component of the exchange spin current is written as

$$j_\beta^{M_z} = \frac{1}{2i}\frac{D}{M_s}[\psi^*(\boldsymbol{r},t)\nabla_\beta\psi(\boldsymbol{r},t) - \psi(\boldsymbol{r},t)\nabla_\beta\psi^*(\boldsymbol{r},t)]. \tag{3.113}$$

By introducing creation and annihilation operators $(b_q^\dagger, b_q)$ of spin-wave excitations (magnons) with the frequency ω_q and the wave number q by $\psi = M_+ = \sqrt{2/M_s}\sum_q b_q e^{i\boldsymbol{q}\cdot\boldsymbol{r}}$ and $\psi^* = M_- = \sqrt{2/M_s}\sum_q b_q^\dagger e^{-i\boldsymbol{q}\cdot\boldsymbol{r}}$, the exchange spin current is expressed as

$$j_x^{M_z} = \sum_{p,q} v_q n_q, \tag{3.114}$$

where $v_q = \partial\omega_q/\partial q = 2Dq$ is the spin-wave group velocity and $n_q = \langle b_q^\dagger b_q\rangle$ is the number of spin waves. Equation (3.114) means that, when the numbers of excited spin waves are different between q and $-q$ in the k space, a nonzero net exchange spin current is carried by the spin waves: a spin-wave spin current. Such a spin-wave property was observed in Ref. 12.

3.6 Topological Spin Current

In the previous section, we showed that exchange interaction among spins can be rewritten by introducing an equilibrium exchange spin current. In this section, another type of an equilibrium spin current is discussed, which is a topological spin current. Topological spin currents are driven by topological band structure and classified into bulk and surface topological spin currents.

3.6.1 Bulk Topological Spin Current

Electrons in crystals are confined onto electron-band manifolds and their motions are sometimes affected by this confinement. This contribution can be argued in terms of Berry's phase. We here go over a standard method to treat this problem [13]; a method combining the equations of motion and the Boltzmann equation for semiclassical electrons in a band. This method considers a wave-packet of electrons and tracks its motion, assuming that its position and momentum is defined with moderate accuracy without violating the uncertainty principle. We obtain the following equations assuming that the band index n does not change because interband transition of electrons does not happen under a weak external perturbation

$$\frac{d\boldsymbol{r}}{dt} = \frac{\partial\varepsilon_n(\boldsymbol{k})}{\partial\boldsymbol{k}}, \tag{3.115}$$

$$\frac{d\boldsymbol{p}}{dt} = -e(\boldsymbol{E} + \frac{1}{c}\frac{d\boldsymbol{r}}{dt} \times \boldsymbol{B}). \tag{3.116}$$

A simple semiclassical motion of an electron is determined without knowledge of wave functions. However, exact motions in solids must be modified reflecting the electron-band curvature, which is represented by the connection of a wave function in k space. This is due to the band structure which causes the Hilbert space to be projected into electron-band manifolds. The operator for the position of the electrons $\boldsymbol{r} = (x_j) = (x, y, z)$ is canonically conjugate with the wavenumber vector $\boldsymbol{k} = (k_j)$ and satisfies the commutation relation

$$[x_i, k_j] = i\delta_{i,j}, \tag{3.117}$$

thus

$$\boldsymbol{r} = i\nabla_{\boldsymbol{k}} \tag{3.118}$$

holds. Due to the curvature of the electron-band manifold, this position operator should be generalized into the gauge covariant derivative using the "vector potential" in $\boldsymbol{k}$ space expressing the connection due to the curvature

$$a_{nj}(\boldsymbol{k}) = -i\langle n\boldsymbol{k}|\nabla_k|n\boldsymbol{k}\rangle \tag{3.119}$$

and becomes

$$\boldsymbol{r} = i\nabla_{\boldsymbol{k}} - \boldsymbol{a}_n(\boldsymbol{k}). \tag{3.120}$$

Therefore, a nontrivial noncommutation relationship holds

$$[x_i, x_j] = i\varepsilon_{ij}\frac{\partial}{\partial k_i}a_{nj}(\boldsymbol{k}). \tag{3.121}$$

This noncommutation relationship modifies the equation of motion as

$$\begin{aligned}\dot{x}_\mu &= -\,i[x_\mu, \mathcal{H}] = -i[x_\mu, k_\nu]\frac{\partial}{\partial k_\mu}\mathcal{H} - i[x_\mu, x_\nu]\frac{\partial}{\partial x_\nu}\mathcal{H} \\ &= \frac{\partial}{\partial k_\mu}\varepsilon_n(\boldsymbol{k}) + \frac{\partial}{\partial k_\mu}a_{n\nu}(\boldsymbol{k})\frac{\partial}{\partial x_\nu}V(\boldsymbol{r}).\end{aligned} \tag{3.122}$$

The second term on the right-hand side is the new term called anomalous velocity. This term generates an electric current (topological current), even under thermal equilibrium. Anomalous velocity is one mechanism for the spin Hall effects and the anomalous Hall effect.

Taking this effect into consideration, the semiclassical equation of motion of electrons in solids becomes [13]

$$\frac{d\boldsymbol{x}}{dt} = \frac{\partial\varepsilon_n(\boldsymbol{k})}{\partial\boldsymbol{k}} + \frac{d\boldsymbol{k}}{dt} \times \boldsymbol{b}_n(\boldsymbol{k}), \tag{3.123}$$

$$\frac{d\boldsymbol{k}}{dt} = -e(-\frac{\partial\phi(\boldsymbol{r})}{\partial\boldsymbol{r}} + \frac{d\boldsymbol{r}}{dt} \times \boldsymbol{B}(\boldsymbol{r})). \tag{3.124}$$

In these equations, the duality of r and k is obvious; using the field $\boldsymbol{b}_n(\boldsymbol{k})$ instead of the magnetic field $\boldsymbol{B}(\boldsymbol{r})$ and using $\varepsilon_n(\boldsymbol{k})$ instead of the electrostatic potential $\phi(\boldsymbol{r})$ results in a dual relationship between the equations of motion for $\boldsymbol{r}$ and $\boldsymbol{k}$. The second term on the right-hand side is the anomalous velocity, and the Hall current can be obtained by adding the states occupied by electrons

$$j_x = -e \sum_{k,n} f(\varepsilon_n(\boldsymbol{k})) b_{nz}(\boldsymbol{k}) \dot{k}_y. \tag{3.125}$$

If an electric field exists in the y-direction only, $\dot{k}_y = -eE_y$, thus

$$\sigma_{xy} = e^2 \sum_{\boldsymbol{k},n} f(\varepsilon_n(\boldsymbol{k})) b_{nz}(\boldsymbol{k}). \tag{3.126}$$

In paramagnetic or diamagnetic metals, when the gauge field $\boldsymbol{b}_n(\boldsymbol{k})$ is spin dependent, electrons with different spins have different velocities; a charge current is converted into a spin current of conduction electrons. Such a spin current is a bulk topological spin current.

3.6.2 Surface Topological Spin Current

The spin currents discussed above are basically those that flow in bulk. Finally, the other type of spin current is introduced very quickly: a surface (edge) spin current, which is limited near surfaces (edges) of a three (two)-dimensional system and flows along the surfaces (edges).

This surface spin current is known to appear in topological insulators. In topological insulators, bulk is insulating but surface or edge is electrically conducting due to the surface or edge state: an electronic state localized at the surface/edge. In such a system, the spin degeneracy of the surface (edge) state is lifted except for the $k = 0$ point and the surface (edge) states of wave vector k and $-k$ have opposite spin, a situation which means the state accompanies a spin current in an equilibrium state even without external perturbation.

3.7 Thermal Spin Current - Spin Seebeck Effect

The Seebeck effect refers to the generation of an electric voltage as a result of a temperature gradient in conductors, which was discovered by T. J. Seebeck in 1820s (see Figure 3.7(a)) [3]. The Seebeck effect plays an important role in the field of electronics; it is widely used in temperature (infrared)-measuring devices, heat flux sensors, and thermoelectric generators.

In the field of spintronics, the spin version of the Seebeck effect, the spin Seebeck effect (SSE), has attracted much attention [15–39]. The SSE refers to the generation of "spin voltage" as a result of a temperature gradient in magnetic materials. Here, a spin voltage is the potential for electron's spins, which drives a spin current; when a conductor is attached to a part of a magnet with a finite spin voltage, the spin voltage injects a spin current into the conductor (see Figure 3.7(b)). In 2008, Uchida *et al.* originally discovered the SSE in a ferromagnetic metal $Ni_{81}Fe_{19}$ film [15] by means of the spin-detection technique based on the inverse spin Hall effect (ISHE) [7–9, 40] in a Pt film. In 2010, using the same experimental method, Jaworski *et al.* also observed this phenomenon in a ferromagnetic semiconductor GaMnAs in a low temperature region [18, 24] and Uchida *et al.* revealed that the SSE appears even in magnetic insulators, such as YIG [19], $LaY_2Fe_5O_{12}$ (La:YIG)

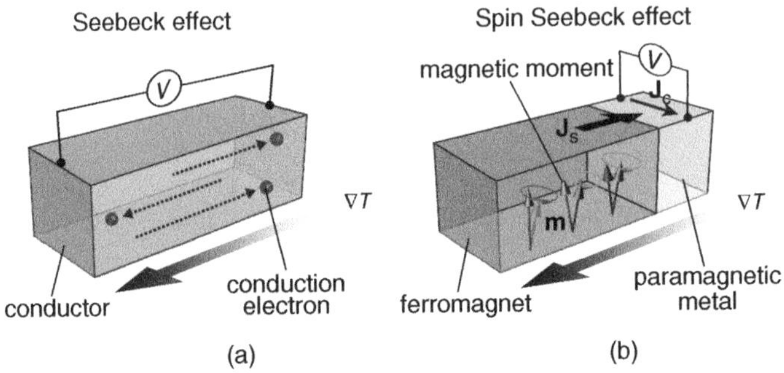

Figure 3.7 (a) A schematic illustration of the Seebeck effect. When a temperature gradient ∇T is applied to a conductor, an electric field **E** (electric voltage V) is generated along the ∇T direction. (b) A schematic illustration of the spin Seebeck effect (SSE). When ∇T is applied to a ferromagnet, a spin voltage is generated via magnetization (**M**) dynamics, which pumps a spin current $\mathbf{J}_s$ into an attached paramagnetic metal. In the paramagnetic metal, this spin current is converted into an electric field $\mathbf{E}_{ISHE}$ due to the inverse spin Hall effect (ISHE)

[17], and (Mn,Zn)Fe_2O_4 [21]. The SSE was observed also in Co_2MnSi [25], $NiFe_2O_4$ [32], Fe_3O_4 [34], and various garnet ferrites [28, 35]. These observations indicate that the SSE is a universal phenomenon in magnetic materials.

The discovery of the SSE in magnetic insulators provides a crucial piece of information for understanding the physics of the SSE. The conventional Seebeck effect requires itinerant charge carriers or conduction electrons, and therefore exists only in metals and semiconductors. It is natural to assume that the same holds for the SSE. Originally, the SSE was phenomenologically formulated in terms of the thermal excitation of conduction electrons [15]. However, the observation of the SSE in insulators upsets this conventional interpretation; conduction electrons are not necessary for the SSE [17]. This is direct evidence that the spin voltage generated by the SSE is associated with thermal magnetization dynamics (see Figure 3.7(b)). Based on this idea, various theoretical models have been proposed [16, 20, 22, 23, 29, 33, 36, 39].

3.7.1 Sample Configuration and Measurement Mechanism

The observation of the SSE exploited the ISHE in two different device structures: One is a *longitudinal* configuration [19, 27], in which a spin current *parallel* to a temperature gradient is measured. This configuration consists of the simplest straightforward structure, although it is applicable only to insulators, as mentioned below. The other setup is a *transverse* configuration [15, 17, 18, 27], in which a spin current flowing *perpendicular* to a temperature gradient is measured. This configuration has a more complicated device structure than the longitudinal one, but it has been used for measuring the SSE in various magnetic materials ranging from metals and semiconductors to insulators.

Figure 3.8(a) shows a schematic illustration of the longitudinal SSE device. The device structure is very simple, which consists of a ferromagnet (F) slab or film covered with a paramagnetic metal (PM) film. When a temperature gradient ∇T is applied to the F layer in the direction perpendicular to the PM/F interface (z direction), a spin voltage is thermally generated and injects a spin current with the spatial direction $\mathbf{J}_s$ and the spin-polarization

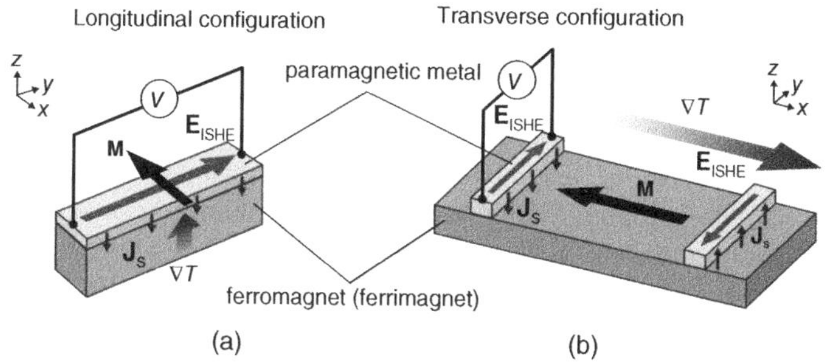

Figure 3.8 *Schematic illustrations of the longitudinal SSE (a) and transverse SSE (b)*

vector $\boldsymbol{\sigma}$, parallel to the magnetization **M** of F, into the PM film along the ∇T direction (see Figure 3.8(a)). In the PM wire, this injected spin current is converted into an electric field $\mathbf{E}_{ISHE}$ due to the ISHE. When **M** is along the x direction, $\mathbf{E}_{ISHE}$ is generated in the PM along the y direction because of the relation [7]

$$\mathbf{E}_{\text{ISHE}} = \frac{\theta_{\text{SH}}\rho}{A}\left(\frac{2e}{\hbar}\right)\mathbf{J}_s \times \boldsymbol{\sigma}, \tag{3.127}$$

where θ_{SH}, ρ, and A are the spin Hall angle of PM, the electric resistivity of PM, and the contact area between F and PM, respectively. Therefore, by measuring $\mathbf{E}_{ISHE}$ in the PM film, the longitudinal SSE can be detected electrically. Here, we note that, if ferromagnetic metals are used as the F layer, the ISHE signal not only will be suppressed by short-circuit currents [40] in the F layer due to the electric conduction of F but will also be contaminated by the signal of the anomalous Nernst effect [41] in F. By using an insulator such as YIG, these artifacts are eliminated. Furthermore, we note that, although the SSE in a Pt/F hybrid system might be contaminated by the anomalous Nernst effect because of the magnetic proximity effect of Pt at the Pt/F interface [42], recent experiments confirmed that such a contribution is negligible in the Pt/YIG [31], Pt/$NiFe_2O_4$ [32], and Pt/Fe_3O_4 [34] systems.

Figure 3.8(b) shows a schematic illustration of the transverse SSE device, which is composed of a rectangular F with one or several PM wires attached onto the top surface of F. The typical length of the F layer along the x direction is $\sim$ 10 mm, much longer than the spin-diffusion length [43]. In the transverse configuration, to generate the ISHE voltage induced by the SSE along the PM-wire direction, the F layer has to be magnetized along the ∇T direction (x direction) (see Equation (3.127) and note that $\mathbf{J}_s$ is parallel to the z direction also in the transverse configuration). Therefore, the anomalous Nernst effect in the F layer vanishes owing to the collinear orientation of ∇T and **M**, enabling the pure detection of the transverse SSE in various magnetic materials.

3.7.2 Longitudinal Spin Seebeck Effect

Here, we show the data for the longitudinal SSE in a Pt/YIG sample [19]. The sample consists of a single-crystalline or polycrystalline YIG slab and a Pt film fabricated on a top surface of YIG. The length, width, and thickness of the YIG slab are 6, 2, and 1 mm, respectively. The Pt/YIG sample was sandwiched between two heat baths, the temperatures of which were stabilized to 300 K and 300 K + ΔT to generate a temperature gradient along the z direction. To detect the ISHE induced by the longitudinal SSE, we measured the electric voltage V between the ends of the Pt film while applying a magnetic field **H** in the x-y plane at an angle θ to the y direction.

Figure 3.9(a) shows V in the single-crystalline Pt/YIG sample as a function of ΔT at $H = 1$ kOe. When **H** is applied along the $\theta = 90°$ direction, the magnitude of V was observed to be proportional to ΔT. The sign of the V signal for finite values of ΔT is clearly reversed by reversing the temperature gradient. Since YIG is an insulator, thermoelectric phenomena in the YIG slab, such as the conventional Seebeck and Nernst effects, do not exist at all. As also shown in Figure 3.9(a), the V signal disappears when **H** is applied along the $\theta = 0$ direction. We also found that the sign of V at $\theta = 90°$ in the Pt/YIG sample is reversed by reversing **H**, indicating that the V signal is affected by the magnetization direction of the YIG layer (see Figure 3.9(b)). These results are consistent with the feature of the ISHE induced by the longitudinal SSE (see Equation (3.127) and Figure 3.8(a)).

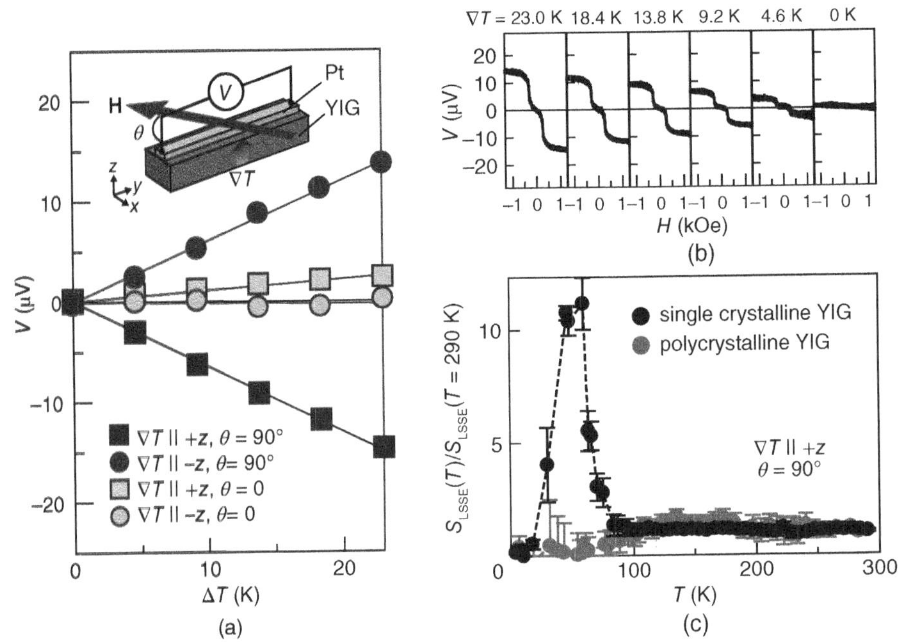

Figure 3.9 *(a) ΔT dependence of V in the Pt/YIG sample at $H = 1$ kOe, measured when ∇T was applied along the +z or −z direction. The magnetic field* **H** *was applied along the x direction ($\theta = 90°$) or the y direction ($\theta = 0$). (b) H dependence of V in the Pt/YIG sample for various values of ΔT at $\theta = 90°$, measured when ∇T was along the +z direction. All the data shown in this figure were measured at room temperature. (c) Temperature T dependences of $S_{LSSE}(T)/S_{LSSE}(T = 290$ K$)$ for the single-crystalline and polycrystalline Pt/YIG samples at $H = 1$ kOe, measured when ∇T and* **H** *were applied along the +z and x directions, respectively. $S_{LSSE}(T)$ denotes the ISHE voltage induced by the longitudinal SSE per unit temperature difference: $S_{LSSE}(T) = V(T)/\Delta T$. The vertical axis is normalized by $S_{LSSE}(T)$ at 290 K*

The critical difference between the SSEs in single-crystalline and polycrystalline samples becomes apparent in the temperature dependence [27]. Figure 3.9(c) shows $V/\Delta T(\equiv S_{LSSE})$ as a function of the temperature T in the single-crystalline and polycrystalline Pt/YIG samples at $H = 1$ kOe. In the entire temperature range (4.2–290 K), the clear SSE signals were observed in both the samples for **H** along the x direction. Notable is that, in the single-crystalline Pt/YIG sample, the magnitude of $V/\Delta T$ is dramatically enhanced at approximately $T = 50$ K, while the V signal in the polycrystalline Pt/YIG sample does not exhibit strong T dependence (see Figure 3.9(c)). We found that this V-peak position coincides with the temperature at which the thermal conductivity of the single-crystalline YIG becomes maximum owing to the increase in the phonon lifetime, suggesting the importance of phonons in creating the nonequilibrium state that drives the spin current in the Pt contact [20, 24, 26, 27].

3.7.3 Transverse Spin Seebeck Effect

In this subsection, we review the experimental results for the transverse SSE in a Pt/La:YIG sample 17. The sample consists of a 3.9-μm-thick La:YIG film with 15-nm-thick Pt wires attached to the surface. A single-crystalline La:YIG (111) film was grown on a GGG (111) substrate by liquid-phase epitaxy. Here, the La:YIG layer has an 8×4 mm^2 rectangular surface. Two (and later more) Pt wires were then sputtered on the top of the La:YIG film. The distance of the Pt wires from the center of the La:YIG layer is 2.8 mm and the wires are electrically well insulated. The length and width of the Pt wires are 4 and 0.1 mm, respectively.

Figure 3.10(a) shows V between the ends of the Pt wires placed near the lower- and higher temperature ends of the La:YIG layer as a function of ΔT at $H = 100$ Oe, measured

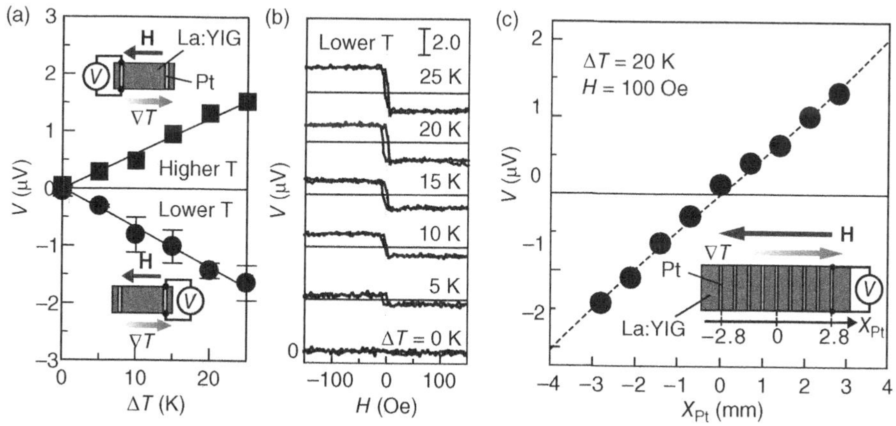

Figure 3.10 *(a) ΔT dependence of V in the Pt/La:YIG sample at $H = 100$ Oe, measured when the Pt wires were attached near the lower temperature (300 K) and higher temperature (300 K + ΔT) ends of the La:YIG layer. (b) H dependences of V in the Pt/La:YIG sample for various values of ΔT. (c) Dependence of V on x_{Pt}, the displacement of the Pt wire from the center of the La:YIG layer along the x direction, in the Pt/La:YIG sample at $\Delta T = 20$ K and $H = 100$ Oe*

when the magnetic field **H** was applied along the ∇T direction [17]. The magnitude of V is proportional to ΔT in both the Pt wires. Notably, the signs of V are opposite for the lower- and higher temperature ends of the sample. This sign reversal of V is a characteristic behavior of the ISHE voltage induced by the transverse SSE [15, 17, 18, 25, 26]. As shown in Figure 3.10(b), the sign of the V signal at each end of the sample is reversed by reversing **H**. The V signal in the Pt/La:YIG sample has been shown to vanish when **H** is along the Pt-wire direction [17], a situation consistent with Equation (3.127). The signal also disappears when the Pt wires are replaced by Cu wires with weak spin-orbit interaction [17]. All the data shown above confirm that the observed V signal is due to the transverse SSE in the Pt/La:YIG sample.

The spatial dependence of the thermally generated spin voltage can be measured by changing the position of the Pt strip [17]. As shown in Figure 3.10(c), the V signal induced by the transverse SSE has a quasi-linear spatial dependence. Here, the signal changes its sign at both the ends of the sample and vanishes at the center of the sample.

3.7.4 Thermoelectric Coating based on Spin Seebeck Effect

The SSE in magnetic insulators can be applied directly to the design of thermospin generators and, in combination with the ISHE, thermoelectric generators, allowing new approaches to improving thermoelectric generation efficiency. In general, the efficiency is improved by suppressing the energy loss due to heat conduction and Joule dissipation, which are realized respectively by reducing the thermal conductivity κ for the sample part where heat currents flow and by reducing the electric resistivity ρ for the part where charge currents flow. However, in electric conductors, when κ is dominated by the electronic thermal conductivity κ_e, the Wiedemann-Franz law ($\kappa_e \rho =$ constant) limits this improvement. A conventional way to overcome this limitation is to use semiconductor-based thermoelectric materials, where the thermal conductance is usually dominated by phonons while the electric conductance is determined by charge carriers and thus κ and ρ are separated according to the kind of the carriers. The SSE provides another way to overcome the Wiedemann-Franz law; in the spin Seebeck device, the heat and charge currents flow in different parts of the sample: κ is the thermal conductivity of the magnetic insulator, and ρ is the electrical resistivity of the metallic wire, such that κ and ρ in the spin Seebeck device are segregated according to the part of the device elements. Therefore, the SSE in insulators allows us to construct thermoelectric devices operated by an entirely new principle, although the thermoelectric conversion efficiency is still small at present.

In 2012, Kirihara *et al.* proposed a novel concept of thermoelectric technology "spin-thermoelectric (STE) coating," which is driven by the SSE and characterized by a simple film structure, convenient scaling capability, and easy fabrication (see Figure 3.11(a)) [28]. In their experiments, an STE coating with a 60-nm-thick Bi-substituted YIG film was applied on a nonmagnetic substrate by a metal organic decomposition method (see Figure 3.11(b) and (c)). Notably, the thermoelectric conversion driven by the longitudinal SSE was successfully demonstrated under a temperature gradient perpendicular to such an ultrathin STE-coating layer (amounting to only 0.01% of the total sample thickness). The STE coating was found to be applicable even onto glass surfaces with amorphous structures [28]. Such versatile implementation of the thermoelectric function may open opportunities for various applications making full use of omnipresent heat.

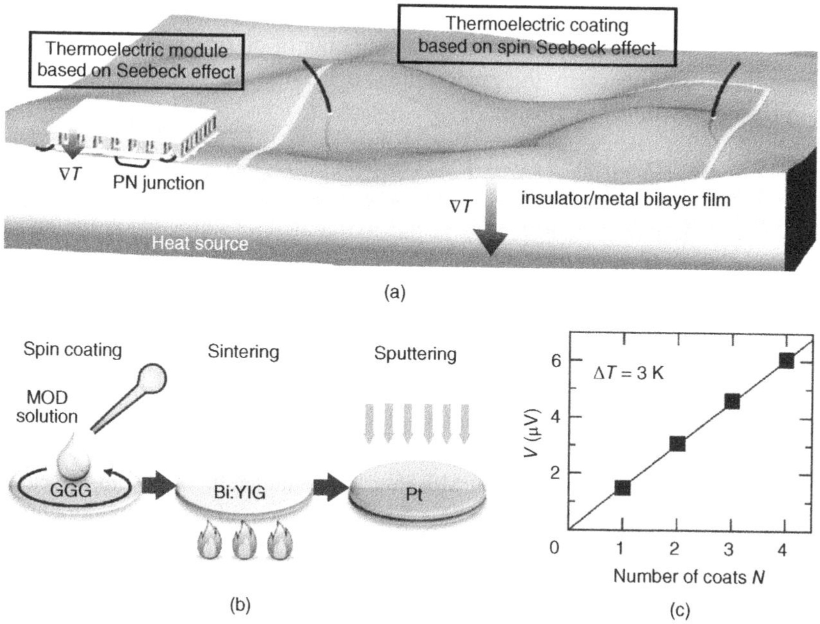

Figure 3.11 (a) Concept of the spin-thermoelectric (STE) coating based on the SSE. The STE coating exhibits a straightforward scaling: a larger film area leads to a larger thermoelectric output. Such a simple film structure can be coated directly onto heat sources with surfaces of various (curved or uneven) shapes. (b) Schematic of the STE-coating process. (c) Thermoelectric voltage V at $\Delta T = 3$ K as a function of the number of coats N, showing that V increases linearly with repeated STE coating

3.7.5 Basic Mechanism of Spin Seebeck Effect

We now present a qualitative picture of the physics underlying the SSE. Since the SSE appears even in magnetic insulators, it cannot be fully expressed in terms of thermal excitation of conduction electrons. The SSE in insulators cannot also be explained by the equilibrium spin pumping [7, 40, 44, 45, 46], since the average spin-pumping current from thermally fluctuating magnetic moments is exactly canceled by the thermal (Johnson-Nyquist) spin-current noise. Therefore, the observed spin voltage requires us to introduce an unconventional nonequilibrium state between magnetic moments in F and electrons in an attached PM. The microscopic mechanism of the SSE was proposed both by Xiao *et al.* [16] and by Adachi *et al.* [22, 33] by means of scattering and linear-response theories, respectively. In this subsection, we review basic concepts of their calculations.

The thermally excited state in the SSEs at the PM/F interface can be described in terms of an effective magnon temperature T_{m}^{*} in F and an effective electron temperature T_{e}^{*} in PM, which can be different from local thermodynamic temperature at a position. The ISHE signal derived from the net spin current is proportional to $T_{\mathrm{m}}^{*} - T_{\mathrm{e}}^{*}$, as shown below. The effective temperatures are related to thermal fluctuations through the fluctuation-dissipation theorem. The fluctuations of the magnetization $\mathbf{m}$ at the PM/F interface are excited by a random magnetic field $\mathbf{h} = \Sigma_j \mathbf{h}^{(j)}$ $(j = 0, 1, \ldots)$, which represents the thermal disturbance from various sources (such as lattice, contacts, etc.). If we, respectively, denote $T^{(j)}$ and $\alpha^{(j)}$ as the temperature and the damping of dissipative source j, due to the fluctuation-dissipation theorem, the random field $\mathbf{h}$ satisfies the following equal-position time-correlation function:

$$\langle h_i^{(j)}(t) h_{i'}^{(j')}(t') \rangle = \left(\frac{2k_B T^{(j)} \alpha^{(j)}}{\gamma M_s V_a} \right) \delta_{jj'} \delta_{ii'} \delta(t - t'), \tag{3.128}$$

where k_B is the Boltzmann constant, γ is the gyromagnetic ratio, M_s is the saturation magnetization, $\alpha = \alpha^{(0)} + \alpha^{(1)} + \cdots$ is the effective damping parameter, $T_m^* = (\alpha^{(0)}T^{(0)} + \alpha^{(1)}T^{(1)} + \cdots)/\alpha$ is the effective magnon temperature, and V_a is the magnetic coherence volume in F, which depends on the magnon temperature and the spin-wave stiffness constant D. Assuming the dissipative sources 0 and 1 are the F lattice and the PM contact respectively, then $T^{(0)}, \alpha^{(0)}, T^{(1)}$, and $\alpha^{(1)} = \gamma \hbar g_r / 4\pi M_s V_a$ represent the bulk lattice temperature, the bulk Gilbert damping parameter, the electron temperature in the PM contact ($T^{(1)} = T_e^*$), and the damping enhancement due to the spin pumping with g_r being the real part of the mixing conductance for the PM/F interface. The net thermal spin current across the PM/F interface is given by the sum of a fluctuating thermal spin-pumping current $\mathbf{J}_{sp}$ from F to PM proportional to T_m^* and a Johnson-Nyquist spin-current noise $\mathbf{J}_{fl}$ from PM to F proportional to T_e^* [16, 47, 48]:

$$\mathbf{J}_s = \mathbf{J}_{sp} + \mathbf{J}_{fl} = \frac{M_s V_a}{\gamma} [\alpha^{(1)} \mathbf{m} \times \dot{\mathbf{m}} + \gamma \mathbf{m} \times \mathbf{h}^{(1)}]. \tag{3.129}$$

The DC component along the magnetization equilibrium direction reduces to

$$J_s \equiv \langle \mathbf{J}_s \rangle_{DC} = 2\alpha^{(1)} k_B (T_m^* - T_e^*). \tag{3.130}$$

Therefore, when $T_m^* \neq T_e^*$, spin currents are generated across the PM/F interface. At an equilibrium state ($T_m^* = T_e^*$), no spin current is generated at the PM/F interface since the spin-pumping current, $\mathbf{J}_{sp}$, is canceled out by the spin-current noise, $\mathbf{J}_{fl}$. Equations (3.127) and (3.130) indicate that the magnitude and sign of the ISHE voltage induced by the SSE are determined by those of $T_m^* - T_e^*$ at the PM/F under a temperature gradient. As demonstrated by the previous studies, this effective temperature difference $T_m^* - T_e^*$ is induced by the magnon-mediated [16, 17, 22, 23, 39] and the phonon-mediated [20, 24, 26] processes; when nonequilibrium magnons and phonons travel around F with feeling a temperature gradient, T_m^* in F and/or T_e^* in PM are modulated by interacting with the traveling magnons and phonons, respectively. Due to these processes, magnons in F and/or electrons in PM in a lower temperature (higher temperature) region feel the temperature information in a higher temperature (lower temperature) region. Therefore, the resultant T_m^* and/or T_e^* in the lower temperature (higher temperature) region increases (decreases).

An important clue to calculate the effective temperature distribution was provided by Sanders and Walton in 1977 [49]. They discussed the effective magnon temperature (T_m^*) and phonon temperature (T_p^*) distributions in a magnetic insulator, especially YIG, placed in a uniform temperature gradient and solved a simple heat-rate equation of the coupled magnon-phonon system under a situation similar to the transverse SSE configuration. The solution of the heat-rate equation yields a hyperbolic-sine $T_m^* - T_p^*$ profile with a decay length λ_m. In a magnetic insulator with weak magnetic damping, such as YIG and La:YIG, a rapid equilibration of magnons is prevented and the resultant λ_m was shown to reach several millimeters. Making use of these results and assuming that T_p^* in F is equal to T_e^*

in an attached PM, Xiao *et al.* formulated the magnon-mediated SSE and their calculation reproduces the magnitude and spatial distribution of the SSE-induced ISHE signal observed in the transverse Pt/La:YIG system [16, 17]. The magnon-mediated SSE was formulated also by means of a many-body technique using nonequilibrium Green's functions by Adachi *et al.* [22] and numerical calculation based on a stochastic LLG equation by Ohe *et al.* [23].

The temperature-dependent behavior of the SSE [20, 24] and the spatial distribution of the thermally generated spin voltage in ferromagnetic metals [15] may be explained by the phonon-based mechanism [20, 26], where the effective temperature difference $T^*_{\mathrm{m}} - T^*_{\mathrm{e}}$ is generated by nonequilibrium phonons. The phonon-mediated thermal spin pumping was formulated also by Adachi *et al.* using a linear-response theory.

3.8 Concluding Remarks

Electrons have electrical charges and spins as internal degrees of freedom. Just as the flow of electrical charge constitutes an electric current, a spin current has an important impact on spintronics, magnetics, and condensed-matter physics. A spin current, however, cannot be treated the same as an electrical current. This is because the spin current normally decays within a distance of some μm owing to spin-orbit interactions since a spin is not a conserved quantity unlike a charge, which is conserved. With the microfabrication technology that has been developed in the present century, nanodevices using magnetic materials have been successfully fabricated, therefore, spin currents have been controlled.

The conservation law of angular momentum should be built into the unified picture of electronics and spintronics. Applications of the electrical control of magnetism as well as methods of transferring electrical signals by spin currents are expected to be developed further. We anticipate that establishing the principles for the interconversion of electricity, magnetism, heat, and even mechanics [50] will provide an opportunity to extend spintronics greatly, which can lead to a new paradigm of physics and its application.

References

1. P. A. M. Dirac, *The Principles of Quantum Mechanics*; 4th edition, Oxford University Press, Oxford, 1982.
2. J. D. Jackson, *Classical Electrodynamics*; 3rd edition, Wiley, New York, 1998.
3. N. W. Ashcroft and N. D. Mermin, *Sold State Physics*, Brooks Cole, New York, 1976.
4. C. Kittel, *Introduction to Solid State Physics*; 8th edition, Wiley, New York, 2004.
5. H. Ibach and H. Lüth, *Solid-State Physics: An Introduction to Principles of Materials*; 4th edition, Springer, Berlin, 2009.
6. S. Takahashi and S. Maekawa, Spin current in metals and superconductors, *J. Phys. Soc. Jpn.*, **77**, 031009 (2008).
7. E. Saitoh, M. Ueda, H. Miyajima, and G. Tatara, Conversion of spin current into charge current at room temperature: Inverse spin-Hall effect, *Appl. Phys. Lett.*, **88**, 182509 (2006).
8. S. O. Valenzuela and M. Tinkham, Direct electronic measurement of the spin Hall effect, *Nature*, **442**, 176–179 (2006).

9. T. Kimura, Y. Otanu, T. Sato, S. Takahashi, and S. Maekawa, Room-temperature reversible spin Hall effect *Phys. Rev. Lett.*, **98**, 156601 (2007).
10. L. K. Werake and H. Zhao, Observation of second-harmonic generation induced by pure spin currents, *Nat. Phys.*, **6**, 875–878 (2010).
11. S. Chikazumi, *Physics of Ferromagnetism (International Series of Monographs on Physics)*; 2nd edition, Oxford University Press, Oxford, 2009.
12. Y. Kajiwara, K. Harii, S. Takahashi, *et al.*, Transmission of electrical signals by spin-wave interconversion in a magnetic insulator, *Nature*, **464**, 262–266 (2010).
13. D. Xiao, M. C. Chang, and Q. Niu, Berry phase effects on electronic properties, *Rev. Mod. Phys.*, **82**, 1959–2007 (2010).
14. A. H. Morrish, *The Physical Principles of Magnetism*, Robert E. Krieger, New York, 1980.
15. K. Uchida, S. Takahashi, K. Harii, *et al.*, Observation of the spin Seebeck effect, *Nature*, **455**, 778–781 (2008).
16. J. Xiao, G. E. W. Bauer, K. Uchida, *et al.*, Theory of magnon-driven spin Seebeck effect, *Phys. Rev. B*, **81**, 214418 (2010).
17. K. Uchida, J. Xiao, H. Adachi, *et al.*, Spin Seebeck insulator, *Nat. Mater.*, **9**, 894–897 (2010).
18. C. M. Jaworski, J. Yang, S. Mack, *et al.*, Observation of the spin-Seebeck effect in a ferromagnetic semiconductor, *Nat. Mater.*, **9**, 898–903 (2010).
19. K. Uchida, H. Adachi, T. Ota, *et al.*, Observation of longitudinal spin-Seebeck effect in magnetic insulators, *Appl. Phys. Lett.*, **97**, 172505 (2010).
20. H. Adachi, K. Uchida, E. Saitoh, *et al.*, Gigantic enhancement of spin Seebeck effect by phonon drag, *Appl. Phys. Lett.*, **97**, 252506 (2010).
21. K. Uchida, T. Nonaka, T. Ota, and E. Saitoh, Longitudinal spin-Seebeck effect in sintered polycrystalline (Mn,Zn) Fe_2O_4, *Appl. Phys. Lett.*, **97**, 262504 (2010).
22. H. Adachi, J. Ohe, S. Takahashi, and S. Maekawa, Linear-response theory of spin Seebeck effect in ferromagnetic insulators, *Phys. Rev. B*, **83**, 094410 (2011).
23. J. Ohe, H. Adachi, S. Takahashi, and S. Maekawa, Numerical study on the spin Seebeck effect, *Phys. Rev. B*, **83**, 115118 (2011).
24. C. M. Jaworski, J. Yang, S. Mack, *et al.*, Spin-Seebeck effect: a phonon driven spin distribution, *Phys. Rev. Lett.*, **106**, 186601 (2011).
25. S. Bosu, Y. Sakuraba, K. Uchida, *et al.*, Spin Seebeck effect in thin films of the Heusler compound Co_2 MnSi, *Phys. Rev. B*, **83**, 224401 (2011).
26. K. Uchida, H. Adachi, T. An, *et al.*, Long-range spin Seebeck effect and acoustic spin pumping, *Nat. Mater.*, **10**, 737–741 (2011).
27. K. Uchida, T. Ota, H. Adachi, *et al.*, Thermal spin pumping and magnon-phonon-mediated spin-Seebeck effect, *J. Appl. Phys.*, **111**, 103903 (2012).
28. A. Kirihara, K. Uchida, Y. Kajiwara, *et al.*, Spin-current-driven thermoelectric coating, *Nat. Mater.*, **11**, 686–689 (2012).
29. Y. Ohnuma, H. Adachi, E. Saitoh, and S. Maekawa, Spin Seebeck effect in antiferromagnets and compensated ferrimagnets, *Phys. Rev. B*, **87**, 014423 (2013).
30. D. Qu, S. Y. Huang, J. Hu, *et al.*, Intrinsic spin Seebeck effect in Au/YIG, *Phys. Rev. Lett.*, **110**, 067206 (2013).
31. T. Kikkawa, K. Uchida, Y. Shiomi, *et al.*, Longitudinal spin Seebeck effect free from the proximity Nernst effect, *Phys. Rev. Lett.*, **110**, 067207 (2013).

32. D. Meier, T. Kuschel, L. Shen, *et al.*, Thermally driven spin and charge currents in thin $NiFe_2O_4$/Pt films, *Phys. Rev. B*, **87**, 054421 (2013).
33. H. Adachi, K. Uchida, E. Saitoh, and S. Maekawa, Theory of the spin Seebeck effect, *Rep. Prog. Phys.*, **76**, 036501 (2013).
34. R. Ramos, T. Kikkawa, K. Uchida, *et al.*, Observation of the spin Seebeck effect in epitaxial Fe_3O_4 thin films, *Appl. Phys. Lett.*, **102**, 072413 (2013).
35. K. Uchida, T. Nonaka, T. Kikkawa, *et al.*, Longitudinal spin Seebeck effect in various garnet ferrites, *Phys. Rev. B*, **87**, 104412 (2013).
36. H. Adachi and S. Maekawa, Linear-response theory of the longitudinal spin Seebeck effect, *J. Korean Phys. Soc.*, **62**, 1753–1758 (2013).
37. K. S. Tikhonov, J. Sinova, and A. M. Finkel'stein, Spectral non-uniform temperature and non-local heat transfer in the spin Seebeck effect, *Nat. Commun.*, **4**, 1945 (2013).
38. B. F. Miao, S. Y. Huang, D. Qu, and C. L. Chien, Inverse spin Hall effect in a ferromagnetic metal, *Phys. Rev. Lett.*, **111**, 066602 (2013).
39. S. Hoffman, K. Sato, and Y. Tserkovnyak, Landau-Lifshitz theory of the longitudinal spin Seebeck effect, *Phys. Rev. B*, **88**, 064408 (2013).
40. K. Ando, S. Takahashi, J. Ieda, *et al.*, Inverse spin-Hall effect induced by spin pumping in metallic system, *J. Appl. Phys.*, **109**, 103913 (2011).
41. H. B. Callen, The application of Onsager's reciprocal relations to thermoelectric, thermomagnetic, and galvanomagnetic effects, *Phys. Rev.*, **73**, 1349–1358 (1948).
42. S. Y. Huang, X. Fan, D. Qu, *et al.*, Transport magnetic proximity effects in platinum, *Phys. Rev. Lett.*, **109**, 107204 (2012).
43. J. Bass and W. P. Pratt, Jr, Spin-diffusion lengths in metals and alloys, and spin-flipping at metal/metal interfaces: an experimentalist's critical review, *J. Phys. Condens. Matter*, **19**, 183201 (2007).
44. R. H. Silsbee, A. Janossy, and P. Monod, Coupling between ferromagnetic and conduction-spin-resonance modes at a ferromagnetic-normal-metal interface, *Phys. Rev. B*, **19**, 4382–4399 (1979).
45. Y. Tserkovnyak, A. Brataas, and G. E. W. Bauer, Enhanced Gilbert damping in thin ferromagnetic films, *Phys. Rev. Lett.*, **88**, 117601 (2002).
46. S. Mizukami, Y. Ando, and T. Miyazaki, Effect of spin diffusion on Gilbert damping for a very thin permalloy layer in Cu/permalloy/Cu/Pt films, *Phys. Rev. B*, **66**, 104413 (2002).
47. J. Foros, A. Brataas, Y. Tserkovnyak, and G. E. W. Bauer, Magnetization noise in magnetoelectronic nanostructures, *Phys. Rev. Lett.*, **95**, 016601 (2005).
48. J. Xiao, G. E. W. Bauer, S. Maekawa, and A. Brataas, Charge pumping and the colored thermal voltage noise in spin valves, *Phys. Rev. B*, **79**, 174415 (2009).
49. D. J. Sanders and D. Walton, Effect of magnon-phonon thermal relaxation on heat transport by magnons, *Phys. Rev. B*, **15**, 1489–1494 (1977).
50. M. Matsuo, J. Ieda, E. Saitoh, and S. Maekawa, Effects of mechanical rotation on spin currents, *Phys. Rev. Lett.*, **106**, 076601 (2011).

4

Spin Hall Effect and Inverse Spin Hall Effect

Shuichi Murakami
Department of Physics, Tokyo Institute of Technology, Tokyo, Japan

In this chapter, we review basic aspects of the spin Hall effect (SHE) and inverse spin Hall effect. In Section 4.1, we introduce a basic framework and discuss the mechanism of the SHE in semiconductors and in metals. Section 4.2 is devoted to the topological insulators.

4.1 Spin Hall Effect

4.1.1 Introduction

In this chapter, we explain basic properties of the spin Hall effect (SHE) [1–3], inverse spin Hall effect (ISHE), and related effects. First, we explain basic properties of these effects. Schematic pictures of the SHE and ISHE are shown in Figure 4.1. In these effects, time-reversal symmetry is imposed; namely we consider nonmagnetic metals or doped semiconductors with no external magnetic field. If we apply an electric field to the system, the electrons are driven along the electric field. If we assume that the system has the spin–orbit coupling (SOC), the electrons can undergo spin-dependent motion. As shown in Figure 4.1(a). the up-spins and down-spins are deflected in the opposite way, perpendicular to the electric field. As a result there will be a transverse spin current in response to the electric field. We note that it does not violate the time-reversal symmetry, because spin current is invariant under time-reversal symmetry. If the system has no SOC, there cannot be spin-dependent motions of electrons, leading to an absence of SHE. The reciprocal effect to the SHE is the ISHE shown in Figure 4.1(b). In this case, when the spin current

Spintronics for Next Generation Innovative Devices, First Edition. Edited by Katsuaki Sato and Eiji Saitoh.
© 2015 John Wiley & Sons, Ltd. Published 2015 by John Wiley & Sons, Ltd.

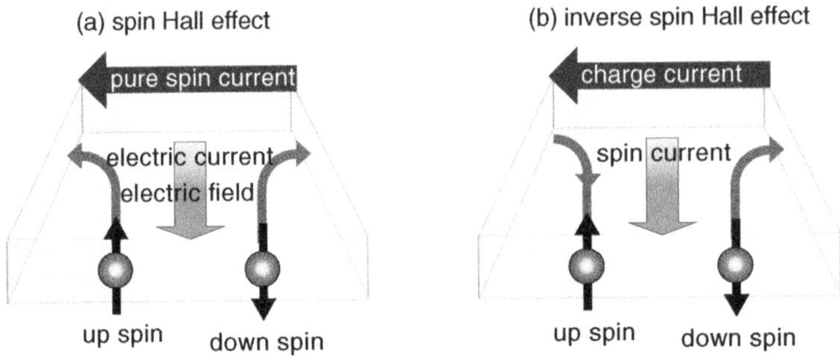

Figure 4.1 *Schematics for (a) spin Hall effect and (b) inverse spin Hall effect*

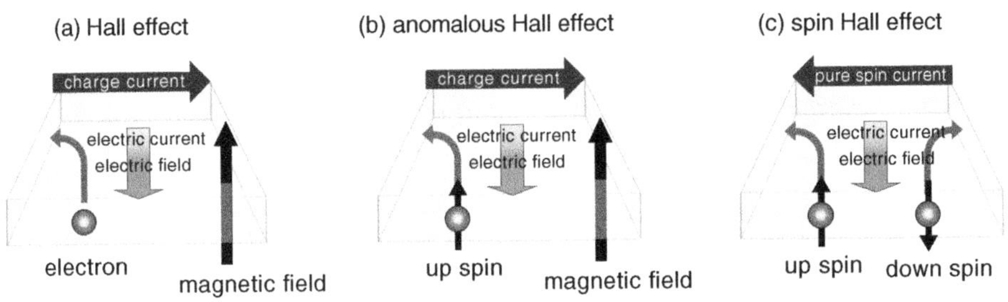

Figure 4.2 *Schematics for (a) Hall effect, (b) anomalous Hall effect, and (c) spin Hall effect*

is injected into the system, an electric field is induced. In the injected spin current, the up-spins and down-spins move in the opposite directions. The SOC causes deflections of the electrons, and the deflections have the same direction for up- and down-spins, causing a transverse charge current. The SHE and ISHE enable us to electrically manipulate or detect spin currents.

Let us compare the SHE with other Hall effect. Thereby one can intuitively understand that this spin-dependent motion in SHE is natural. For simplicity we consider two-dimensional systems. We first consider the Hall effect (Figure 4.2(a)). Here an external magnetic field is applied perpendicular to the plane. By applying the electric field, the electrons will be deflected due to Lorentz force, and eventually the transverse charge current results. This is the Hall effect. Next we replace the magnetic field with the magnetization. It leads to the anomalous Hall effect (AHE) shown in Figure 4.2(b). It is natural to expect such an effect, considering an analogy between the magnetization and the magnetic field. Nevertheless, the mechanism itself might be different from the Hall effect, because there is no Lorentz force acting onto the electrons. There are several mechanisms for the AHE, as will be discussed later, and all the mechanisms require the SOC. If such an AHE response is possible for up-spin electrons, down-spins, if any, will be deflected in the opposite direction. It causes a spin current perpendicular to the applied field, as shown in Figure 4.2 (c). It is the SHE as we discuss here.

The mechanisms for the AHE and the SHE are similar. They are classified into extrinsic and intrinsic mechanisms. In the extrinsic mechanism, impurity scattering becomes spin-dependent with the help of the SOC, leading to the AHE or the SHE. The intrinsic mechanism, on the other hand, does not rely on impurity scattering, but is due to band structure itself. The details of these mechanisms will be discussed later. Historically, the extrinsic mechanism was first proposed in 1971 [1]. On the other hand, the intrinsic mechanism of the SHE was proposed in 2003 for p-type semiconductors like p-GaAs [2] and for n-type semiconductors in two-dimensional heterostructures [3]. After these two proposals for intrinsic SHE, the SHE has been intensively studied, both theoretically and experimentally.

In the SHE, the applied electric field induces a transverse pure spin current. The pure spin current is defined as a spin-dependent electron flow without a charge current. Namely, the up-spin electrons and the down-spin electrons are flowing in the opposite directions with equal amount. A pure spin current is described as a tensor j_j^i, which corresponds to a product between the spin S^i and the velocity v_j. For example, j_x^z represents the current where the spin along the $+z$ direction flows to $+x$, and the spin along $-z$ direction flows to $-x$. It is important to note that the pure spin current is even under time-reversal operation. Therefore, the pure spin current can be induced even when the system is time-reversal invariant. One can see from symmetry that this SHE is possible as a response to electric field. For example, if we impose cubic symmetry and time-reversal symmetry, such as in p-type semiconductors, the spin-current response to the external electric field is restricted to the form

$$j_j^i = \sigma_s \varepsilon_{ijk} E_k, \tag{4.1}$$

where ε_{ijk} is the totally antisymmetric tensor. This is exactly the SHE, shown in Figure 4.2 (c). The coefficient σ_s is called spin Hall conductivity (SHC). When this spin current arrives at the boundaries of the system, spins will accumulate there. Such an incoming spin current to the boundaries due to the SHE will eventually balance with spin relaxation and spin diffusion.

4.1.2 Intrinsic and Extrinsic Hall Effect

Here we explain the intrinsic mechanism of the Hall effect. The intrinsic mechanism stems from the Berry curvature [4] in momentum space. The Berry curvature represents geometric structure of the Bloch wavefunction in momentum space. As we see later, the Berry curvature is responsible for various types of intrinsic Hall effects, such as the SHE [2, 3], the integer quantum Hall effect (QHE) [5, 6] and the AHE [7–10]. Another famous example is the topological insulators, as we discuss in the next section. Here we restrict ourselves to a crystalline system without impurities. The Bloch theorem says that the wavefunctions are written in the form of the Bloch wavefunction

$$\psi(\boldsymbol{r}) = u_{\boldsymbol{k}}(\boldsymbol{r}) e^{i\boldsymbol{k}\cdot\boldsymbol{r}}, \tag{4.2}$$

where $\boldsymbol{k}$ is the Bloch wavevector. By using (4.2) we can rewrite the eigenvalue problem for the state $\psi(\boldsymbol{r})$

$$\hat{H}\psi(\boldsymbol{r}) = E\psi(\boldsymbol{r}) \tag{4.3}$$

into that for the periodic part of the Bloch wavefunction $u_{\boldsymbol{k}}(\boldsymbol{r})$:

$$H_{\boldsymbol{k}} u_{\boldsymbol{k}}(\boldsymbol{r}) = E_{\boldsymbol{k}} u_{\boldsymbol{k}}(\boldsymbol{r}), \qquad H_{\boldsymbol{k}} \equiv e^{-i\boldsymbol{k}\cdot\boldsymbol{r}} \hat{H} e^{i\boldsymbol{k}\cdot\boldsymbol{r}}. \tag{4.4}$$

For example, for $\hat{H} = \frac{(-i\hbar\nabla)^2}{2m} + V(\boldsymbol{r})$ we get

$$H_{\boldsymbol{k}} = \frac{(-i\hbar\nabla + \hbar\boldsymbol{k})^2}{2m} + V(\boldsymbol{r}). \tag{4.5}$$

The dependence of $H_{\boldsymbol{k}}$ on the Bloch wavevector $\boldsymbol{k}$ is then reflected to $\boldsymbol{k}$-dependence of $u_{\boldsymbol{k}}(\boldsymbol{r})$. This dependence is described in the Berry curvature $\boldsymbol{B}_n(\boldsymbol{k})$.

This Berry curvature affects the motion of electrons under an electric field. We define the Berry connection (gauge field) $\boldsymbol{A}_n(\boldsymbol{k})$ and the corresponding Berry curvature (field strength) $\boldsymbol{B}_n(\boldsymbol{k})$ as

$$\boldsymbol{A}_n(\boldsymbol{k}) = i\langle u_{n\boldsymbol{k}}|\nabla_{\boldsymbol{k}}|u_{n\boldsymbol{k}}\rangle \equiv i\int_{\text{unit cell}} u^{\dagger}_{n\boldsymbol{k}}\frac{\partial u_{n\boldsymbol{k}}}{\partial k_i}d^dx, \qquad \boldsymbol{B}_n(\boldsymbol{k}) = \nabla_{\boldsymbol{k}}\times\boldsymbol{A}_n(\boldsymbol{k}), \tag{4.6}$$

where n is the band index. In fact, mathematical properties for the Berry connection $\mathbf{A}_n(\boldsymbol{k})$ and the Berry curvature $\boldsymbol{B}_n(\boldsymbol{k})$ resemble those for the vector potential $\boldsymbol{A}(\boldsymbol{r})$ and magnetic flux density $\boldsymbol{B}(\boldsymbol{r})$, although their physical meanings are very different. One of the resemblances lies in the properties under gauge transformation. $\boldsymbol{A}_n(\boldsymbol{k})$ is a gauge field, and is subject to a gauge transformation. Namely, one can add a $\boldsymbol{k}$-dependent arbitrary phase to the Bloch wavefunction $u_{n\boldsymbol{k}}$: $u'_{n\boldsymbol{k}} = e^{i\theta(\boldsymbol{k})}u_{n\boldsymbol{k}}$. It is a gauge transformation, leading to a transformation of $\boldsymbol{A}_n(\boldsymbol{k})$ as $\boldsymbol{A}'_n(\boldsymbol{k}) = \boldsymbol{A}_n(\boldsymbol{k}) - \nabla_{\boldsymbol{k}}\theta$. Meanwhile, the corresponding Berry curvature $\boldsymbol{B}_n(\boldsymbol{k}) = \nabla_{\boldsymbol{k}}\times\boldsymbol{A}_n(\boldsymbol{k})$ remains unchanged, i.e., gauge invariant. Thus the gauge transformations for the Berry connection and Berry curvature are similar to that of the vector potential and magnetic flux density.

This Berry curvature manifests itself as various Hall effects, as can be seen in the semiclassical dynamics of a wavepacket. A wavepacket is a superposition of various Bloch wavefunctions distributed around a central wavevector $\boldsymbol{k}$. Then the Berry curvature $\boldsymbol{B}_n(\boldsymbol{k})$ contributes to the Boltzmann-type semiclassical equation of motion (EOM) as an additional term [11–13]. The resulting semiclassical EOM for the central position $\boldsymbol{r}$ and the central wavevector $\boldsymbol{k}$ of a wavepacket is

$$\dot{\boldsymbol{x}} = \frac{1}{\hbar}\frac{\partial E_n(\boldsymbol{k})}{\partial \boldsymbol{k}} - \dot{\boldsymbol{k}}\times\boldsymbol{B}_n(\boldsymbol{k}), \tag{4.7}$$

$$\dot{\boldsymbol{k}} = -e\boldsymbol{E}. \tag{4.8}$$

In the right-hand side of (4.7), the first term represents the group velocity, whereas the second term is called anomalous velocity, caused by the Berry curvature. This anomalous velocity is perpendicular to the electric field $\boldsymbol{E}$ and therefore causes various Hall effects depending on systems, among which are the AHE in ferromagnetic metals or the SHE in nonmagnets. As for the AHE, the Hall conductivity calculated from Equations (4.7) and (4.8) is identical with that from linear response theory. In nonmagnets, the Berry curvatures for up-spin and down-spin become opposite to each other, giving rise to the opposite anomalous velocities, due to the time-reversal symmetry. It is nothing but the SHE [2, 3].

From the definition (4.6), the Berry curvature describes how rapidly the Bloch wavefunction changes as a function of the wavevector $\boldsymbol{k}$. One can further rewrite the formula for the Berry curvature as

$$\boldsymbol{B}_n(\boldsymbol{k}) = i\langle \nabla_{\boldsymbol{k}} u_{n\boldsymbol{k}}| \times |\nabla_{\boldsymbol{k}} u_{n\boldsymbol{k}}\rangle = i \sum_{m(\neq n)} \frac{\left\langle n\boldsymbol{k} \left| \frac{\partial H}{\partial \boldsymbol{k}} \right| m\boldsymbol{k} \right\rangle \times \left\langle m\boldsymbol{k} \left| \frac{\partial H}{\partial \boldsymbol{k}} \right| n\boldsymbol{k} \right\rangle}{(E_n(\boldsymbol{k}) - E_m(\boldsymbol{k}))^2}, \tag{4.9}$$

where the sum is taken over all the Bloch eigenstates but for n. Here we assumed that the energy spectrum has no degeneracy. This formula implies that when the nth band comes close with other energy bands, such as in band anticrossings, the Berry curvature is enhanced. It is a guiding principle for enhancing the Berry curvature. Once a Hamiltonian is given, the Berry curvature can be calculated, and one can check that around such band anticrossings the Berry curvature is indeed enhanced. For example, the *ab initio* calculation for platinum [14] shows that it is indeed the case. In platinum the band anticrossing is close to the Fermi energy, which helps to enhance the intrinsic SHE. In particular, when two bands touch each other at a certain wavevector $\boldsymbol{k}$, $\boldsymbol{B}(\boldsymbol{k})$ may diverge there. Such divergence of the Berry curvature $\boldsymbol{B}(\boldsymbol{k})$ occurs only at band degeneracies, but we note that the inverse is not true; the band degeneracy does not necessarily mean divergence of $\boldsymbol{B}(\boldsymbol{k})$

We note that in addition to the intrinsic SHE from the Berry curvature, there is also an extrinsic contribution to the SHE, as we discuss later. Whether the intrinsic or the extrinsic SHE dominates depends on materials, sample purity, sample geometry, and so on. This extrinsic contribution comes from the excitation around the Fermi energy and therefore vanishes for insulators, where the Fermi energy lies in a gap. Furthermore, in insulators, the intrinsic one (4.12) can be shown to be quantized in a unit of e^2/h, which explains the integer QHE [5, 6].

Here we discuss the Hall conductivity calculated from the Kubo formula, to see that it is indeed expressed in terms of the Berry curvature. We restrict ourselves to two-dimensional systems for simplicity. According to the Kubo formula, describing the current response to the applied electric field in the linear-response regime, the Hall conductivity σ_{xy} is written as

$$\sigma_{xy} = -ie^2\hbar \frac{1}{L^2} \sum_n f(E_\alpha) \sum_{\beta(\neq\alpha)} \frac{\langle\alpha|v_x|\beta\rangle\langle\beta|v_y|\alpha\rangle}{(E_\alpha - E_\beta)^2}, \tag{4.10}$$

where $-e = -|e|$ is an electronic charge, $f(E)$ is the Fermi distribution function, and v_i is the velocity operator along i direction. α and β are labels for eigenvectors; $\alpha = (n, \boldsymbol{k})$, $\beta = (m, \boldsymbol{k})$, where $\mathbf{k}$ is a Bloch wavevector. In the Bloch basis, the velocity operator v_i is written as $v_i \equiv \frac{1}{\hbar}\frac{\partial H}{\partial k_i}$. Therefore, we have

$$\sigma_{xy} = -i\frac{e^2}{\hbar}\frac{1}{L^2} \sum_{\boldsymbol{k}} \sum_n f(E_{n\boldsymbol{k}}) \sum_{m(\neq n)} \frac{\langle n\boldsymbol{k}|\frac{\partial H}{\partial k_x}|m\boldsymbol{k}\rangle\langle m\boldsymbol{k}|\frac{\partial H}{\partial k_y}|n\boldsymbol{k}\rangle}{(E_{n\boldsymbol{k}} - E_{m\boldsymbol{k}})^2}. \tag{4.11}$$

We can then use the formula for the Berry curvature, Equation (4.9), to obtain

$$\sigma_{xy} = -\frac{e^2}{\hbar}\frac{1}{L^2} \sum_{\boldsymbol{k}} \sum_n f(E_{n\boldsymbol{k}}) B_{n,z}(\boldsymbol{k}) = -\frac{e^2}{h} \int \frac{d^2k}{2\pi} \sum_n f(E_{n\boldsymbol{k}}) B_{n,z}(\boldsymbol{k}). \tag{4.12}$$

It means that each filled state gives rise to a contribution of $-\frac{e^2}{\hbar}B_{n,z}(\boldsymbol{k})$ to the Hall conductivity. In particular, for zero temperature, it is further simplified as

$$\sigma_{xy} = -\frac{e^2}{h} \sum_n \int_{E_{n\boldsymbol{k}} < E_{\mathrm{F}}} \frac{d^2k}{2\pi} B_{n,z}(\boldsymbol{k}) \tag{4.13}$$

Properties of this formula (4.13) for insulators and those for metals are quite different, and we discuss these two cases separately.

4.1.2.1 *Quantized Intrinsic Hall Effect for Band Insulators*

First, we consider intrinsic Hall conductivity (4.13) for band insulators. In such cases, the Hall conductivity is shown to be quantized as an integer multiple of e^2/h as explained in the following. In band insulators, all the bands are classified into completely occupied bands and completely empty bands. Therefore, the Hall conductivity is expressed as a sum over the occupied bands

$$\sigma_{xy} = -\frac{e^2}{h} \sum_{n:\text{filled}} \mathrm{Ch}_n, \tag{4.14}$$

where Ch_n is defined for respective bands as

$$\mathrm{Ch}_n \equiv \int_{\mathrm{BZ}} \frac{d^2k}{2\pi} B_{n,z}(\boldsymbol{k}) \tag{4.15}$$

and is called the Chern number for the nth band. This number can be mathematically shown to be always an integer, provided the considered nth band is separated from other bands by energy gaps. This property of integer quantization of the Chern number comes from the 2π degree of freedom for phases of wavefunctions, and therefore is universal for any waves described by eigenvalue problems. This integer quantization of the Chern number leads to the quantization of the Hall conductivity as an integer multiple of e^2/h. This is exactly the integer QHE. The Chern number is an example of a topological number or a topological invariant. It means that the Chern number is unchanged even when the Hamiltonian is continuously changed, as long as the gap remains open.

It is noted that the integer QHE usually refers to a two-dimensional electron gas (2DEG) in a strong magnetic field. It is in fact an example of the integer QHE discussed so far, but the integer QHE (in the wider sense) is not limited to the 2DEG in a magnetic field but can be realized in other systems. For example, realizations of such an integer QHE in ferromagnets (without magnetic field) is called quantum anomalous Hall effect (QAHE), or Chern insulators. After a theoretical proposal of the QAHE, systems realizing the QAHE have been sought for. It was now realized in a Cr-doped Bi_2 Se_3 thin film [15]. To achieve such a system, both the spin–orbit coupling and the spin splitting should be large.

4.1.2.2 *Intrinsic and Extrinsic Hall Effect in Metals*

We then consider the intrinsic Hall effect in metals. In Equation (4.13), the integral over the wavevector $\boldsymbol{k}$ is limited to the interior of the Fermi surface. Such an integral is not quantized and therefore the total intrinsic Hall conductivity is not quantized. Furthermore, in addition to Equation (4.13), there will arise an additional contribution to the Hall conductivity. It is an extrinsic Hall conductivity, contributed from impurity scattering. These intrinsic and extrinsic mechanisms have been discussed in the context of the AHE [16]. In contrast to the Hall effect, which can be attributed to Lorentz force within the classical picture, the AHE is a purely quantum-mechanical effect, and is accounted for by including wave nature of electrons.

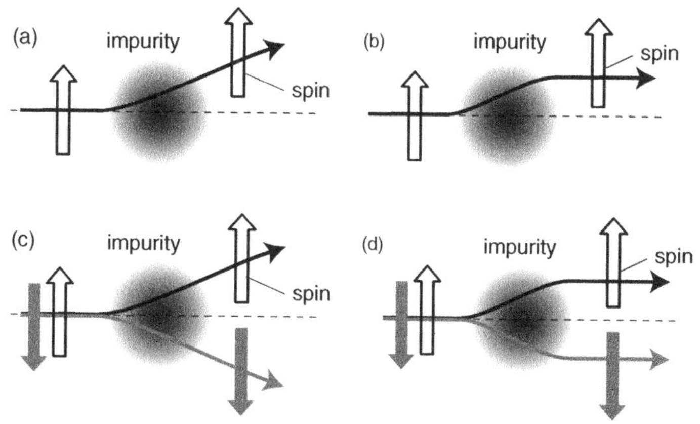

Figure 4.3 *Schematics of the extrinsic mechanisms of the AHE: (a) skew scattering and (b) side jump. Schematics of the extrinsic mechanisms of the SHE: (c) skew scattering and (d) side jump*

Both the intrinsic and extrinsic mechanisms of the AHE require the SOC. In the intrinsic picture, the SOC leads to a spin-dependent Berry curvature, leading to the AHE. On the other hand, in the extrinsic picture, the impurity scattering in the presence of the SOC becomes spin dependent, and it causes the AHE. The extrinsic mechanisms are classified into (i) the skew-scattering mechanism [17], and (ii) the side-jump mechanism [18]. These mechanisms are schematically shown in Figure 4.3(a) and (b). In the skew scattering (Figure 4.3(a)) the wavevector changes at the scattering, whereas the transverse shift at the scattering is attributed to the side jump (Figure 4.3(b)).

The two extrinsic mechanisms for the Hall effect have different scaling rules for the strength of impurity scattering. Strength of impurity scattering, i.e., the impurity concentration or the strength of impurity potentials, is characterized by the conductivity σ_{xx}. When the system becomes cleaner, both σ_{xx} and σ_{xy} increase; we focus how σ_{xy} scales with σ_{xx}. By definition, intrinsic Hall conductivity does not depend on impurity scattering and we have $\sigma_{xy} \propto \sigma_{xx}^0$. On the other hand, the skew scattering leads to a certain Hall angle θ_{H}, which means that σ_{xx} and σ_{xy} are proportional: $\sigma_{xy} \propto \sigma_{xx}$. On the other hand, the side jump leads to the constant value of σ_{xy}, independent of the impurity scattering, and have $\sigma_{xy} \propto \sigma_{xx}^0$, which is similar to the intrinsic contribution. Thus the total Hall conductivity has a crossover behavior depending on the impurity scattering strength. When the system is very clean, the skew scattering is dominant, and σ_{xy} is proportional to σ_{xx}. If the system becomes dirtier, the skew scattering is suppressed, and dominant mechanisms will be the intrinsic one and the side jump. The Hall conductivity then stays almost constant as a function of σ_{xx}. When the system becomes dirtier, the band structure will be lost, and the transport will be of hopping nature. σ_{xy} will decrease again [19].

We have so far explained the extrinsic mechanism for the AHE. The extrinsic mechanisms for the SHE are quite similar to those in the AHE. Namely, the SOC is required both for the intrinsic [2, 3] and for the extrinsic mechanisms [1]. Extrinsic mechanisms are classified into the skew scattering (Figure 4.3(c)) and the side jump (Figure 4.3(d)).

4.1.3 Experimental Observation of Spin Hall Effect in Semiconductors and Metals

Here we briefly explain experimental observations on the SHE. The experimental methods differ between semiconductors and metals, and they are separately argued in the following. For semiconductors, optical methods are useful for measurements of the SHE, such as the Kerr rotation and the spin light-emitting diode (spin LED), as we discuss in the following.

When an electric field is applied to a semiconductor sample, the transverse spin current is induced, and it will eventually reach the edges of the sample. It leads to a steady state of accumulated spins along the edges, with a balance between the incoming spin current, spin diffusion and spin relaxation. Such spin accumulation is measured by Kerr rotation [20]. In the experiment, a laser beam with linear polarization is incident onto the sample, and the rotation angle of the polarization plane at the reflection is measured. The measured rotation angle is proportional to the out-of-plane spin polarization. By scanning the incident beam over the sample, a space-resolved profile of spin accumulation is obtained. As a result, at the edge of n-type semiconductors, spin accumulation is found to distribute uniformly along the edges of the sample [21]. The spin accumulation is the opposite for the opposite sides of the sample, and in the bulk there is no spin accumulation. It is considered to be mainly extrinsic, because the spin splitting in this sample is small.

The SHE in a 2D p-type semiconductor is observed in a p–n junction structure forming a spin LED [22]. An electric field is applied to the p-type semiconductor channel. Then a spin-polarized hole current is induced in the p-type channel due to the SHE, and is injected into the p–n junction, to recombine with electrons. The emitted light is then expected to be circularly polarized due to the spin polarization of the hole current. Indeed the emitted light has a circular polarization around $\sim 1\%$. It is argued that the obtained SHE is mostly intrinsic because the spin splitting in the p-type semiconductor due to SOC is large.

In semiconductors, the carrier concentration is small and therefore the SHE is also small. In contrast, in metals the SHE signal is expected to be much larger. In metals, there are two electrical detection methods for detection of the SHE in metals. One is to use the nonlocal spin injection from a ferromagnet. In this method, a charge current is injected into a circuit including a ferromagnet. The charge current is accompanied by a spin current due to the ferromagnet. It is noted that in contrast to a charge current, a pure spin current can flow into devices which do not form a closed circuit, because of nonconservation of spin current. By utilizing this property of spin current, one can inject pure spin current into devices by spin diffusion, when these devices are in contact with the circuit (but not within the circuit). By the ISHE such an injected spin current will induce a transverse voltage drop, which is to be measured. The first experiment of SHE by the nonlocal spin injection is in Al [23]. From the measured signal, the spin Hall angle is evaluated as $\alpha_{\mathrm{SH}} = \sigma_s/\sigma_c = (1\text{–}3) \times 10^{-4}$ at 4.2 K. It is small even at low temperatures, because of the smallness of the SOC in Al.

The SHE for other metals with larger SOC have also been measured. The SHE in Pt is measured even at room temperature [24], and the spin Hall angle is evaluated as $\alpha_{\mathrm{SH}} = \sigma_s/\sigma_c = 3.7 \times 10^{-3}$. So far the SHE has been measured for various metals. For example, in FePt/Au devices at room temperature [25], the measured spin Hall angle $\alpha_{\mathrm{SH}} = 0.113$ is very large compared with other materials. The mechanism for this large SHE in the FePt/Au device is considered to be extrinsic.

Another method for measurement of SHE in metals is by injection of spins using spin pumping. The nonmagnetic metal, for which the SHE is to be measured, is attached to a ferromagnetic metal. Ferromagnetic resonance is induced by dc magnetic field and transverse ac magnetic field, and magnetization precesses within the ferromagnetic metal. In the magnetization precession, the spin angular momentum dissipates into the nonmagnetic metal, causing a spin injection into it. This is called the spin pumping, The injected spin current is expected to produce the charge current in the nonmagnetic metal due to the ISHE. In the experiments on Pt [26], a voltage drop across Pt due to the ISHE is observed even at room temperature.

As previously noted, the large SHE in platinum can be attributed to the enhanced Berry curvature from the first-principle calculations [14]. The large SHE in platinum is now utilized as a probe of spin current in various experiments. To measure the SHE, one has only to put a Pt bar onto the sample where the spin current is flowing. The spin current is then absorbed into the Pt bar to some extent, and absorbed spin current will induce a transverse voltage drop. The ISHE has been recently applied to the experimental observation of the spin Seebeck effect [27], and to the demonstration of conversions between the spin current in metals and spin-wave spin current in ferromagnetic insulators [28].

4.1.4 Intrinsic Hall Effects for Photons and Magnons

Here we discuss various intrinsic Hall effects for various particles, not limited to electrons. Because the EOM (4.7) and (4.8) for Bloch electrons is based on a wave nature of electrons, they can be applied to other waves, with appropriate modifications. For example, for photons, we can consider a similar EOM for a photonic wavepacket [29–31]

$$\dot{\boldsymbol{r}} = \boldsymbol{\nabla}_k \mathcal{E}_{nk;r} - \dot{\boldsymbol{k}} \times (z|\boldsymbol{\Omega}_{nk}|z), \qquad \dot{\boldsymbol{k}} = -\boldsymbol{\nabla}_r \mathcal{E}_{nkr}, \qquad |\dot{z}) = -i[\dot{\boldsymbol{k}} \cdot \boldsymbol{\Lambda}_{nk}], \tag{4.16}$$

where $|z)$ is a polarization vector, $\mathcal{E}_{nk;r}$ is an eigenfrequency, $\boldsymbol{\Omega}_{nk}$ is a Berry curvature, and $\boldsymbol{\Lambda}_{nk}$ is a gauge field. For a photon in a isotropic uniform medium, $\boldsymbol{\Omega}_{nk} = \mp \boldsymbol{k}/k^3$ corresponds to a left- (right)-circularly polarized light. The EOM for the position $\boldsymbol{r}$ involves a term which is proportional to the Berry curvature, and therefore photons will have the intrinsic Hall effect, similar to the electronic case. In order to induce the Hall effect, one needs to have a nonzero $\dot{\boldsymbol{k}}$, which is proportional to a force exerted onto the particle. For electrons it is the electric field, and for photons it is a gradient of the refractive index. Thus one can realize the Hall effect of photons by using an interface between two media with different refractive index. Thus, at an interface between two media, the anomalous velocity term involving the Berry curvature induces a transverse shift, which is perpendicular to both the interface normal and the incident wavevector $\boldsymbol{k}$. This is in fact the so-called Imbert shift, which has been theoretically proposed in 1955 [32], and later experimentally observed [33, 34]. Thus this Imbert shift can be interpreted as an intrinsic Hall effect by the Berry curvature, and from this one can design the system to have a large Imbert shift, by introducing periodicity to make a photonic crystal with large Berry curvature [29].

Let us turn to magnons, which are low-energy excitations in magnets such as ferromagnets. One can also realize intrinsic Hall effect due to Berry curvature. An EOM is formulated as

$$\dot{\boldsymbol{r}} = \frac{1}{\hbar}\frac{\partial \varepsilon_{n\boldsymbol{k}}}{\partial \boldsymbol{k}} - \dot{\boldsymbol{k}} \times \boldsymbol{\Omega}_n, \quad \hbar\dot{\boldsymbol{k}} = -\nabla U(\boldsymbol{r}), \tag{4.17}$$

where $\Omega_n(\boldsymbol{k})$ is the Berry curvature for magnons in the nth band, and $\varepsilon_{n\boldsymbol{k}}$ is an eigenenergy [35]. $U(\boldsymbol{r})$ is a potential for the magnons. It is however difficult to externally apply a potential $U(\boldsymbol{r})$ at our disposal, in order to induce an intrinsic Hall effect. Instead, temperature gradient can be used as a driving force. From this semiclassical EOM, we can derive the following formula for thermal Hall conductivity:

$$\kappa^{xy} = -\frac{k_{\mathrm{B}T}^2}{\hbar V}\sum_{n,\boldsymbol{k}} c_2(\rho(\varepsilon_{n\boldsymbol{k}}))\Omega_{n,z}(\boldsymbol{k}), \tag{4.18}$$

where $c_2(\rho) = (1+\rho)\left(\log\frac{1+\rho}{\rho}\right)^2 - (\log\rho)^2 - 2\,\mathrm{Li}_2(-\rho)$, $\rho(\varepsilon)$ is the Bose distribution function, and $\mathrm{Li}_2(z)$ is the polylogarithm function. This thermal Hall conductivity is a ratio of a transverse heat-current response to the applied temperature gradient. In this sense, the Berry curvature of magnons appears as a thermal Hall effect of magnons.

For photons and magnons, one can engineer the band structure by introducing artificial periodicity into the system. For photons it is theoretically proposed [36, 37], and experimentally shown [38] that in some photonic crystals with broken time-reversal symmetry, a photonic band gap is opened between bands, and the sum of the Chern number below the gap becomes nonzero. It then follows that there should be a chiral edge modes within the photonic gap, which is similar to that of the integer QHE. For magnons a similar kind of artificial crystals can be theoretically proposed [39, 40]. In some magnonic crystal, which is a periodic structure of ferromagnets, a magnonic gap is opened, and the sum of the Chern numbers below the gap becomes nonzero. Therefore an existence of topological chiral edge modes is proposed. Experimental realization of such a magnonic crystal is yet to be seen.

4.2 Topological Insulators

As the SHE has been investigated theoretically and experimentally, a new topological phase related with the SHE was theoretically proposed [41–43]. It is called a topological insulator (TI), and it was later confirmed experimentally as we explain later. It is shown schematically in Figure 4.4 for two-dimensional (2D) TI [41–43] and for three-dimensional (3D) TI [44–46]. In the bulk, the energy band is gapped and there are no states on the Fermi energy; namely it is a nonmagnetic insulator in the bulk. Meanwhile, states on the edge/surface of the 2D/3D TI are gapless. These edge/surface states are characteristic; they consist of counterpropagating states with opposite spins, carrying a pure spin current. Such a property of edge/surface states is called helical. Such helical states exhibit novel transport properties, since backscattering by nonmagnetic impurities is suppressed. As such, the TIs in total show novel metallic conduction carried by the helical edge/surface states [41–43]. We note that the time-reversal symmetry is preserved in the TI. It means that it is nonmagnetic, and no magnetic field is applied. The pair of helical edge states are degenerate because time-reversal operation transforms these two states into each other.

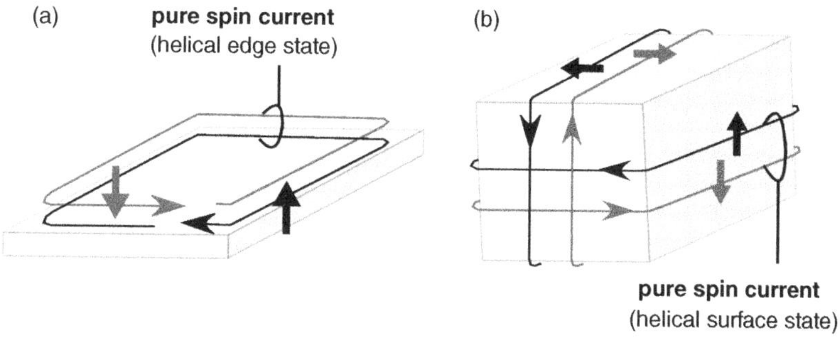

Figure 4.4 *Schematic of the topological insulators for (a) 2D and (b) 3D*

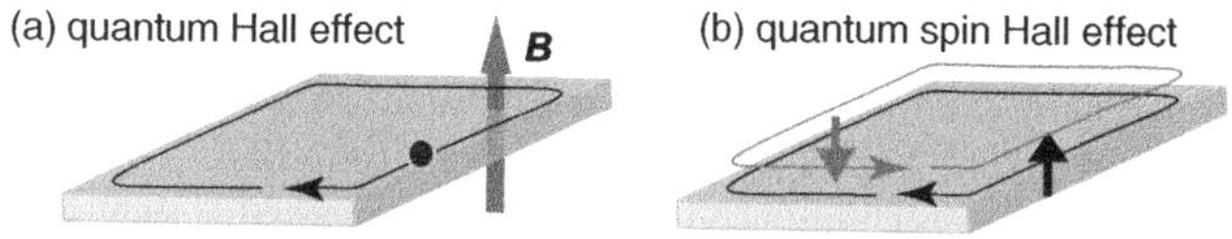

Figure 4.5 *Edge states in (a) quantum Hall effect and (b) quantum spin Hall effect (2D TI)*

This edge/surface spin current circulates around the system in equilibrium. This equilibrium spin current is persistent and cannot be taken out. If an electric field is applied, this circulating spin current no longer balances between the opposite edges/surfaces, and the spin current will flow in the transverse way. In that sense, the TIs can be regarded as an insulator variant of the SHE, and therefore can be called as quantum spin Hall system. This is similar to the relation that the QHE is an insulator variant of the Hall effect. The QHE and the quantum spin Hall effect (QSHE) are similar in that they are insulating in the bulk. Meanwhile, the QHE has chiral (i.e., one-way) edge states, while the QSHE has helical edge/surface states (Figure 4.5).

In theoretical characterization of TIs, we introduce special momenta called time-reversal invariant momenta (TRIM). They are defined as momenta satisfying $\boldsymbol{k} \equiv -\boldsymbol{k}$ (mod $\boldsymbol{G}$). Namely, $\boldsymbol{k}$ and $-\boldsymbol{k}$ differ by a reciprocal lattice vector, meaning that $\boldsymbol{k}$ and $-\boldsymbol{k}$ are equivalent. Specifically, in 2D there are 4 TRIM $\boldsymbol{k} = 0, \boldsymbol{G}_1/2, \boldsymbol{G}_2/2, (\boldsymbol{G}_1 + \boldsymbol{G}_2)/2$, where $\boldsymbol{G}_1, \boldsymbol{G}_2$ are primitive reciprocal vectors. We write them as $\boldsymbol{k} = \boldsymbol{\Gamma}_i (i = 1, 2, 3, 4)$. In 3D, there are 8 TRIM $\boldsymbol{k} = \frac{1}{2}(n_1\mathbf{G}_1 + n_2\boldsymbol{G}_2 + n_3\boldsymbol{G}_3)(n_1, n_2, n_3 = 0, 1)$ where $\boldsymbol{G}_1, \boldsymbol{G}_2, \boldsymbol{G}_3$ are primitive reciprocal vectors. We write them as $\boldsymbol{k} = \boldsymbol{\Gamma}_i (i = 1, 2, \ldots, 8)$.

4.2.1 Two-Dimensional Topological Insulators

The following picture for 2D TIs is useful. Suppose we consider an up-spin system and down-spin system separately. The up-spin system is assumed to be the quantum Hall system with $\mathrm{Ch}_\uparrow = 1$, and the down-spin system is assumed to be the quantum Hall system with $\mathrm{Ch}_\downarrow = -1$. This is realized when a magnetic field along $+z$ is applied for up spins, and a magnetic field along $-z$ is applied for down spins. Their superposition is nothing but the quantum spin Hall system, or the two-dimensional TI (Figure 4.6) [41–43]. Such a

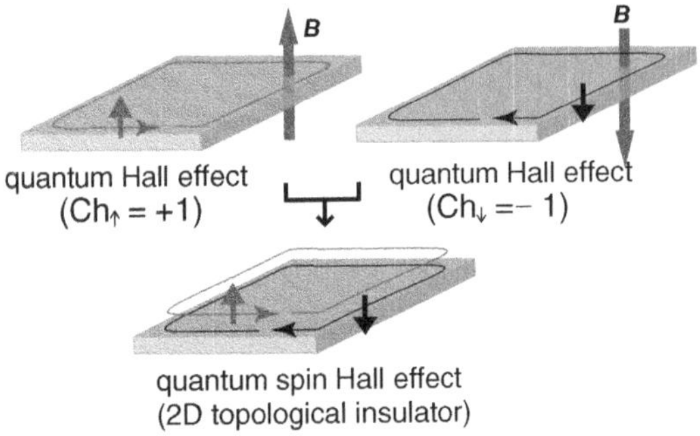

Figure 4.6 *Quantum spin Hall system as a superposition of two quantum Hall systems*

spin-dependent magnetic field cannot be realized as such. Nonetheless, it can be realized physically by the SOC itself. Therefore the TIs require strong SOC.

Here the up-spins and the down-spins do not necessarily mean that they are perpendicular to the plane, but the spins can point to general directions, depending on materials and on details of the edges/surfaces. Meanwhile, a universal property for the 2D TI is that the spins flowing in the clockwise way and those in the counterclockwise way have the opposite directions. This follows from the time-reversal symmetry. They are degenerate because of the time-reversal symmetry, due to Kramers theorem. Therefore, this pair of degenerate states is called a Kramers pair. If the time-reversal symmetry is broken by external magnetic field, time-reversl symmetry is violated, and the topologicall insulator phase is broken.

4.2.1.1 Band Structure of Edge States of 2D Topological Insulators

Edge states appear in the band structure, only when the system geometry has edges, e.g. a semi-infinite plane, or a strip. For a semi-infinite plane, the band structures for normal insulators (NIs) and TIs are compared in Figure 4.7. The Fermi energy is assumed to be within the bulk gap. The wavenumber along the edge direction is a good quantum number, because of the translational symmetry.

First we note that the edge states in the TI lie across the gap, connecting between the valence and the conduction bands, as shown in Figure 4.7(b) [41–43]. Because of the time-reversal symmetry, the dispersion of the edge states is symmetric with respect to $k = 0$, and the states which are transformed to each other by flipping the sign of k have opposite spins. We here note that the time-reversal transformation changes the wavenumber k into $-k$, and changes the spin directions. On the other hand, the NI may not have any edge states within the bulk gap. Even if the edge states exist, they do not run across the gap; they come from the conduction (or valence) band, and enters the same band again as shown in Figure 4.7(a). We note that the NI (Figure 4.7(a)) and the TI (Figure 4.7(b)) cannot be transformed to each other as long as the bulk gap remains open. Thus they are considered to be different phases. One might think that the crossing at $k = 0$ in Figure 4.7(b) is lifted by some hybridization. Nevertheless it never occurs because of the time-reversal symmetry; namely, the degeneracy at $k = 0$ is the Kramers degeneracy, protected by the time-reversal symmetry [47, 48].

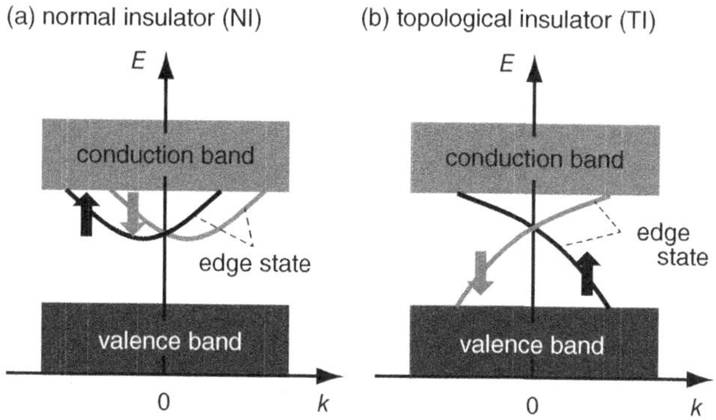

Figure 4.7 Schematic band structures for (a) 2D NI and (b) 2D TI

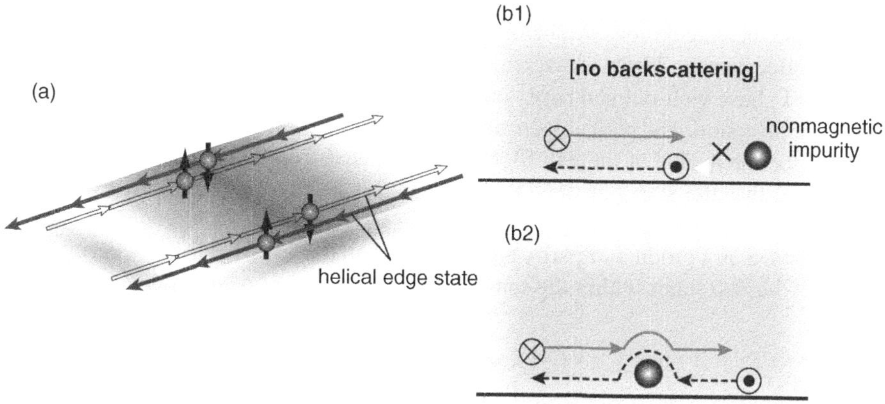

Figure 4.8 Impurity scattering at the edge of 2D TI

Since the electron velocity is expressed as $v = \frac{1}{\hbar}\frac{dE}{dk}$, the edge states (Figure 4.7(b)) in the real space is shown as in Figure 4.8(a). From this figure, an important property of the edge states of 2D TIs is seen; the edge states are immune to backscattering by nonmagnetic impurities [49, 50] (Figure 4.8(b1) and (b2)). Because the two edge states forming a Kramers pair have opposite spins, backscattering will require a spin flip. Nevertheless, nonmagnetic impurities cannot induce spin flip at the scattering. This protection of edge states against elastic scattering is called topological protection. This absence of backscattering has been experimentally confirmed in a transport measurement in HgTe quantum well, which is a 2D TI [51, 52].

4.2.1.2 Z_2 Topological Numbers

We now discuss how to characterize topological insulators. We introduce the Z_2 topological number (Z_2 topological invariant) ν here [41, 47, 48]. The Z_2 topological number can take only two values, $\nu = 0$ and $\nu = 1$; $\nu = 0$ represents the NI, and $\nu = 1$ represents the TI. As we have seen, the Z_2 topological number ν expresses information on the edge states. It is important to note that this Z_2 topological number is calculated from bulk wavefunctions. In this sense, the Z_2 topological number represents a bulk "topological order." If the bulk has a topological order, i.e., if $\nu = 1$, then the edge states should exist and traverse across the bulk gap, irrespective of the details of the edge, which is a highly nontrivial property of the TIs.

We show how to calculate the Z_2 topological number ν. As follows from the previous explanations, one method is to calculate the band structure with a geometry with edges; the dispersion of the edge states clearly show whether it is a NI or a TI. Alternatively, the Z_2 topological number can also be calculated from the bulk, as we explain below. We note that one cannot distinguish between TI and NI only from the bulk band structure. It is because the information of the Z_2 topological number is encoded in the eigenstates, not in the eigenenergies.

For simplicity we restrict ourselves to systems with inversion symmetry [48]. The calculation method of Z_2 topological numbers for systems without inversion symmetry is omitted here because the calculations are involved [47]. We consider the Bloch wavefunctions at the TRIM. Because the TRIM is inversion symmetric ($\boldsymbol{k} \equiv -\boldsymbol{k}$), the Bloch eigenstates at the TRIM $\boldsymbol{k} = \boldsymbol{\Gamma}_i$ have well-defined parity eigenvalues ± 1 : $Pu_{\boldsymbol{k}}(\boldsymbol{r}) = u_{\boldsymbol{k}}(-\boldsymbol{r}) = \pm u_{\boldsymbol{k}}(\boldsymbol{r})$, where P is a space inversion. The parity eigenvalue is ± 1 because P satisfies $P^2 = 1$. At each TRIM $\boldsymbol{\Gamma}_i$, the parity eigenvalue of the mth Bloch state, counted from lower energy, is denoted as $\xi_m(\boldsymbol{\Gamma}_i)$. Note that the states at the TRIM $\boldsymbol{k} = \boldsymbol{\Gamma}_i$ are doubly degenerate due to Kramers theorem, and they share the same parity: $\xi_{2m} = \xi_{2m-1}$ Then the Z_2 topological number ν can be obtained as a product of parity eigenvalues at all the TRIM below the Fermi energy, noting that the two states within the same Kramers pair are counted only once:

$$(-1)^{\nu} = \prod_{i=1}^{4} \prod_{m=1}^{N} \xi_{2m}(\boldsymbol{\Gamma}_i), \tag{4.19}$$

where we assume that there are $2N$ states below the Fermi energy. Since each term is either $+1$ or -1 the product is also ± 1, which corresponds to $\nu = 0$ and $\nu = 1$, respectively. In real materials, by *ab initio* calculations, one can easily calculate the parity eigenvalues for each TRIM, and the Z_2 topological number can be calculated.

Suppose the valence and the conduction bands at some TRIM have the opposite parity eigenvalues. We also assume that by controlling some parameter, their energies will gradually change their order. At this change the Z_2 topological number ν changes between 1 and 0. This is called band inversion, and it accompanies the topological phase change. Conversely, the change of the Z_2 topological number requires the band inversion, and closing of the bulk gap. As long as the bulk gap remains open, the topological number never changes.

4.2.2 Three-Dimensional Topological Insulators

In three-dimensional (3D) TIs [44, 48], there are 8 TRIM. To define the Z_2 topological numbers, we define an index at each TRIM $\boldsymbol{k} = \boldsymbol{\Gamma}_i$ as

$$\delta_i = \prod_{m=1}^{N} \xi_{2m}(\boldsymbol{\Gamma}_i). \tag{4.20}$$

In three dimensions, we define four Z_2 topological numbers $\nu_0, \nu_1, \nu_2, \nu_3$. These Z_2 topological numbers $\nu_0, \nu_1, \nu_2, \nu_3$ take the values 0 or 1, and are written as $\nu_0; (\nu_1\nu_2\nu_3)$. These Z_2 topological numbers are defined by

$$(-1)^{\nu_0} = \prod_{i=1}^{8} \delta_i, \quad (-1)^{\nu_j} = \prod_{i=(n_1n_2n_3),n_j=1} \delta_i. \tag{4.21}$$

Namely, ν_0 is defined from the indices δ_i at all the TRIM, whereas the others, $\nu_j (j = 1, 2, 3)$, are defined as a product of indices δ_i for four TRIM on the plane in the reciprocal space $\boldsymbol{k} = \frac{1}{2}(n_1\boldsymbol{G}_1 + n_2\boldsymbol{G}_2 + n_3\boldsymbol{G}_3), n_j = 1$. As a result there are 16 possible phases labeled by $\nu_0; (\nu_1\nu_2\nu_3)$.

These Z_2 topological numbers lead to properties of the surface Fermi surface in the following way. First we project the bulk Brillouin zone (BZ) onto the surface; by this projection the eight TRIM are projected onto the four surface TRIM in the surface BZ. We make the product of two indices δ_i and δ_j for two TRIM, which are projected onto the same surface TRIM point $\boldsymbol{\Lambda}_a$. Let π_a denote the product $\pi_a = \delta_i\delta_j$. Thereby four indices $\pi_a (= \pm 1)$ for the four surface TRIM $\boldsymbol{\Lambda}_a$ is obtained. From these four indices π_a the surface Fermi surface obeys the following rules:

- The number of surface Fermi surfaces running between two TRIM with the same indices π_a is even.
- The number of surface Fermi surfaces running between two TRIM with the different indices π_a is odd.

In various TIs, these rules are confirmed through comparison between the calculation of Z_2 topological numbers and the experimentally observed surface Fermi surfaces.

From these rules of the surface Fermi surfaces, if $\nu_0 = 1$, there exists at least one surface Fermi surface. Namely, every surface of any crystallographic orientation supports gapless surface states and is metallic. From this property, systems with $\nu_0 = 1$ are called strong topological insulators (STIs) and systems with $\nu_0 = 0$ are called weak topological insulators (WTIs) [44]. In a STI every surface has gapless surface states, whereas in a WTI some surfaces do not have gapless surface states. In fact, a WTI can be regarded as a layered structure of 2D TIs with weak interlayer coupling. On the other hand, in STI three-dimensionality is essential.

Let us consider an example of Bi_2Se_3 and Bi_2Te_3. In these materials the *ab initio* calculation for bulk band structure gives the Z_2 topological number of 1;(000), meaning that it is a STI [53]. Among the indices on the 8 TRIM, only the index at the TRIM $\boldsymbol{k} = 0$ is -1 while others are $+1$. In the surface BZ for the (111) surface, where these materials cleave, the surface TRIM are one $\bar{\Gamma}$ point and three $\bar{\text{M}}$ points. The index π_a is -1 for $\bar{\Gamma}$ and $+1$ for $\bar{M}$ points. Among various possibilities of the topology of the surface Fermi surface, the simplest possibility is to have a single surface Fermi surface encircling the point $\boldsymbol{k} = 0(\bar{\Gamma})$. The surface states obtained from experiments and from the *ab initio* calculation agree with this expectation. The dispersion of the surface states forms a Dirac cone, a linearly dispersed band with degeneracy at $\boldsymbol{k} = 0$ as shown in Figure 4.9(a). This dispersion corresponds to a variant of the edge states of the 2D TIs (Figure 4.7(b)). Namely, if we rotate Figure 4.7(b) around the vertical axis, the surface Dirac cone (Figure 4.9) is obtained. This emergence of the surface Fermi surface around $\boldsymbol{k} = 0$ is traced back to the parity eigenvalue -1 at $\boldsymbol{k} = 0$,

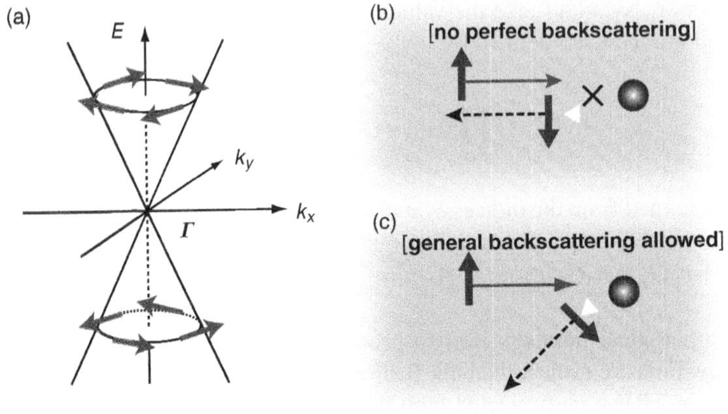

Figure 4.9 (a) Surface Dirac cone on the surface of a 3D TI. (b) and (c) Scattering by non-magnetic impurities on the surface of 3D TIs

which is different from other TRIM in the 3D BZ. This different index at $\boldsymbol{k} = 0$ comes from a band inversion at $\boldsymbol{k} = 0$ due to the SOC.

Up to the linear order in $\boldsymbol{k}$, the Hamiltonian for this Dirac cone is

$$H = \lambda(\boldsymbol{\sigma} \times \boldsymbol{k})_z = \lambda \begin{pmatrix} 0 & k_y + ik_x \\ k_y - ik_x & 0 \end{pmatrix}, \tag{4.22}$$

where $\boldsymbol{k} = (k_x, k_y)$ is a wavevector along the surface, z represents a coordinate along the surface normal, and λ is a constant. The energy eigenvalues are $E = s\lambda k, s = \pm 1$, forming a Diracc cone. The eigenstates are

$$\Psi_{sk} \propto \begin{pmatrix} 1 \\ -is(k_x + ik_y)/k \end{pmatrix} \tag{4.23}$$

and therefore the spins are given by $\langle \Psi_{sk} | \boldsymbol{\sigma} | \Psi_{sk} \rangle = s\left(\frac{k_y}{k}, -\frac{k_x}{k}, 0\right)$, perpendicular to the wavevector $\boldsymbol{k}$. The velocity is $\boldsymbol{v} = \frac{1}{\hbar}\frac{\partial E}{\partial \boldsymbol{k}}$, which is always perpendicular to the spins, as shown in Figure 4.9(a). The speed is a constant λ. Therefore, the velocity and the spin of an electron within the Dirac cone is locked to each other.

Dirac-cone dispersions are realized not only in the surfaces of 3D TIs, but also in graphene, for example. Nevertheless, there are two important differences between the graphene Dirac cone and the TI Dirac cone. The Dirac cones in graphene are degenerate for up- and down spins, and there are two Dirac points (K, K') in the BZ. In contrast, the Dirac cone on the surface of a TI does not have spin degeneracy; a state at each $\boldsymbol{k}$ has a fixed spin direction, and it leads to novel transport properties. We note that the band structure with only one Dirac cone within the Brillouin zone, e.g., on the surface of Bi_2Se_3, cannot be realized in two-dimensional systems, because it is theoretically shown that the number of Dirac cones in the Brillouin zone for two-dimensional systems should be even [54]. Hence surface states with a single Dirac cone are unique to TI surfaces.

In 2D TIs, the edge states consist of two opposite spins propagating in opposite directions, and therefore backscattering by nonmagnetic impurities is forbidden by time-reversal symmetry. On the other hand, in 3D TIs, perfect backscattering, which changes the wavevector by 180°, is forbidden (Figure 4.9(b)). Meanwhile, other backscattering is suppressed but not forbidden (Figure 4.9(c)). Therefore the surface transport is diffusive and it shows relatively good conduction similar to graphene.

4.2.3 Experiments on Topological Insulators

TIs require time-reversal symmetry; the systems should be nonmagnetic insulators. In addition the Z_2 topological numbers should be nontrivial. Nevertheless from the bulk band structure one cannot tell TIs from NIs. Therefore it is desirable to have a criterion for choosing candidate materials for TIs, before resorting to numerical calculations. In TIs, band inversions due to the SOC are required, because systems without the SOC are NIs. In some materials one can consider how band inversions occur by increasing the SOC from zero. When the SOC is absent, the Z_2 topological numbers are trivial (zero). By increasing the SOC, some band inversions occur, and the parity eigenvalues of the bands involved in the band inversion tell us how the Z_2 topological numbers change upon band inversion. Because the magnitude of the SOC is some hundreds of meV at maximum, the bulk gap of a TI is at most of that order of magnitude, and it should belong to narrow-gap semiconductors. Robustness of the TIs against various perturbations such as temperature and disorder is determined by the size of the bulk gap. As long as the energy scales of the perturbations are smaller than the bulk gap, the properties of the TI remain. Therefore, it is vital to look for TIs with large bulk gaps.

4.2.3.1 Two-Dimensional Topological Insulators: Bi Thin Film and HgTe Quantum Wells

Bismuth (Bi) is a semimetal with large SOC, having three tiny electron pockets and a tiny hole pocket. Although it is not an insulator, it is quite close to an insulator. Therefore, it is probable that bismuth can be made insulating by some external perturbations. One is to make bismuth into a thin film. Because of the finite-size effect, a thin film tends to open a gap compared with bulk 3D crystals. Indeed, some *ab initio* calculations show that it should become insulating in a bilayer film. From these considerations, Bi bilayer was theoretically proposed to be a 2D TI [55]. The predicted nontrivial topological number and the existence of the topological edge states are confirmed by *ab initio* calculations [56]. Nevertheless, in reality, another thin-film structure is more stable than the bulk-like structure, and it is not easy to experimentally confirm this TI in Bi bilayer in bulk-like structure. Nonetheless, it is recently confirmed by experiments using STM [57].

On the other hand, CdTe/HgTe/CdTe quantum well has been intensively studied both theoretically and experimentally [51–53, 58]. After a theoretical prediction as a 2D TI [53], it was confirmed to be a TI by conductance measurement [51, 52]. CdTe is a cubic semiconductor similar to GaAs, and is a NI but not a TI. Meanwhile in HgTe, due to the strong SOC the valence and conduction bands are inverted, compared with other cubic semiconductors. Nevertheless, because of the cubic symmetry, the bulk gap is closed at $\boldsymbol{k} = 0$ in HgTe, and it is called zero-gap semiconductor. It is not a TI because the bulk gap is not opened due to cubic symmetry. One way to make it into a TI is to make a quantum-well structure, by sandwiching the HgTe layer by CdTe layers, in order to break cubic symmetry. In this quantum well, there are two states, E1 and H1, which are close to the Fermi energy at $\boldsymbol{k} = 0$. Their parities are the opposite at $\boldsymbol{k} = 0$. The Fermi energy lies between them, and which of the two states lies below the Fermi energy depends on the width of the HgTe layer d. Therefore, at a critical width $d = d_c = 60$Å a topological phase transition occurs due to a band inversion at $\boldsymbol{k} = 0$. It was proposed to be a 2D TI when $d > d_c$, and a NI when $d < d_c$ [53].

The Fermi energy in this quantum well can be controlled by gate voltage. The physics of TIs appear only when the Fermi energy lies in the bulk gap, namely between the E1 and the H1 levels. The transport through the quantum well is then contributed only by the edge states on both sides (Figure 4.8(a)) of the quantum well, and the total conduction shows two-channel conductance $2e^2/h$. This expectation is confirmed in experiments, showing the existence of helical edge states [51]. Another experimental evidence for the helical edge states is nonlocal conduction, by using multiterminal geometry. By changing voltage and current probes in various ways, nonlocal conductances become e^2/h times a simple fraction, as predicted from the Landauer–Büttiker formalism [52].

4.2.3.2 Three-Dimensional Topological Insulator $Bi_{1-x}Sb_x$

We explained that bismuth (Bi) is close to a TI phase, and making a ultrathin film leads to a TI phase. Another way to drive Bi to a TI phase is to dope Sb into Bi to make $Bi_{1-x}Sb_x$. By increasing x, the carrier pockets vanish and it was predicted to become TI at $0.07 < x < 0.22$ [59]. The indices are $+1$ at the L points and -1 at the other TRIM. Hence in a (111) surface of $Bi_{1-x}Sb_x$, the indices π_a at the surface TRIM are $+1$ at the Γ point, and -1 at the three M points. Hence from the previous arguments, there should be an odd number of surface Fermi surfaces between the Γ point and the M points.

Indeed, by angle-resolved photoemission spectroscopy (ARPES), there are 5 [60] or 3 [61] surface Fermi surfaces, which confirmed that $Bi_{1-x}Sb_x$ is the first experimental material for the 3D TI. Furthermore by spin-resolved ARPES [61], each surface state has a fixed spin direction and is free from spin degeneracy. This further guarantees that it is a 3D STI. Scattering by impurities can be probed by scanning tunneling spectroscopy (STS). A Fourier transform of the STS image tells us about the momentum change at impurity scattering. The measured STS image is consistent with the results with each band having no spin degeneracy, and with suppressed perfect backscattering, which confirms again that it is a 3D STI [62].

4.2.3.3 Three-Dimensional Topological Insulators: Bi_2Se_3, Bi_2Te_3

Bi_2Se_3 and Bi_2Te_3 are 3D TIs which have been studied most [63–66]. These materials have bulk gaps larger than that of $Bi_{1-x}Sb_x$. In Bi_2Se_3 the gap is as large as 0.3 eV. The surface states and their spin states are measured by ARPES [64–66] and spin-resolved ARPES [64]. As a result, the surface Fermi surface is found to be a circle enclosing $\boldsymbol{k} = 0$ as shown in Figure 4.9. They have fixed spin directions as shown in Figure 4.9 as arrows. As previously noted, the Z_2 topological numbers are calculated to be 1;(000) [63], in agreement with the measured surface states. Moreover, STS measurement confirms absence of perfect backscattering in the surface states [67]. In Bi_2Se_3 the surface Fermi surface deviates from a circle to a shape like a snowflake, as the Fermi energy deviates from the Dirac point [68].

This is called a Fermi-surface warping, and it agrees with a theoretical consideration on crystallographic symmetry [69]. In reality it is not easy to realize TI phases in these materials, because Bi_2Se_3 tends to be n-doped and Bi_2Te_3 tends to be p-doped by extrinsic reasons. Hence, in order to realize TIs, the carrier density should be controlled to bring the Fermi energy inside the bulk gap. By this reason, transport measurements often suffer from considerable contributions from bulk carriers.

4.3 Summary

Since the proposals for the extrinsic SHE in 1971 and the intrinsic SHE in 2003, there have been a vast number of studies on SHE, both in theories and in experiments. Unexpectedly, an insulating variant of the SHE has led to a discovery of new topological phases called topological insulators, which have grown up to a big research field in condensed matter physics. These new phenomena are caused by the spin–orbit coupling; the spin–orbit coupling enters the Hamiltonian in a unique way, similar to the gauge coupling, and directly modulates the phases of electronic wavefunctions. In reality, explicit forms of the spin–orbit coupling depend on materials and system geometry such as surfaces and heterostructures. From such a large variety of the spin–orbit coupling, we expect there will be more phenomena due to spin–orbit coupling which are to be discovered in the future.

Acknowledgment

This work is partially supported by PRESTO, Japan Science and Technology Agency (JST), Japan.

References

1. M. I. Dyakonov and V. I. Perel, *JETP Lett.* **13**, 467 (1971); *Phys. Lett* **A35**, 459 (1971).
2. S. Murakami, N. Nagaosa, and S.-C. Zhang, *Science* **301**, 1348 (2003).
3. J. Sinova, , D. Culcer, Q. Niu, *et al.*, *Phys. Rev. Lett.* **92**, 126603 (2004).
4. M. V. Berry, *Proc. R. Soc. London* **A392**, 45 (1984).
5. D. J. Thouless, M. Kohmoto, M. P. Nightingale, and M. den Nijs, *Phys. Rev. Lett.* **49**, 405 (1982).
6. M. Kohmoto, *Ann. Phys.* **160**, 343 (1985).
7. K. Ohgushi, S. Murakami, and N. Nagaosa, *Phys. Rev. B* **62**, R6065 (2000).
8. Y. Taguchi, Y. Oohara, H. Yoshizawa, *et al.*, *Science* **291**, 2573 (2001).
9. T. Jungwirth, Q. Niu, and A. H. MacDonald, *Phys. Rev. Lett.* **88**, 207208 (2002).
10. Z. Fang, N. Nagaosa, K. S. Takahashi, *et al.*, *Science* **302**, 92 (2003).
11. E. N. Adams and E. I. Blount, *J. Phys. Chem. Sol.* **10**, 286 (1959).
12. E. I. Blount, Formalisms of band theory. In: F. Seitz, D. Turnbull, (Eds.), *Solid State Physics* (Academic Press, New York, 1962) p. 305.
13. G. Sundaram and Q. Niu, *Phys. Rev. B* **59**, 14915 (1999).
14. G. Y. Guo, S. Murakami, T. W. Chen, and N. Nagaosa, *Phys. Rev. Lett.* **100**, 096401 (2008).

15. C.-Z. Chang, J. Zhang, X. Feng, *et al.*, *Science* **340**, 167 (2013).
16. R. Karplus and J. M. Luttinger, *Phys. Rev.* **95**, 1154 (1954).
17. J. Smit, *Physica* **21**, 877 (1955); *Physica* **24**, 39 (1958).
18. L. Berger, *Phys. Rev. B* **2**, 4559 (1970).
19. S. Onoda, N. Sugimoto, and N. Nagaosa, *Phys. Rev. Lett.* **97**, 126602 (2006).
20. Y. K. Kato and D. D. Awschalom, *J. Phys. Soc. Jpn.* **77**, 031006 (2008).
21. Y. K. Kato, R. C. Myers, A. C. Gossard, and D. D. Awschalom, *Science* **306**, 1910 (2004).
22. J. Wunderlich, B. Kaestner, J. Sinova, and T. Jungwirth, *Phys. Rev. Lett.* **94**, 047204 (2005).
23. S. O. Valenzuela and M. Tinkham, *Nature* **442**, 176 (2006).
24. T. Kimura, Y. Otani, T. Sato, *et al.*, *Phys. Rev. Lett.* **98**, 156601 (2007).
25. T. Seki, Y. Hasegawa, S. Mitani, *et al.*, *Nature Mater.* **7**, 125 (2008).
26. E. Saitoh, M. Ueda, H. Miyajima, and G. Tatara, *Appl. Phys. Lett.* **88**, 182509 (2006).
27. K. Uchida, S. Takahashi, K. Harii, *et al.*, *Nature* **455**, 778 (2008).
28. Y. Kajiwara, K. Harii, S. Takahashi, *et al.*, *Nature* **464**, 262 (2010).
29. M. Onoda, S. Murakami, and N. Nagaosa, *Phys. Rev. Lett.* **93**, 083901 (2004); *Phys. Rev. E* **74**, 066610 (2006).
30. A. V. Dooghin , N. D. Kundikova, V. S. Liberman, and B. Ya. Zel'dovich, *Phys. Rev. A* **45**, 8204 (1992).
31. V. S. Liberman and B. Ya. Zel'dovich, *Phys. Rev. A* **46**, 5199 (1992).
32. F. I. Fedorov, *Dokl. Akad. Nauk SSSR* **105**, 465 (1955).
33. C. Imbert, *Phys. Rev. D* **5**, 787 (1972).
34. O. Hosten, P. Kwiat, *Science* **319**, 787 (2008).
35. R. Matsumoto and S. Murakami, *Phys. Rev. Lett.* **106**, 197202 (2011); *Phys. Rev. B* **84**, 184406 (2011).
36. F. D. M. Haldane and S. Raghu, *Phys. Rev. Lett.* **100**, 013904 (2008).
37. S. Raghu and F. D. M. Haldane, *Phys. Rev. A* **78**, 033834 (2008).
38. Z. Wang, Y. Chong, J. D. Joannopoulos, and M. Soljačić, *Nature* **461**, 772 (2008).
39. R. Shindou, J. I. Ohe, R. Matsumoto, *et al.*, *Phys. Rev. B* **87**, 174402 (2013).
40. R. Shindou, R. Matsumoto, S. Murakami, and J. Ohe, *Phys. Rev. B* **87**, 174427 (2013).
41. C. L. Kane and E. J. Mele, *Phys. Rev. Lett.* **95**, 146802 (2005).
42. C. L. Kane and E. J. Mele, *Phys. Rev. Lett.* **95**, 226801 (2005).
43. B. A. Bernevig and S.-C. Zhang, *Phys. Rev. Lett.* **96**, 106802 (2006).
44. L. Fu, C. L. Kane, and E. J. Mele, *Phys. Rev. Lett.* **98**, 106803 (2007).
45. J. E. Moore and L. Balents, *Phys. Rev. B* **75**, 121306 (2007).
46. R. Roy, *Phys. Rev. B* **79**, 195322 (2009).
47. L. Fu and C. L. Kane, *Phys. Rev. B* **74**, 195312 (2006).
48. L. Fu and C. L. Kane, *Phys. Rev. B* **76**, 045302 (2007).
49. C. Wu, B. A. Bernevig, and S.-C. Zhang, *Phys. Rev. Lett.* **96**, 106401 (2006).
50. C. Xu and J. E. Moore, *Phys. Rev. B* **73**, 045322 (2006).
51. M. König, S. Wiedmann, C. Brüne, *et al.*, *Science* **318**, 766 (2007).
52. A. Roth, C. Brüne, H. Buhmann, *et al.*, *Science* **325**, 294 (2009).
53. B. A. Bernevig, T. L. Hughes, and S.-C. Zhang, *Science* **314**, 1757 (2006).
54. H. B. Nielsen and M. Ninomiya, *Nucl. Phys.* **B185**, 20 (1981).
55. S. Murakami, *Phys. Rev. Lett.* **97**, 236805 (2006).

56. M. Wada, S. Murakami, F. Freimuth, and G. Bihlmayer, *Phys. Rev. B* **83**, 121310(R) (2011).
57. C. Sabater, D. Gosálbez-Martínez, J. Fernández-Rossier, *et al.*, *Phys. Rev. Lett.* **110**, 176802 (2013).
58. M. König, H. Buhmann, L. W. Molenkamp, *et al.*, *J. Phys. Soc. Jpn.* **77**, 031007 (2008).
59. L. Fu and C. L. Kane, *Phys. Rev. B* **79**, 161408 (2009).
60. D. Hsieh, D. Qian, L. Wray, *et al.*, *Nature* **452**, 970 (2008), D. Hsieh, Y. Xia, L. Wray, *et al.*, *Science* **323**, 919 (2009).
61. A. Nishide, A.A. Taskin, Y. Takeichi, *et al.*, *Phys. Rev. B* **81**, 041309 (2010).
62. P. Roushan, J. Seo, C. V. Parker, *et al.*, *Nature* **460**, 1106 (2009).
63. H. Zhang, C-X. Liu, X-L. Qi, *et al.*, *Nat. Phys.* **5**, 438 (2009).
64. D. Hsieh, Y. Xia, D. Qian, *et al.*, *Nature* **460**, 1101 (2009).
65. D. Hsieh, Y. Xia, D. Qian, *et al.*, *Phys. Rev. Lett.* **103** (2009) 146401.
66. Y. Xia, D. Qian, D. Hsieh, *et al.*, *Nat. Phys.* **5**, 398 (2009).
67. T. Zhang, P. Cheng, X. Chen, *et al.*, *Phys. Rev. Lett.* **103**, 266803 (2009).
68. Y. L. Chen, J. G. Analytis, J.-H. Chu, *et al.*, *Science* **325**, 178 (2010).
69. L. Fu, *Phys. Rev. Lett.* **103**, 266801 (2009).

5

Spin Torque (Domain Wall Drive, Magnetization Reversal)

Akinobu Yamaguchi
Laboratory of Advanced Science and Technology for Industry, University of Hyogo, Hyogo, Japan

5.1 Introduction

Magnetization reversal, namely change of spin structure, is an important property of magnetism. The study on magnetization reversal in an artificial low-dimensional magnet has stimulated the discovery of novel phenomena in thin films and multilayers, and suggested the potential application of thin-films systems to magnetic storage media and sensor technology [1]. One of them is current-induced magnetization reversal [2–5] or current-driven magnetic domain wall (DW) displacement [6–13]. The displacement of the DW, which exists between neighboring domains, within each of which magnetic moments align, occurs magnetization reversal. In a DW, the direction of magnetic moments gradually changes as schematically shown in Figure 5.1(a). When an electric current flows through a DW, there are basically two possibilities: reflection or transmission [6, 13]. Magnetization reversal induced by an electric current through the s–d interaction has been theoretically predicted and experimentally demonstrated [6–12]. The idea of switching spin structure using the s–d interaction was firstly discussed by Berger in 1978 [6]. When a spin-polarized current passes through a ferromagnetic conductor, the transfer of spin angular momentum from the conduction electrons exerts a torque on the magnetic moment. As a consequence, the electric current can displace the DW as schematically illustrated in Figure 5.1(b). Since the discovery of the spin-transfer torque (STT), it has received lots of attention not only

Spintronics for Next Generation Innovative Devices, First Edition. Edited by Katsuaki Sato and Eiji Saitoh.
© 2015 John Wiley & Sons, Ltd. Published 2015 by John Wiley & Sons, Ltd.

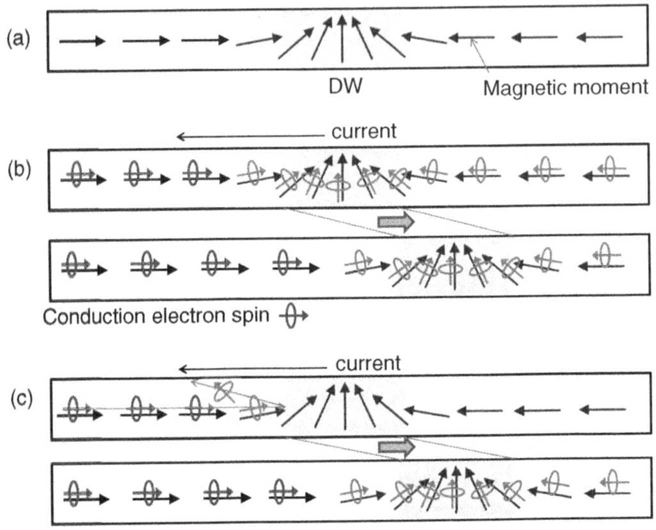

Figure 5.1 (a) Schematic depiction of a magnetic domain wall (DW) structure in a ferromagnetic wire. Frame format of the DW displacement induced (b) by the adiabatic spin angular momentum from conduction electron spins to the magnetic moments within the DW and (c) by the momentum transfer mechanism

due to its potential application such as magnetic random access memory but also its exotic physics.

In the vortex-typed DW displacement induced by the STT, the magnetic vortex plays an important role on the magnetization reversal process. The magnetic vortex is characterized by two binary properties: a chirality of the in-plane rotating magnetization and a polarity of the vortex core magnetization direction. The study on the vortex gyration confined in a disk attracts considerable attention in the physics of magnetization reversal [14–18].

In this chapter, we review a DW displacement and a vortex core gyration induced by the contributions of the STT and the Oersted field. We also distinguish the polarity and chirality of the vortex core in a $Ni_{81}Fe_{19}$ disk using a current-induced rectifying effect. This manuscript is organized as follows. The studies on the current-induced DW displacement are described in Section 5.2. The underlying mechanisms responsible for the experimental setup are presented. Section 5.3 describes the experimental results of the DW displacement and the rectifying spectra of the vortex gyration. We explain the analysis of the rectifying spectra. In Section 5.4, the conclusions are summarized.

5.2 Experiment: Current-Driven DW Displacement in a Magnetic Nanowire

A designed special L-shaped magnetic wire with a round corner was schematically illustrated in Figure 5.2(a). One end of the L-shaped magnetic wire is connected to a diamond-shaped pad which acts as a DW injector, and the other end is sharpen pointed to prevent the nucleation of a DW from this end [6]. The wires of 10-nm-thick $Ni_{81}Fe_{19}$ were fabricated

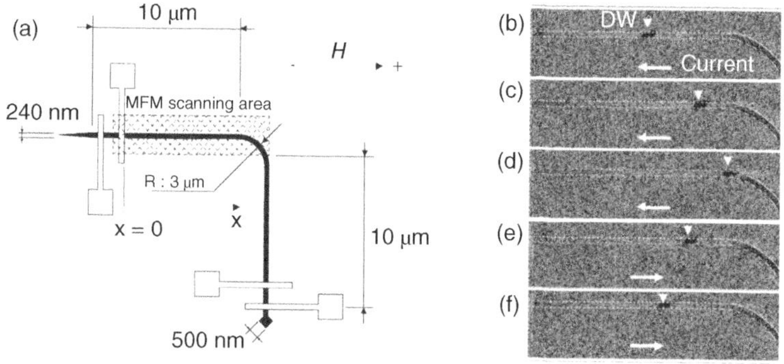

Figure 5.2 *(a) Schematic illustration of a top view of the L-shaped wire made of 10-nm-thick* $Ni_{81}Fe_{19}$*. One end of the wire is connected to a diamond-shaped pad which acts as a DW injector, and the other end is sharply pointed to prevent a nucleation of a DW from this end. The wire has four Cu electrodes. Magnetic force microscope (MFM) observations were performed for the hatched area at room temperature. (b)–(f) Successive MFM images with one pulse applied between each consecutive image. The current density and the pulse duration were* 7.0×10^{11} *A/m*^{2} *and 0.5μs, respectively. The white arrows show the direction of the pulsed current*

onto thermally oxidized Si substrates by means of e-beam lithography and a lift-off method. The width of the wire is 240 nm. The 20-nm-thick Cu electrodes are connected for electrical transport measurements. Magnetic force microscope (MFM) observations were performed for the hatched area in Figure 5.2(a) at room temperature.

A single DW can be introduced into the MFM scanning area from the diamond-shaped pad by manipulation of an external magnetic field. The MFM observations were carried out in the absence of a magnetic field. The existence of the single DW in the MFM scanning area was confirmed as shown in Figure 5.2(b). The DW is imaged as a dark contrast which corresponds to the stray field from a negative magnetic charge, indicating that a tail-to-tail DW is realized. Pinning effect due to a local structural defect enables us to observe the DW using MFM because a stray field from the MFM probe was too small to change the magnetic structure and position of the DW.

Figures 5.2(b)–(f) show successive MFM images with one pulsed current applied between each consecutive image. The current density and the pulse duration were 7.0×10^{11} A/m^2 and 0.5μs, respectively [6]. The direction of the pulsed current is represented as a white-colored arrow. As shown in Figures 5.2(b)–(f), the direction of the DW displacement is opposite to the current direction. The fact that both head-to-head and tail-to-tail DWs are displaced opposite to the current direction indicates that the DW displacement is not caused by a magnetic field due to the current but spin-transfer mechanism. The critical current density below which the DW cannot be driven by the current was observed to be about 6.7×10^{11} A/m^2. It has been suggested theoretically that the intrinsic and extrinsic critical current densities are proportional to the product of the hard-axis magnetic anisotropy and pinning, respectively. Recently, it has been experimentally shown that the critical current density depends on the cross-sectional shape of wires, composition of materials. The Joule heating due to the high current density plays an important role on the current-driven DW displacement [6, 19–21]. The sample

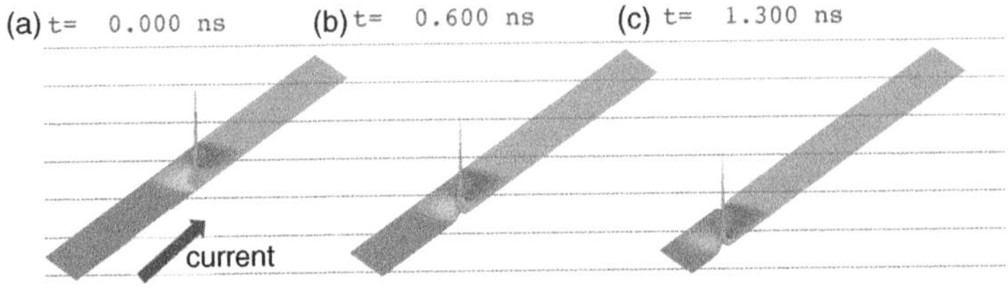

Figure 5.3 Results of micromagnetics simulation for the current-induced DW displacement at t = (a) 0.0, (b) 0.6, and (c) 1.3 ns after the application of the dc current. The size of the calculated model was the same as the sample for the experiment except for the length of the wire

temperature increase induced by the Joule heating decreases the saturation magnetization and crystalline anisotropy. The reduction of the saturation magnetization and crystalline anisotropy results in the decrease of not only the intrinsic pinning but also the extrinsic pinning [12, 13]. As a result, the critical current density, required for the DW displacement, can be decreased with increasing the sample temperature.

Another important physical mechanism for the current-driven DW displacement is the internal spin structure of a DW. The DW has a complicated internal spin structure as shown in Figure 5.3. Spin torque depends on the spin structure, which can be considered to be the fact that the analytical theories do not agree well with experimental results. In particular, the vortex core displaced perpendicular to the dc current while the DW proceeded in the opposite direction of the current as shown in Figures 5.3(a)–(c). Thus, the vortex core plays an important role in the magnetization switching process. Therefore, in the following section, we focus on vortex gyration, which is one of the typical spin structures and is suitable for the investigation.

5.3 EXPERIMENT: Electrical Spectroscopy of Vortex State and Gyration in a Magnetic Disk

Rectifying effect provides highly sensitive detection of the information and dynamics of microscopic spin structure [17, 18, 22–25]. These rectifying spectra can precisely detect the vortex dynamics. The electrical measurement allows us to determine the vortex properties, as well as the strength of the driving torque, accurately and directly through developed analytical calculation [18]. The resonant properties were probed using a microwave rectifying effect. A radio-frequency (rf) current is used to generate small amplitude excitation of magnetization in a magnet with confined structure.

A vortex in a micron- or nanostructured magnetic thin-film element forms when the in-plane magnetization curls around a vortex core. The vortex core magnetization turns out of the plane to minimize the exchange energy. The vortex states are characterized by two binary properties, a chirality and a polarity as shown in Figure 5.4. The chirality describes direction of the in-plane rotating magnetization. The clockwise and counterclockwise direction correspond to the parameters $C = +1$ and -1, respectively. The polarity

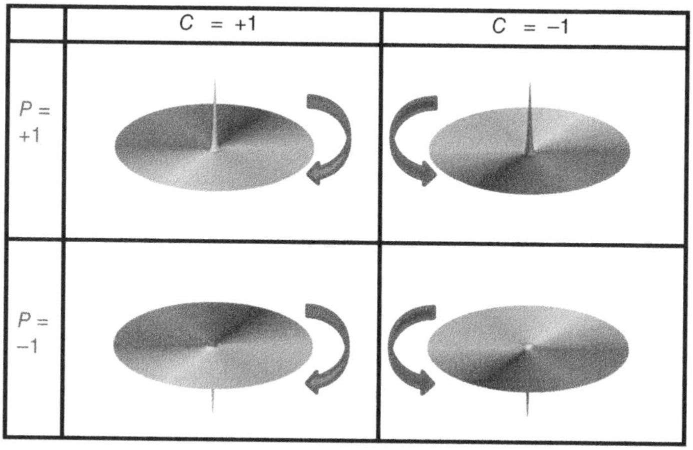

Figure 5.4 *Four vortex states with the Boolean parameters (P, C) = (±1, ±1)*

represents direction of the vortex core magnetization, denoting up ($P = +1$) or down ($P = -1$). Thus, an understanding of the stability and dynamical behavior of the magnetic vortex provides comprehension of current-driven DW displacement because the vortex core is the starting point in the current-driven DW displacement as shown in Figure 5.3.

The rectifying voltage spectrum induced by the vortex core gyration can be calculated using the following modified Thiele equation including the spin torque term [14]

$$\mathbf{G}(P) \times (\mathbf{u} - \dot{r}) = -\frac{\delta U}{\delta \mathbf{r}} - \alpha D\dot{r} + \beta D\mathbf{u}, \tag{5.1}$$

where $\mathbf{G}(P) = -PG_0\mathbf{e}_z = -P\frac{2\pi M_S}{\gamma}\mathbf{e}_z$ is the gyrovector. M_S, γ, and P denote the saturation magnetization, gyromagnetic ratio and spin polarization, respectively; $\mathbf{e}_i$ ($i = x, y, z$) indicates the unit vector. The potential energy combined with the Zeeman, magnetostatic, and exchange energies, in the system is given by $U(r) = \frac{1}{2}\kappa r^2 + \mathbf{q} \cdot (\mathbf{H} \times \mathbf{r})$, where $\mathbf{H}$ is the magnetic field vector and $\mathbf{q} = Cq_0\mathbf{e}_z = \pi CM_SLr_{\text{disk}}\mathbf{e}_z$ defines the stiffness coefficient vector. Here, C, L and r_{disk} are the chirality, thickness and radius of disk, respectively. $\mathbf{u}$ is the spin–torque interaction term of the current and the magnetization, which is given by $\mathbf{u} = u_0e^{i\omega t}e_x$. Here, $u_0 = -\frac{\mu_B J}{eM_S}$, where J is the current density, μ_B is Bohr magnton and e is the electron charge. Here, we assumed that the current is injected to the x-axis. α is the Gilbert damping constant, β is the nonadiabatic contribution to the spin-transfer torque, and D is the diagonal element of the damping tensor, which is given by $D = \frac{G_0}{2}\ln\left(\frac{r_{\text{disk}}}{\xi}\right)$ with vortex core radius ξ.

We firstly obtain the oscillation amplitude of the vortex core [18],

$$\begin{pmatrix} X \\ Y \end{pmatrix} = \begin{pmatrix} X' + iX'' \\ P\left(Y' + iY''\right) \end{pmatrix} \tag{5.2}$$

with

$$X'(\omega) = u_0 \frac{\tilde{\beta}\tilde{\kappa}(\omega_r^2 - \omega^2) + (1 + \tilde{\alpha}\tilde{\beta})\alpha_*\omega^2}{(1 + \tilde{\alpha}^2)\{(\omega_r^2 - \omega^2)^2 + (\alpha_*\omega)^2\}} + h \frac{Cq_0\{\tilde{\kappa}(\omega_r^2 - \omega^2) + \alpha_*\tilde{\alpha}\omega^2\}}{(1 + \tilde{\alpha}^2)\{(\omega_r^2 - \omega^2)^2 + (\alpha_*\omega)^2\}}$$

$$X''(\omega) = u_0\omega \frac{-\tilde{\beta}\tilde{\kappa}\alpha_* + (1 + \tilde{\alpha}\tilde{\beta})(\omega_r^2 - \omega^2)}{(1 + \tilde{\alpha}^2)\{(\omega_r^2 - \omega^2)^2 + (\alpha_*\omega)^2\}} + h\omega \frac{Cq_0\{-\tilde{\kappa}\alpha_* + \tilde{\alpha}(\omega_r^2 - \omega^2)\}}{(1 + \tilde{\alpha}^2)\{(\omega_r^2 - \omega^2)^2 + (\alpha_*\omega)^2\}}$$

$$Y'(\omega) = u_0 \frac{\tilde{\kappa}(\omega_r^2 - \omega^2) + (\tilde{\alpha} - \tilde{\beta})\alpha_*\omega^2}{(1 + \tilde{\alpha}^2)\{(\omega_r^2 - \omega^2)^2 + (\alpha_*\omega)^2\}} - h \frac{Cq_0\alpha_*\omega^2}{(1 + \tilde{\alpha}^2)\{(\omega_r^2 - \omega^2)^2 + (\alpha_*\omega)^2\}} \quad (5.3)$$

$$Y''(\omega) = u_0\omega \frac{-\tilde{\kappa}\alpha_* + (\tilde{\alpha} - \tilde{\beta})(\omega_r^2 - \omega^2)}{(1 + \tilde{\alpha}^2)\{(\omega_r^2 - \omega^2)^2 + (\alpha_*\omega)^2\}} - h\omega \frac{Cq_0(\omega_r^2 - \omega^2)}{(1 + \tilde{\alpha}^2)\{(\omega_r^2 - \omega^2)^2 + (\alpha_*\omega)^2\}}.$$

Here, X' (X'') and Y' (Y'') are the x and y component of the real part (imaginary part) of the oscillation amplitude of the vortex core position, respectively. Additionally, we introduce the following reduced parameters to simplify the notation:

$$\tilde{\alpha} = \alpha\frac{D}{G_0}, \quad \tilde{\beta} = \beta\frac{D}{G_0}, \quad \tilde{\kappa} = \frac{\kappa}{G_0}, \quad \tilde{q_0} = \frac{q_0}{G_0}, \quad \alpha_* = \frac{2\tilde{\alpha}\tilde{\kappa}}{1 + \tilde{\alpha}^2}, \quad \omega_r = \frac{\tilde{\kappa}}{\sqrt{1 + \tilde{\alpha}^2}}. \quad (5.4)$$

As seen in Equation (5.3), the polarity P is included only in the y component. This is because the polarity-dependent gyroforce $\mathbf{G}(P) \times \mathbf{u}$ points in the direction of the y-axis. Thus, the vortex core position can be analytically traced.

The rectification occurs when rf current flows across a magnet and the magnetoresistance oscillates because of the magnetization precession induced by the STT or rf magnetic field. As in homodyne detection, a dc output voltage is the result of rectification due to the combination of an rf current and an oscillating resistance. The rectification can provide highly sensitive detection of spin dynamics in a nanoscale magnet. Here, the electrical resistance arising from the anisotropic magnetoresistance (AMR) effect in the $Ni_{81}Fe_{19}$ disk depends on the position of the vortex core. When the direction of the electrical current is parallel to the magnetization, AMR increases. On the other hand, AMR decreases when the direction of the current is perpendicular to the magnetization. Considering the relationship between the vortex core position and the resistance change, we calculate the rectifying spectrum based on the approximation that the resistance oscillates when the vortex core gyrates near the center of the disk. Additionally, considering that the rf-Oersted field dominantly drives the vortex core dynamics in the present electrical measurement system with a micron-scale radius, we approximate the rectifying spectrum in the following form [18]:

$$V_{dc} = -\frac{I_0 a\tilde{q}_0^2 h(\omega_r^2 - \omega^2)}{(1 + \tilde{\alpha}^2)\{(\omega_r^2 - \omega^2)^2 + (\alpha_*\omega)^2\}} H_y + \frac{I_0 a \frac{\tilde{q}_0^2}{\tilde{\kappa}} h\alpha_*\omega^2}{(1 + \tilde{\alpha}^2)\{(\omega_r^2 - \omega^2)^2 + (\alpha_*\omega)^2\}} PH_x. \quad (5.5)$$

The first term is proportional to the y component of the external field. This spectrum shape correlates with the dispersion function. The second term is proportional to the polarity and the x component of the external field. This second term provides the Lorentzian-typed spectrum. According to Equation (5.5), it is possible to detect the sign of the polarity by the application of the static field in the x direction. For detection of chirality, we focus on the external field dependence of the spectrum amplitude. The asymmetric electrode structure enables us to detect chirality even if the contribution of the spin torque is negligible because the current and the proportionality constant of AMR are dependent on the chirality and electrode structure. Therefore, we can detect both polarity and chirality by using analytical calculation and electrical spectrum measurement with the asymmetric electrode.

Schematic illustration of the measurement circuit and photo-image of magnetic disk are shown in Figure 5.5. A $Ni_{81}Fe_{19}$ disk with a diameter of 3 μm and a thickness of 30 nm was directly fabricated onto a polished MgO substrate by means of e-beam lithography and lift-off technique. The coplanar wave guide structure made from Cr (5 nm)/Au (80 nm) was connected to the disk, and the center conductive strip line was placed on the disk. Here, the center of the electrode gap was shifted to the $+x$ direction as shown in Figure 5.5. A sinusoidal continuous wave rf current with a power of 5.0×10^{-5} W was subsequently injected into the disk by a signal generator in the frequency range from 50 to 150 MHz. A rectified dc voltage between the electrodes was measured through a bias tee, which separates the dc and rf components of the current. The coordinate system used in this study is also shown in Figure 5.5. In order to control the vortex polarity and chirality, we designed the circular disk with two tags on both sides [26]. All measurements were performed at room temperature.

The sense of direction of the static magnetic field is defined as positive along the $+x$ $(+y)$ direction. For clarity, the measured spectra are shifted vertically.

Figure 5.6(a) shows the external field dependence of the rectifying voltage spectra at the angle parallel to the electrode for $(P, C) = (+1, -1)$ and $(-1, -1)$, respectively. As seen in Figure 5.6(a), the result indicates that the sign of the rectifying spectra is dependent on the direction of the static field and polarity. These results correlate well with the analytical calculations shown in Equation (5.5).

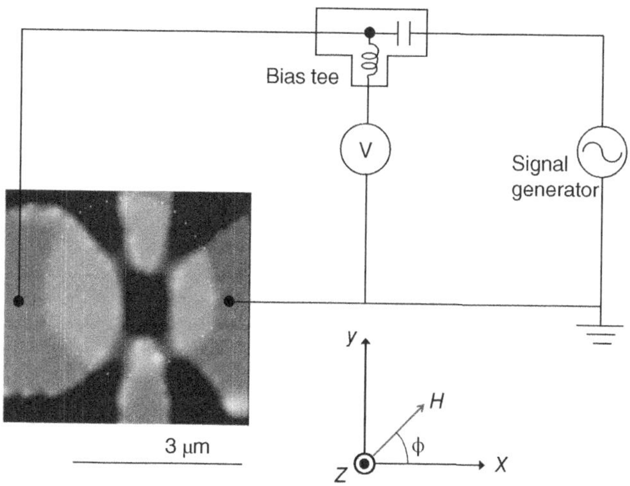

Figure 5.5 Schematic diagram of measurement circuit and atomic force microscope image of the disk. 30-nm-thick $Ni_{81}Fe_{19}$ disk with two tags were patterned directly onto a polished MgO substrate. The diameter of the disk was 3 μm. An external magnetic field H was applied in the substrate plane at an angle φ from the x-axis parallel to the horizontal electrode

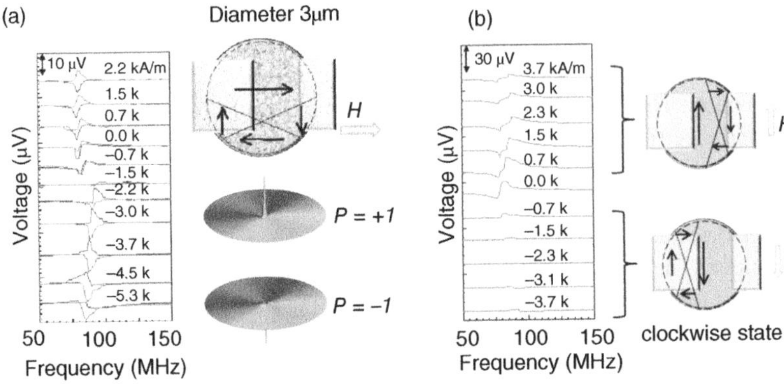

Figure 5.6 *(a) External magnetic field dependence of rectifying voltage spectra at the angle* $\phi = 0^\circ$ *for* $(P, C) = (+1, -1)$ and $(-1, -1)$. *The red and blue solid curves correspond to the spectra in the case of* $(P, C) = (+1, -1)$ and $(-1, -1)$, *respectively. (b) Field dependence of rectifying spectra for the sample with asymmetric electrodes at the angle* $\phi = 90^\circ$ *for* $(P, C) = (+1, -1)$

Next, to detect chirality, we measured the rectifying spectra in the field perpendicular to the electrode. Figure 5.6(b) shows the external magnetic field dependence of the rectifying spectra for $(P, C) = (-1, +1)$ and $(-1, -1)$, respectively. As shown in Figure 5.6(b), we were able to detect the resonant signals from the vortex gyration above the field $H = -0.7k$ A/m. The asymmetry in the external field dependence can be explained as follows; when the static field is applied in the direction perpendicular to the electrode, the vortex core position displaces in the x-direction. The rectifying spectra were only observed while the vortex core gyrated between the electrodes. On the other hand, when the vortex displaces under the electrode, the rectifying spectrum vanishes because the magnetoresistance does not oscillate. As described the above, we can evaluate the ratio of the driving torques such as the spin transfer torques and Oersted field torque by fitting the rectifying spectrum. In particular, the driving force of the Oersted field from the rf current is dominant in this case. Thus, this rectifying spectroscopy of vortex gyration enables us not only to detect both polarity and chirality but also to distinguish the respective contribution of driving torques such as the STT and Oersted field. In addition, the spectroscopy also provides the detection of ultrafast magnetic vortex core switching induced by the topological inverse Faraday effect [27].

5.4 Conclusion

This chapter reviewed a real-space observation of the current-driven magnetic domain wall displacement in a nanoscale ferromagnetic wire and a highly sensitive electric detection of the vortex core dynamics which plays an important role in the current-driven domain wall displacement. Our investigation sheds light on understanding vortex dynamics and current-driven domain wall displacement, and helps develop nonvolatile magnetic random access memory or microwave devices applications.

Acknowledgements

We thank Prof. Nakatani for micromagnetics simulation. We also acknowledge support from the JST PRESTO program.

References

1. B. Hillebrands and K. E. Ounadjela, *Spin Dynamics in Confined Magnetic Structures*, Springer, Berlin, 2003.
2. J. C. Slonczewski, Current-driven excitation of magnetic multilayers, *J. Magn. Magn. Mater.*, **159**, L1–L7 (1996).
3. L. Berger, Emission of spin waves by a magnetic multilayer traversed by a current, *Phys. Rev. B*, **54**, 9353–9358 (1996).
4. J. A. Katine, F. J. Albert, R. A. Buhrman, *et al.*, Current-driven magnetization reversal and spin-wave excitations in Co/Cu/Co pillars, *Phys. Rev. Lett.*, **84**, 3149–3152 (2000).
5. J. Grollier, V. Cros, A. Hamzic, *et al.*, Spin-polarized current induced switching in Co/Cu/Co pillars, *Appl. Phys. Lett.* **78**, 3663–3665 (2001).
6. L. Berger, Low-field magnetoresistance and domain drag in ferromagnets, *J. Appl. Phys.*, **49**, 2156–2161 (1978); Exchange interaction between ferromagnetic domain wall and electric current in very thin metallic films, *J. Appl. Phys.*, **55**, 1954–1956 (1988); Exchange interaction between electric current and magnetic domain wall containing Bloch lines, *ibid*, **63**, 1663–1669 (1988).
7. P. P. Feritas and L. Berger, Observation of s–d exchange force between domain walls and electric current in very thin Permalloy films, *J. Appl. Phys.* **57**, 1266–1269 (1985); C. Y. Hung and L. Berger, Exchange forces between domain wall and electric current in permalloy films of variable thickness, *J. Appl. Phys.* **63**, 4276–4278 (1988).
8. A. Yamaguchi, T. Ono, S. Nasu, *et al.*, Real-space observation of current-driven domain wall motion in submicron magnetic wires, *Phys. Rev. Lett.* **92** 077205-1–077205-4 (2004); A. Yamaguchi, S. Nasu, H. Tanigawa, *et al.*, Effect of Joule heating in current-driven domain wall motion, *Appl. Phys. Lett.* **86**, 012511-1-012511-3–(2005).
9. N. Vernier, D. A. Allwood, D. Atkinson, *et al.*, Domain wall propagation in magnetic nanowires by spin-polarized current injection, *Europhys. Lett.*, **65**, 526–532 (2004).
10. C. K. Lim, T. Devolder, C. Chappert, *et al.*, Domain wall displacement induced by subnanosecond pulsed current, *Appl. Phys. Lett.*, **84**, 2820–2822 (2004).
11. G. S. D. Beach, C. Knutson, C. Nistor, *et al.*, Nonlinear domain-wall velocity enhancement by spin-polarized electric current, *Phys. Rev. Lett.*, **97**, 057203-1–057203-4 (2006).
12. M. Kläui, P.-O. Jubert, R. Allkenspach, *et al.*, Direct observation of domain-wall configurations transformed by spin currents, *Phys. Rev. Lett.*, **95**, 026601-1–026601-4 (2005).
13. M. Yamanouchi, D. Chiba, F. Matsukura, and H. Ohno, Current-induced domain-wall switching in a ferromagnetic semiconductor structure, *Nature*, **428**, 539–541 (2004); M. Yamanouchi, D. Chiba, F. Matsukura, T. Dietl, and H. Ohno, Velocity of domain-wall motion induced by electrical current in the ferromagnetic semiconductor

(Ga, Mn)As, *Phys. Rev. Lett.*, **96**, 096601-1–096601-4 (2006); M. Yamanouchi, J. Ieda, F. Matsukura, *et al.*, Universality classes for domain wall motion in the ferromagnetic semiconductor (Ga, Mn)As, *Science*, **317**, 1726–1729 (2007).
14. M. Hayashi, L. Thomas, Ya B Bazaliy, *et al.*, Dependence of current and field driven depinning of domain walls on their structure and chirality in permalloy nanowires, *Phys. Rev. Lett.*, **97**, 207205-1–207205-4 (2006).
15. Y. Togawa, T. Kimura, K. Harada, *et al.*, Current-excited magnetization reversal under in-plane magnetic field in a nanoscaled ferromagnetic wire, *Appl. Phys. Lett.*, **92**, 012505-1–012505-3 (2008).
16. T. Koyama, D. Chiba, K. Ueda, *et al.*, Observation of the intrinsic pinning of a magnetic domain wall in a ferromagnetic nanowire, *Nat. Mat.*, **10**, 194–197 (2011).
17. G. Tatara and H. Kohno, Theory of current-driven domain wall motion: Spin transfer versus momentum transfer, *Phys. Rev. Lett.*, **92**, 086601-1–086601-4 (2004); G. Tatara, H. Kohno, and J. Shibata, Microscopic approach to current-driven domain wall dynamics, *Phys. Rep.*, **468**, 213–301 (2008).
18. A. Zhang and Z. Li, Roles of nonequilibrium conduction electrons on the magnetization dynamics of ferromagnets, *Phys. Rev. Lett.* **93**, 127204-1–127204-4 (2004).
19. A. Thiaville, Y. Nakatani, J. Miltat, and Y. Suzuki, Micromagnetic understanding of current-driven domain wall motion in patterned nanowires, *Europhys. Lett.*, **69**, 990–996 (2005).
20. J. Shibata, G. Tatara, and H. Kohno, Effect of spin current on uniform ferromagnetism: Domain nucleation, *Phys. Rev. Lett.*, **94**, 076601-1–076601-4 (2005).
21. K. Yu. Guslienko, W. Scholz, R. W. Chantrel, and V. Novosad, Vortex-state oscillations in soft magnetic cylindrical dots, *Phys. Rev. B*, **71**, 144407-1–144407-8 (2005).
22. B. Van Waeyenberge, A. Puzic, H. Stoll, *et al.*, Magnetic vortex core reversal by excitation with short bursts of an alternating field, *Nature*, **444**, 461–464 (2006); M. Bolte, G. Meier, B. Krüger, *et al.*, Time-resolved X-ray microscopy of spin-torque-induced magnetic vortex gyration, *Phys. Rev. Lett.*, **100**, 176601-1–176601-4 (2008).
23. S. Kasai, Y. Nakatani, K. Kobayashi, *et al.*, Current-driven resonant excitation of magnetic vortices, *Phys. Rev. Lett.*, **97**, 107204-1–107204-4 (2006); S. Kasai, K. Nakano, K. Kondo, *et al.*, Three-terminal device based on the current-induced magnetic tunnel junction, *Appl. Phys. Exp.*, **1**, 091302-1–091302-3 (2008).
24. A. Yamaguchi, K. Motoi, H. Miyajima, *et al.*, Current manipulation of a vortex confined in a micron-sized $Fe_{19}Ni_{81}$ disk, *Appl. Phys. Lett.*, **95**, 122506-1–122506-3 (2009).
25. M. Goto, H. Hata, A. Yamaguchi, *et al.*, Electric spectroscopy of vortex states and dynamics in magnetic disks, *Phys. Rev. B*, **84**, 064406-1–064406-9 (2011).
26. C. You, I. Sung and B. Joe, Analytic expression for the temperature of the current-heated nanowire for the current-induced domain wall motion, *Appl. Phys. Lett.*, **89**, 222513-1–222513-3 (2006).
27. H. Fangohr, D. S. Chernyshenko, M. Franchin, *et al.*, Joule heating in nanowires, *Phys. Rev. B*, **84**, 054437-1–054437-12 (2011).
28. A. Yamaguchi, A. Hirohata, T. Ono, and H. Miyajima, Temperature estimation in a ferromagnetic Fe–Ni nanowire involving a current-driven domain wall motion, *J. Phys: Condens. Matter.*, **24**, 024201-1–024201-9 (2012).

29. A. A. Tulapurkar, Y. Suzuki, A. Fukushima, *et al.*, Spin-torque diode effect in magnetic tunnel junctions, *Nature*, **438**, 339–342 (2005).
30. J. C. Sankey, P. M. Braganca, A. G. F. Garcia, *et al.*, Spin-transfer-driven ferromagnetic resonance of individual nanomagnets, *Phys. Rev. Lett.*, **96**, 227601-1–227601-4 (2006).
31. A. Yamaguchi, H. Miyajima, T. Ono, *et al.*, Rectification of radio frequency current in ferromagnetic nanowire, *Appl. Phys. Lett.*, **90**, 182507-1–182507-3 (2007); A. Yamaguchi, K. Motoi, A. Hirohata, H. Miyajima, *et al.*, Broadband ferromagnetic resonance $Ni_{81}Fe_{19}$ wires using a rectifying effect, *Phys. Rev. B*, **78**, 104401-1–104401-11 (2008); A. Yamaguchi, K. Motoi, A. Hirohata, and H. Miyajima, Anomalous Hall voltage rectification and quantized spin-wave excitation induced by simultaneous application of dc and rf currents in a single-layered $Ni_{81}Fe_{19}$ nanoscale wire, *Phys. Rev. B*, **79**, 224409-1–224409-11 (2009).
32. H. J. Juretschke, Electromagnetic theory of dc effects in ferromagnetic resoance, *J. Appl. Phys.*, **31**, 1401–1406 (1960).
33. M. V. Costache, M. Sladkov, C. H. van der Wal, and B. J. van Wees, On-chip detection of ferromagnetic resonance of a single submicron Permalloy strip, *Appl. Phys. Lett.*, **89**, 192506-1–192506-3 (2006).
34. N. Mecking, Y. S. Gui, and C.-M, Hu, Microwave photovoltage and photoresistance effects in ferromagnetic microstrips, *Phys. Rev. B*, **76**, 224430 (2007); Y. S. Gui, N. Mecking, X. Zhou, *et al.*, Realiztion of a room-temperature spin dynamo: The spin rectification effect, *Phys. Rev. Lett.*, **98**, 107602-1–107602-4 (2007).
35. T. Taniuchi, M. Oshima, H. Akinaga, and K. Ono, Vortex-chirality control in mesoscopic disk magnets observe by photoelectron emission microscopy, *J. Appl. Phys.*, **97**, 10J904-1–10J904-3 (2005).
36. K. Taguchi, J. Ohe, and G. Tatara, Ultrafast magnetic vortex cores switching driven by the topological inverse Faraday effect, *Phys. Rev. Lett.*, **109**, 127204-1–127204-5 (2012).

6
Spin Pumping

Matthias Althammer,[1] *Mathias Weiler,*[1,2] *Hans Huebl,*[1] *and Sebastian T.B. Goennenwein*[1]
[1] *Walther-Meißner-Institut für Tieftemperaturforschung, Bayerische Akademie der Wissenschaften, Garching, Germany*
[2] *National Institute of Standards and Technology, Boulder, CO, USA*

This chapter is devoted to *spin pumping* [1–4]. Spin pumping is the Onsager reciprocal of the spin transfer torque (STT) effect [5–12], the latter being discussed in more detail *in Chapter 5 by Yamaguchi*. Both STT and spin pumping are related to the transfer of spin angular momentum between a ferromagnet[1] (F) and an appropriate reservoir. In STT, a spin (polarized) current injected from the reservoir into the ferromagnet transfers angular momentum onto the magnetization **M** of the latter, thereby inducing magnetization dynamics or even a reorientation of the magnetization in F. The situation is inverted for spin pumping, where nonequilibrium magnetization dynamics (magnons) in F act as the source for an angular momentum flow, from the ferromagnet into the reservoir.

With a concise definition of spin currents and their characteristics given in *Chapter 3 by Uchida et al.*, we here simply state that a directed flow of (spin) angular momentum is nothing else than a spin current [13, 14]. Thus, one can think of the spin-pumping process as of a spin current, which carries a certain fraction of the angular momentum due to the nonequilibrium magnetization excitations, out from the ferromagnet into the adjacent angular momentum reservoir. In other words, in the spin pumping process, the excited, precessing magnetization in F relaxes back toward thermal equilibrium by sourcing a pure spin current into the reservoir. In many cases, the reservoir is a "normal" (nonferromagnetic) metal (N).

Two different experimental detection schemes for the spin-pumping process in F/N hybrid structures are well established [1, 15–24]. Both of them are described in more detail in this chapter. In one approach, one takes advantage of the fact that the outflow

[1] For the sake of simplicity, we here use the term "ferromagnet" in the sense of "exhibiting long range magnetic order with a finite spontaneous magnetization." We thus will refer to both ferrimagnets and ferromagnets simply as "ferromagnets."

Spintronics for Next Generation Innovative Devices, First Edition. Edited by Katsuaki Sato and Eiji Saitoh.
© 2015 John Wiley & Sons, Ltd. Published 2015 by John Wiley & Sons, Ltd.

of spin current from F provides a channel for the relaxation of the excited magnetization in F. One thus detects spin pumping spectroscopically, via the corresponding increase in (Gilbert) damping of coherent magnetization precession in F. In the other approach, the spin current is detected electrically, via the inverse spin Hall effect in the normal metal, which converts the spin current into a charge current. In this electrically detected spin-pumping scheme, the normal metal thus acts as both an angular momentum sink and as a spin current detector.

6.1 Spin Pumping and Magnetization Damping

We start by discussing the spin-pumping scheme and its relation to magnetization damping, with the conceptual idea summarized in Figure 6.1. Consider a ferromagnetic (F) slab with its macroscopic magnetization **M**, as sketched in panel (a). In thermal equilibrium, **M**

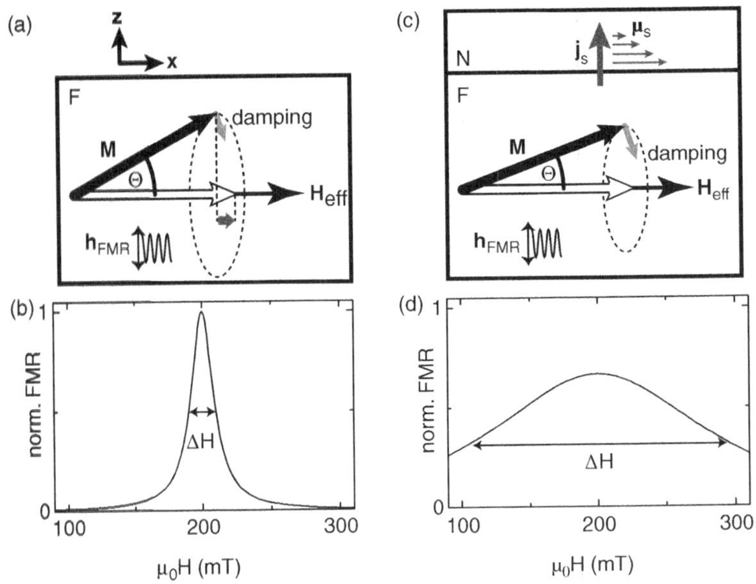

Figure 6.1 *(a) When the magnetization **M** (solid black arrow) of a ferromagnet (F) is driven out of its thermodynamic equilibrium orientation (open arrow) via resonant excitation by a microwave magnetic field* $\mathbf{h}_{FMR}$*, it precesses with a nonzero cone angle Θ. Note that the excited magnetization has nonequilibrium angular momentum, with the horizontal black arrow indicating the corresponding longitudinal component. (b) The resonant absorption of microwave radiation sketched in (a) results in characteristic high-frequency magnetic susceptibility, which can be detected in ferromagnetic resonance (FMR) experiments. The magnetization relaxation mechanisms determine the FMR line width ΔH. (c) In ferromagnet/normal metal (F/N) heterostructures, an additional magnetization damping mechanism arises via the emission of excess angular momentum across the interface, from F into N. This flow of angular momentum can be described as a so-called pure spin current* $\mathbf{j}_s$ *with (longitudinal) spin polarization* **s**. *Thus, exciting a resonant magnetization precession (FMR) in F allows to "pump" a spin current from F into N. (d) The spin-pumping mechanism constitutes an additional magnetization damping mechanism, broadening the FMR line width*

(open white arrow) aligns along the effective magnetic field $\mathbf{H}_{\text{eff}}$ in the ferromagnet [25, 26]. $\mathbf{H}_{\text{eff}}$ hereby comprises the externally applied magnetic field, as well as internal magnetic fields parameterizing demagnetization, magnetic anisotropy, and exchange energies. Upon the application of continuous wave microwave radiation with a frequency ω appropriate to excite ferromagnetic resonance (FMR), $\mathbf{M}$ is driven into a steady-state, resonant precessional motion around $\mathbf{H}_{\text{eff}}$ [25, 26]. In a simple picture, the precessing magnetization (full black arrow in Figure 6.1(a)) then encloses a nonzero angle Θ with $\mathbf{H}_{\text{eff}}$, the so-called precession cone angle. In a more elaborate approach, the magnetization dynamics and the FMR are modeled using the Landau–Lifshitz–Gilbert (LLG) equation [25, 26]

$$\frac{d\mathbf{M}}{dt} = -\gamma\mu_0(\mathbf{M} \times \mathbf{H}_{\text{eff}}) + \frac{\alpha}{|\mathbf{M}|}\left(\mathbf{M} \times \frac{d\mathbf{M}}{dt}\right), \tag{6.1}$$

with the gyromagnetic ratio γ, the vacuum permeability μ_0, and the so-called Gilbert damping parameter α describing viscous magnetization damping. More complex, non-Gilbert-like contributions to magnetization relaxation have been omitted in Equation (6.1) [25, 27, 28]. Solving the LLG equation yields the FMR condition (the relation between the microwave frequency and the external magnetic field magnitude required for resonant magnetization excitation), the time-dependent magnetization motion, as well as the dynamic magnetic susceptibility. The latter determines the resonant microwave absorption or dispersion properties of the ferromagnet observed in FMR experiments [25, 26]. Figure 6.1(b) exemplarily shows the FMR absorption of a ferromagnetic film with $\mathbf{M}$ in the film plane. For fixed microwave frequency, the FMR condition is met for a particular value of the external magnetic field H_{res}, at which the microwave absorption peaks with a characteristic FMR line width ΔH. If the magnetization damping is purely Gilbert type, and in the absence of inhomogeneous line width broadening, one can infer the Gilbert damping parameter α directly from the FMR line width. However, for a robust determination of α, FMR experiments as a function of microwave frequency are mandatory, with the external magnetic field orientation carefully adjusted to suppress two-magnon scattering. More details about FMR and magnetization damping processes can be found, e.g., in [25, 27, 28].

Coming back to the simple FMR scenario sketched in Figure 6.1(a), the steady-state balance between magnetization excitation provided by the microwave radiation (acting to increase Θ) and magnetization damping (acting to reduce Θ) determines the precession cone angle Θ [29]. In a Gedanken experiment, let us now abruptly switch off the microwave radiation. The excited, precessing magnetization (full black arrow in Figure 6.1(a)) now must change its angular momentum state in order to relax back along its equilibrium orientation ($\mathbf{M}||\mathbf{H}_{\text{eff}}$, open white arrow). The longitudinal angular momentum change required for this magnetization relaxation is depicted by the small black arrow in Figure 6.1(a).[2] In an isolated ferromagnetic slab, the angular momentum transfer required for magnetization relaxation is dominantly provided by the ferromagnet's lattice, via the magnetization damping mechanisms parameterized by the Gilbert α in Equation (6.1). In contrast, in a ferromagnet/normal metal (F/N) hybrid structure, the spin-pumping mechanisms opens an additional channel for magnetization relaxation. As sketched in Figure 6.1(c), the excess angular momentum now also can flow out or be "pumped" across the interface, from the ferromagnet into the normal metal – hence the name "spin pumping" [1–4, 18, 19]. Please

[2] Note, however, that the precessing magnetization has both longitudinal (along $\mathbf{H}_{\text{eff}}$) and transverse (perpendicular to $\mathbf{H}_{\text{eff}}$) nonequilibrium angular momentum components.

note that in the spin-pumping process, both longitudinal and transverse angular momenta are transferred, with the latter providing the dominant contribution to damping for small Θ.

As we have just seen, the additional angular momentum relaxation channel provided by spin pumping in F/N hybrids increases the magnetization damping in these structures. Spin pumping thus manifests itself as an increase in FMR line width, arising upon covering a ferromagnetic specimen with a normal metal layer, as sketched in Figure 6.1(d). Following this notion, FMR experiments have been extensively used to study the spin-pumping process [1, 15–20, 22–24]. Note also that as per definition, spin pumping is an interface-related process, such that a measurable increase in magnetization damping owing to spin pumping is to be expected only in thin ferromagnetic layers.

The spin current density (the spin current per interface area) pumped across the F/N interface by the spin-pumping mechanism is given by[3] [2–4]

$$\mathbf{j}_s^{\text{pump}} = \frac{\hbar}{4\pi}\left\{\text{Re}(g^{\uparrow\downarrow})\left[\mathbf{m}\times\frac{d\mathbf{m}}{dt}\right] - \text{Im}(g^{\uparrow\downarrow})\frac{d\mathbf{m}}{dt}\right\}. \tag{6.2}$$

Hereby, the spin mixing conductance $g^{\uparrow\downarrow}$ quantifies the available spin transport channels at the interface, $\mathbf{m} = \mathbf{M}/|\mathbf{M}|$ is the magnetization orientation unit vector, and $\hbar$ is the reduced Planck constant. The second term in the brackets on the right-hand side of Equation (6.2) results in a slight modification of the effective gyromagnetic ratio of the ferromagnet [2]. It is usually ignored in the discussion of spin pumping, and we will do likewise here. In contrast, the first term in the brackets on the right-hand side of Equation (6.2) has the same form as the Gilbert damping term in Equation (6.1). Thus, pumping a spin current across the F/N interface is formally equivalent to an additional Gilbert-like damping channel, increasing the FMR line width. The additional Gilbert-like damping arising from spin pumping can thus be expressed as an increase of α by α_{SP}, with [30, 31]

$$\alpha_{\text{SP}} = g\,\mu_{\text{B}}\frac{\text{Re}(g^{\uparrow\downarrow})}{4\pi M_{\text{S}}}\,\frac{1}{t_{\text{F}}}\,\eta \tag{6.3}$$

with the Landé g-value g, the Bohr magneton μ_{B}, the saturation magnetization of the ferromagnet M_{S}, the thickness of the ferromagnet t_{F}, and the backflow parameter $0 \leq \eta \leq 1$ accounting for a possible spin current backflow into F.

The magnitude of the spin mixing conductance $\text{Re}(g^{\uparrow\downarrow})$ has been studied in a variety of different F/N hybrid structures. The ferromagnetic layer hereby consisted either of an electrically conductive (metallic or semiconducting) ferromagnet [21, 32, 33], or of an electrically insulating ferromagnet [17, 20, 22–24, 33–36]. Irrespective of the particular ferromagnet used, the spin mixing conductance into typical 3D transition metals such as Au or Pt was found to be of the order of $\text{Re}(g^{\uparrow\downarrow}) \approx 10^{19}\ \text{m}^{-2}$ [17, 20, 33, 36]. In other words, generating a large spin current via spin pumping primarily requires a strong magnetization excitation, namely, a large precession cone angle Θ, while the particular material combination used appears to be of lesser importance.

[3] For the F/N double layer system considered here, we neglect the spin-dependent transmission $t^{\uparrow\downarrow}$ [2–4]. In other words, we assume that the generated spin current is fully absorbed in the normal metal and is not transmitted to a third layer.

6.2 Electrically Detected Spin Pumping

The spin currents arising from spin pumping can not only be detected via magnetization damping, but also via electrical measurements. One hereby takes advantage of the inverse spin Hall effect (ISHE), according to which a charge current density

$$\mathbf{j}_c = -\frac{2e}{\hbar}\Theta_{SH}\,\mathbf{j}_s \times \mathbf{s}, \tag{6.4}$$

will be induced by a spin current $\mathbf{j}_s$ in electrical conductors with nonzero spin orbit coupling [37–39], and e is the electronic charge. The spin current spin polarization $\mathbf{s}$ represents the orientation of the spins flowing in the direction $\mathbf{j}_s$, and the spin Hall angle Θ_{SH} quantifies the spin-to-charge current conversion efficiency. As evident from Equation (6.4), the ISHE is a Hall-type effect in the sense that $\mathbf{j}_c, \mathbf{j}_s$, and $\mathbf{s}$ are mutually orthogonal [38]. Equation (6.4) furthermore implies that materials with large Θ_{SH} enable an efficient electrical spin current detection. Since the ISHE arises from spin–orbit coupling, large Θ_{SH} are expected in heavy elements. In experiment, materials such as platinum (Pt), tantalum (Ta), tungsten (W), gold (Au), or alloys such as CuBi are used as ISHE spin current detectors, with $|\Theta_{SH}| < 0.4$ reported to date [36, 40–44]. The spin Hall effect in exchange-coupled (ferromagnetic) materials [45], and the impact of spin accumulation and spin backflow [30, 31] are currently vividly investigated in this context.

According to Equation (6.2) and Figure 6.1(a) and (c), the spin current $\mathbf{j}_s$ has both AC and DC components. The DC spin current component stems from the longitudinal nonequilibrium magnetization components, the AC spin current components from the transverse ones. So far, the DC part of $\mathbf{j}_s$ has been exploited in most electrically detected spin-pumping experiments, such that we only consider this component here. As sketched in Figure 6.2, the pumped DC spin current $\mathbf{j}_s$ propagates across the F/N interface, i.e., along the F/N interface normal. Equation (6.4) then implies that $\mathbf{j}_c$ lies in the N film plane, in the direction perpendicular to $\mathbf{s}$. The latter is determined by the magnetization orientation in F, with $\mathbf{s} || -\mathbf{M}$ in typical ferromagnets [26]. Note also that thin film F/N hybrid structures are usually used in electrically detected spin-pumping experiments, to avoid electrical short-circuiting of the generated voltage. This is important, since the spin current pumped into N decays on the

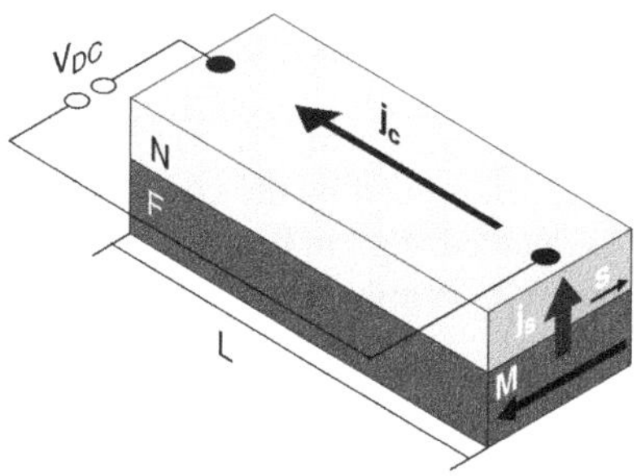

Figure 6.2 In DC electrically detected spin-pumping experiments, the DC spin current pumped into the normal metal is converted into a charge current via the inverse spin Hall effect, and then detected, either as a charge current, or via the corresponding open-circuit voltage V_{DC}

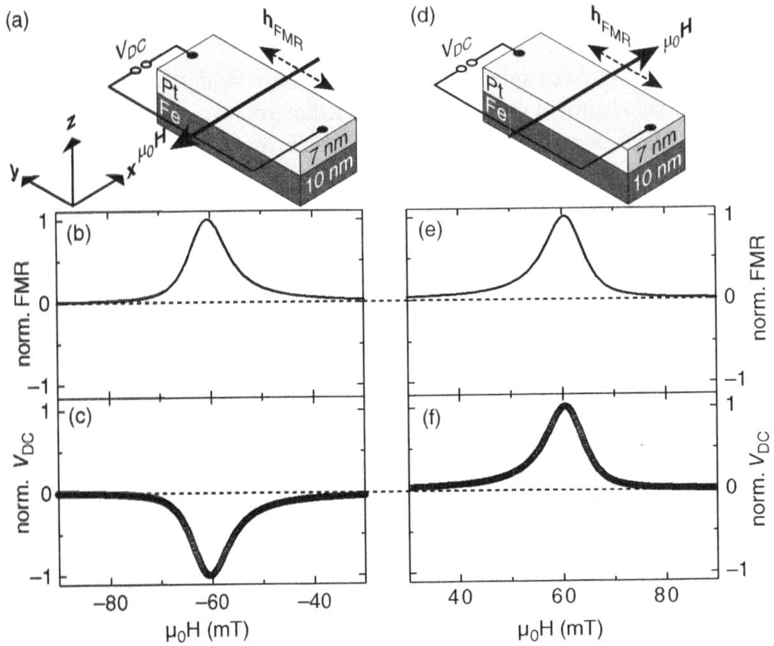

Figure 6.3 *Electrically detected spin-pumping measurements in a Co/Pt thin film hybrid structure. If the external magnetic field **H** is applied along* $-\mathbf{x}$ *as shown in panel (a), one observes the FMR and* V_{DC} *signals depicted in panels (b) and (c). Inverting the magnetic field orientation as sketched in panel (d), the FMR signal is unaltered (panel (e)), but* V_{DC} *changes sign as expected from Equation (6.4) (panel (f)). Note also that the FMR and* V_{DC} *signals have identical line shape*

length scale of the spin diffusion length λ_{sd}, which is of the order of a few nm to a few 100 nm in typical metals [1, 18, 19, 36, 38, 46].

As discussed in Section 6.1, a spin current will only be pumped into N upon (resonant) excitation of the magnetization in F. We thus expect an ISHE charge current only in FMR. In an electrically detected spin-pumping experiment, either the ISHE charge current density $j_c = |\mathbf{j}_c|$, or the corresponding open circuit electric field $E_c = V_{DC}/L = j_c/\sigma_{tot}$ is detected. V_{DC} hereby is the voltage appearing in FMR between two electrical contacts separated by the distance L (see Figures 6.2 and 6.3), and $\sigma_{tot} = (\sigma_F t_F + \sigma_N t_N)(t_F + t_N)^{-1}$ is the total conductivity of the F/N hybrid, determined by the conductivities σ_i and layer thicknesses t_i of F and N, respectively. As detailed, e.g., in Refs. [15, 16, 33, 36], the magnitude of the spin-pumping spin Hall voltage V_{DC} is given by

$$\frac{V_{DC}}{L} = \frac{2e}{\hbar}\Theta_{SH}\frac{j_s \eta \lambda_{sd} \tanh\left(\frac{t_N}{2\lambda_{sd}}\right)}{\sigma_F t_F + \sigma_N t_N}. \tag{6.5}$$

Typical experimental data from electrically detected spin-pumping experiments in a Fe/Pt thin film bilayer are shown in Figure 6.3 [47]. The data were taken at room temperature, with the sample in the antinode of the linearly polarized microwave magnetic field of a commercial microwave cavity operating at a fixed frequency of 9.3 GHz. The FMR signal of the Fe layer, recorded with the external magnetic field applied antiparallel to the x-axis as sketched in Figure 6.3(a), is shown as a solid line in Figure 6.3(b). The DC voltage V_{DC}, recorded between the two contacts along the **y** direction (cf. Figure 6.3(a)), exhibits an extremum with identical line shape and magnetic field position as the FMR (Figure 6.3(c)). Inverting the static external magnetic field orientation (Figure 6.3(d)) leaves the FMR absorption unaltered (Figure 6.3(e)). In contrast, the extremum in V_{DC} changes sign upon magnetic field inversion, as evident from comparing Figure 6.3(c) and (f), respectively. This is the behaviour expected for electrically detected spin pumping according to Equation (6.4). The inversion of magnetic field orientation results in an inversion of the spin quantization axis **s**, and thus also an inversion of the ISHE charge current flow direction (viz. the sign of the corresponding open circuit electric field). This characteristic evolution of V_{DC} with magnetic field orientation, governed by ISHE physics, represents an important consistency check for the interpretation of experimental data in terms of electrically detected spin pumping [32, 48–50].

It is important to note, however, that microwave rectification phenomena [49–54] can also give rise to DC electric fields, with qualitative and quantitative features very similar to those of electrically detected spin pumping. In microwave rectification, the resonant magnetization precession in FMR results in a modulation of the sample's resistivity tensor at the microwave frequency. These resistivity modulations rectify microwave-frequency charge currents, induced in the sample by the microwave irradiation [51]. Since different components of the resistivity tensor (anisotropic magnetoresistance, anomalous Hall effect, etc.) can contribute to the ensuing rectification voltages, arguments based on symmetry or line shape can be ambiguous [49–51, 54]. Careful experimental arrangements and a critical assessment of the experimental results thus are mandatory in electrically detected spin-pumping measurements.

Another, often neglected, aspect of electrically detected spin-pumping experiments is the presence of a static magnetic proximity polarization in the normal metal, very close to the F/N interface. Such a magnetic proximity effect is well established for paramagnetic metals close to the Stoner instability (in particular Pt and Pd) in contact with a metallic ferromagnet (e.g., Co or Fe) [55, 56]. From the point of view of spin pumping, the presence of an additional, magnetically ordered layer may provide an additional magnetization damping channel [20]. Furthermore, microwave rectification or magneto-caloric effects such as the anomalous Nernst effect may then play a role in the proximity polarized metal. Note also that for F/Pt hybrids made from an electrically insulating ferromagnet such as yttrium iron garnet ($Y_3Fe_5O_{12}$, YIG), experimental evidence both for and against a static magnetic polarization in the Pt has been put forward [57–60]. It thus appears that more experimental and theoretical work will be necessary to understand and resolve the impact of magnetically polarized paramagnetic metal layers on the spin-pumping process.

In the context of Figure 6.3, we have discussed the influence of the magnetization orientation (the spin current spin polarization **s**) on the pumped spin current. In F1/N/F2 trilayers, one can furthermore study the impact of the spin current flow direction $\mathbf{j}_s$ on the ISHE voltage V_{DC}. Figure 6.4 shows experimental data from a thin film trilayer structure, consisting of an iron bottom layer, a platinum interlayer, and a cobalt top layer [47]. The data were again taken with the sample at room temperature, using microwave radiation with 9.3 GHz. The FMR and the electrically detected spin-pumping signal owing to the iron layer appear

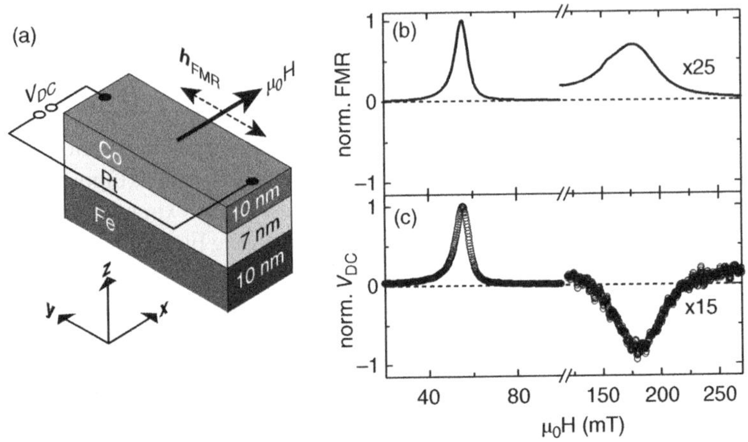

Figure 6.4 In a Fe/Pt/Co trilayer structure as sketched in (a), the FMR absorption signal (panel b) shows two distincitive peaks, attributed to the FMR of Fe at 60 mT and of Co at 180 mT. The sign of the electrically detected spin-pumping signal V_{DC} (panel c) reflects the layer stacking sequence – or more precisely the direction of spin current flow

at $\mu_0 H_{res} \approx 60$ mT, as already discussed in Figure 6.3. The FMR line of the cobalt layer is at $\mu_0 H_{res} \approx 180$ mT, being much broader and less intense than the iron one. Nevertheless, a corresponding DC voltage again is clearly resolved. Interestingly, however, the polarity of V_{DC} at the cobalt FMR is of inverted as compared to its iron counterpart. Since all data were taken for the same orientation of the external magnetic field, **s** is the same for both the cobalt and the iron layer, such that the sign inversion of V_{DC} cannot be due to the spin current spin polarization. The spin current propagation direction, on the other hand, is opposite for the two ferromagnetic layers. As evident from Figure 6.4(a), the spin current pumped from the Co layer flows along −**z** into Pt, while the spin current pumped from Fe flows along +**z** into Pt. According to Equation (6.4), the electrically detected spin-pumping signals V_{DC} of these two magnetic layers (with parallel **M**) thus should be of opposite polarity, in full agreement with experiment.

6.3 A Broader View on Spin Pumping

So far, we have discussed spin-pumping experiments in which the magnetization was driven out of equilibrium via conventional FMR. In other words, the spin current across the F/N interface was excited by microwave photons (viz. GHz frequency electromagnetic waves). According to the spin-pumping scheme (cf. Equation (6.3)), however, the microscopic mechanism used to excite the magnetization is not important. This implies that magnetization dynamics driven by thermal gradients, by acoustic strain fields, by charge currents, etc., also should give rise to spin-pumping spin currents. Indeed, thermally driven spin pumping is invoked to explain the so-called spin Seebeck effect [14, 61–64], and high-frequency elastic deformations (acoustic phonons) also allow us to pump spin currents by virtue of the magnon–phonon interaction [65, 66].

The spin-pumping scheme – i.e., magnetization relaxation via a transfer of nonequilibrium angular momentum across a ferromagnet/normal metal interface – thus can be seen as a powerful and versatile tool for the generation of pure spin currents.

References

1. R. Urban, G. Woltersdorf, and B. Heinrich. Gilbert damping in single and multilayer ultrathin films: Role of interfaces in nonlocal spin dynamics. *Physical Review Letters*, **87**:217204, Nov 2001.
2. Yaroslav Tserkovnyak, Arne Brataas, and Gerrit E. W. Bauer. Enhanced gilbert damping in thin ferromagnetic films. *Physical Review Letters*, **88(11)**:117601, Feb 2002.
3. Yaroslav Tserkovnyak, Arne Brataas, and Gerrit E. W. Bauer. Spin pumping and magnetization dynamics in metallic multilayers. *Physical Review B*, **66**:224403, Dec 2002.
4. Yaroslav Tserkovnyak, Arne Brataas, Gerrit E. W. Bauer, and Bertrand I. Halperin. Nonlocal magnetization dynamics in ferromagnetic heterostructures. *Reviews of Modern Physics*, **77(4)**:1375–1421, Dec 2005.
5. S. I. Kiselev, J. C. Sankey, I. N. Krivorotov, N. C. Emley, R. J. Schoelkopf, R. A. Buhrman, and D. C. Ralph. Microwave oscillations of a nanomagnet driven by a spin-polarized current. *Nature*, **425**:380–383, 2003.
6. D. C. Ralph and M. D. Stiles. Spin transfer torques. *Journal of Magnetism and Magnetic Materials*, **320(7)**:1190 – 1216, 2008.
7. Ioan Mihai Miron, Gilles Gaudin, Stephane Auffret, Bernard Rodmacq, Alain Schuhl, Stefania Pizzini, Jan Vogel, and Pietro Gambardella. Current-driven spin torque induced by the rashba effect in a ferromagnetic metal layer. *Nature Materials*, **9**:230–234, 2010.
8. D. Fang, H. Kurebayashi, J. Wunderlich, K. Vyborny, L. P. Zarbo, R. P. Campion, A. Casiraghi, B. L. Gallagher, T. Jungwirth, and A. J. Ferguson. Spin-orbit-driven ferromagnetic resonance. *Nature Nanotechnology*, **6**:413, 2011.
9. Luqiao Liu, Takahiro Moriyama, D Ralph, and R Buhrman. Spin-torque ferromagnetic resonance induced by the spin Hall effect. *Physical Review Letters*, **106(3)**:036601, January 2011.
10. Arne Brataas, Andrew D Kent, and Hideo Ohno. Current-induced torques in magnetic materials. *Nature Materials*, **11(5)**:372, April 2012.
11. Tomas Jungwirth, Jörg Wunderlich, and Kamil Olejník. Spin Hall effect devices. *Nature Materials*, **11(5)**:382, April 2012.
12. Satoru Emori, Uwe Bauer, Sung-Min Ahn, Eduardo Martinez, and Geoffrey S. D. Beach. Current-driven dynamics of chiral ferromagnetic domain walls. *Nature Materials*, **12**:611, 2013.
13. Axel Hoffmann. Pure spin currents. *physica status solidi (c)*, **4(11)**:4236–4241, November 2007.
14. Gerrit E W Bauer, Eiji Saitoh, and Bart J van Wees. Spin caloritronics. *Nature Materials*, **11(5)**:391, April 2012.
15. O Mosendz, J E Pearson, F Y Fradin, G E W Bauer, S D Bader, and A Hoffmann. Quantifying spin Hall angles from spin pumping: Experiments and theory. *Physical Review Letters*, **104(4)**:046601, January 2010.

16. O. Mosendz, V. Vlaminck, J. E. Pearson, F. Y. Fradin, G. E. W. Bauer, S. D. Bader, and A. Hoffmann. Detection and quantification of inverse spin hall effect from spin pumping in permalloy/normal metal bilayers. *Physical Review B*, **82**:214403, Dec 2010.
17. B. Heinrich, C. Burrowes, E. Montoya, B. Kardasz, E. Girt, Young-Yeal Song, Yiyan Sun, and Mingzhong Wu. Spin pumping at the magnetic insulator (YIG)/normal metal (Au) interfaces. *Physical Review Letters*, **107(6)**:066604, August 2011.
18. M. V. Costache, M. Sladkov, S. M. Watts, C. H. van der Wal, and B. J. van Wees. Electrical detection of spin pumping due to the precessing magnetization of a single ferromagnet. *Physical Review Letters*, **97**:216603, Nov 2006.
19. E. Saitoh, M. Ueda, H. Miyajima, and G. Tatara. Conversion of spin current into charge current at room temperature: Inverse spin-hall effect. *Applied Physics Letters*, **88(18)**:182509, 2006.
20. Yiyan Sun, Houchen Chang, Michael Kabatek, Young-Yeal Song, Zihui Wang, Michael Jantz, William Schneider, Mingzhong Wu, E Montoya, B Kardasz, B Heinrich, Suzanne G E te Velthuis, Helmut Schultheiss, and Axel Hoffmann. Damping in yttrium iron garnet nanoscale films capped by platinum. *Physical Review Letters*, **111(10)**:106601, September 2013.
21. A. Azevedo, L. H. Vilela-Leão, R. L. Rodríguez-Suárez, A. F. Lacerda Santos, and S. M. Rezende. Spin pumping and anisotropic magnetoresistance voltages in magnetic bilayers: Theory and experiment. *Physical Review B*, **83**:144402, Apr 2011.
22. S. M. Rezende, R. L. Rodríguez-Suárez, and A. Azevedo. Magnetic relaxation due to spin pumping in thick ferromagnetic films in contact with normal metals. *Physical Review B*, **88**:014404, Jul 2013.
23. S. M. Rezende, R. L. Rodríguez-Suárez, M. M. Soares, L. H. Vilela-Leão, D. Ley Domínguez, and A. Azevedo. Enhanced spin pumping damping in yttrium iron garnet/pt bilayers. *Applied Physics Letters*, **102(1)**:–, 2013.
24. C Burrowes, B Heinrich, B Kardasz, E A Montoya, E Girt, Yiyan Sun, Young-Yeal Song, and Mingzhong Wu. Enhanced spin pumping at yttrium iron garnet/Au interfaces. *Applied Physics Letters*, **100(9)**:092403, 2012.
25. S. V. Vonsovskii. *Ferromagnetic Resonance*. Pergamon Press, New York, 1960.
26. A. H. Morrish. *The Physical Principles of Magnetism*. IEEE Press, New York, 2001.
27. B. Heinrich. *Ultrathin Magnetic Structures III*, chapter Spin Relaxation in Magnetic Metallic Layers and Multilayers, page 143. Springer, New York, 2005.
28. M Sparks. *Ferromagnetic Relaxation Theory*. McGraw-Hill, New York, 1964.
29. Y. Guan, W. E. Bailey, E. Vescovo, C.-C. Kao, and D. A. Arena. Phase and amplitude of element-specific moment precession in. *Journal of Magnetism and Magnetic Materials*, **312(2)**:374 – 378, 2007.
30. HuJun Jiao and Gerrit E W Bauer. Spin backflow and ac voltage generation by spin pumping and the inverse spin Hall effect. *Physical Review Letters*, **110(21)**:217602, May 2013.
31. Yan-Ting Chen, Saburo Takahashi, Hiroyasu Nakayama, Matthias Althammer, Sebastian T. B. Goennenwein, Eiji Saitoh, and Gerrit E. W. Bauer. Theory of spin hall magnetoresistance. *Physical Review B*, **87**:144411, Apr 2013.
32. K Ando and E Saitoh. Inverse spin-Hall effect in palladium at room temperature. *Journal of Applied Physics*, **108(11)**:113925, 2010.

33. F. D. Czeschka, L. Dreher, M. S. Brandt, M. Weiler, M. Althammer, I.-M. Imort, G. Reiss, A. Thomas, W. Schoch, W. Limmer, H. Huebl, R. Gross, and S. T. B. Goennenwein. Scaling behavior of the spin pumping effect in ferromagnet-platinum bilayers. *Physical Review Letters*, **107**:046601, Jul 2011.
34. Y. Kajiwara, K. Harii, S. Takahashi, J. Ohe, K. Uchida, M. Mizuguchi, H. Umezawa, H. Kawai, K. Ando, K. Takanashi, S. Maekawa, and E. Saitoh. Transmission of electrical signals by spin-wave interconversion in a magnetic insulator. *Nature*, **464(7286)**:262, March 2010.
35. C. Sandweg, Y. Kajiwara, A. Chumak, A. Serga, V. Vasyuchka, M. Jungfleisch, E. Saitoh, and B. Hillebrands. Spin pumping by parametrically excited exchange magnons. *Physical Review Letters*, **106(21)**:216601, May 2011.
36. Mathias Weiler, Georg Woltersdorf, Matthias Althammer, Hans Huebl, and Sebastian T. B. Goennenwein. Spin pumping and spin currents in magnetic insulators. In Mingzhong Wu and Axel Hoffmann, editors, *Recent Advances in Magnetic Insulators – From Spintronics to Microwave Applications*, volume 64 of *Solid State Physics*, chapter 5, pages 123–156. Academic Press, 2013.
37. M. I. D'yakonov and V. I. Perel. Possibility of orienting electron spins with current. *JETP Letters*, **13**:467–469, 1971.
38. J. E. Hirsch. Spin hall effect. *Physical Review Letters*, **83(9)**:1834–1837, Aug 1999.
39. S. O. Valenzuela and M. Tinkham. Direct electronic measurement of the spin hall effect. *Nature*, **442(7099)**:176–179, July 2006.
40. M. Morota, Y. Niimi, K. Ohnishi, D. H. Wei, T. Tanaka, H. Kontani, T. Kimura, and Y. Otani. Indication of intrinsic spin hall effect in 4*d* and 5*d* transition metals. *Physical Review B*, **83**:174405, May 2011.
41. Luqiao Liu, Chi-Feng Pai, Y. Li, H. W. Tseng, D. C. Ralph, and R. A. Buhrman. Spin-torque switching with the giant spin hall effect of tantalum. *Science*, **336(6081)**:555–558, 2012.
42. Chi-Feng Pai, Luqiao Liu, Y. Li, H. W. Tseng, D. C. Ralph, and R. A. Buhrman. Spin transfer torque devices utilizing the giant spin hall effect of tungsten. *Applied Physics Letters*, **101(12)**:122404, 2012.
43. Y. Niimi, Y. Kawanishi, D. H. Wei, C. Deranlot, H. X. Yang, M. Chshiev, T. Valet, A. Fert, and Y. Otani. Giant spin hall effect induced by skew scattering from bismuth impurities inside thin film cubi alloys. *arXiv eprint*, arXiv:1208.6208, 2012. accepted for publication in Phys. Rev. Lett.
44. Kevin Garello, Ioan Mihai Miron, Can Onur Avci, Frank Freimuth, Yuriy Mokrousov, Stefan Blügel, Stéphane Auffret, Olivier Boulle, Gilles Gaudin, and Pietro Gambardella. *arXiv:1301.357*, 2013.
45. B. F. Miao, S. Y. Huang, D. Qu, and C. L. Chien. Inverse spin Hall effect in a ferromagnetic metal. *Physical Review Letters*, **111(6)**:066602, August 2013.
46. Y. K. Kato, R. C. Myers, A. C. Gossard, and D. D. Awschalom. Observation of the spin hall effect in semiconductors. *Science*, **306(5703)**:1910–1913, 2004.
47. Franz Dominik Czeschka. *Spin currents in metallic nanostructures*. PhD thesis, Technische Universität München, 2011.
48. K. Ando, S. Takahashi, J. Ieda, Y. Kajiwara, H. Nakayama, T. Yoshino, K. Harii, Y. Fujikawa, M. Matsuo, S. Maekawa, and E. Saitoh. Inverse spin-hall effect induced by spin pumping in metallic system. *Journal of Applied Physics*, **109(10)**:103913, 2011.

49. Lihui Bai, P. Hyde, Y. S. Gui, C.-M. Hu, V. Vlaminck, J. E. Pearson, S. D. Bader, and A. Hoffmann. Universal method for separating spin pumping from spin rectification voltage of ferromagnetic resonance. *Physical Review Letters*, **111**:217602, Nov 2013.
50. Z. Feng, J. Hu, L. Sun, B. You, D. Wu, J. Du, W. Zhang, A. Hu, Y. Yang, D. M. Tang, B. S. Zhang, and H. F. Ding. Spin hall angle quantification from spin pumping and microwave photoresistance. *Physical Review B*, **85**:214423, Jun 2012.
51. H. J. Juretschke. Electromagnetic theory of dc effects in ferromagnetic resonance. *Journal of Applied Physics*, **31(8)**:1401, 1960.
52. L. Berger. Generation of dc voltages by a magnetic multilayer undergoing ferromagnetic resonance. *Physical Review B*, **59(17)**:11465, May 1999.
53. Y. Gui, N. Mecking, X. Zhou, Gwyn Williams, and C-M Hu. Realization of a room-temperature spin dynamo: The spin rectification effect. *Physical Review Letters*, **98(10)**:107602, March 2007.
54. Lihui Bai, Z. Feng, P. Hyde, H. F. Ding, and C.-M. Hu. Distinguishing spin pumping from spin rectification in a pt/py bilayer through angle dependent line shape analysis. *Applied Physics Letters*, **102(24)**:–, 2013.
55. F. Wilhelm, P. Poulopoulos, G. Ceballos, H. Wende, K. Baberschke, P. Srivastava, D. Benea, H. Ebert, M. Angelakeris, N. K. Flevaris, D. Niarchos, A. Rogalev, and N. B. Brookes. Layer-resolved magnetic moments in Ni/Pt multilayers. *Physical Review Letters*, **85(2)**:413–416, July 2000.
56. Focko Meier, Samir Lounis, Jens Wiebe, Lihui Zhou, Swantje Heers, Phivos Mavropoulos, Peter H. Dederichs, Stefan Blgel, and Roland Wiesendanger. Spin polarization of platinum (111) induced by the proximity to cobalt nanostripes. *Physical Review B*, **83(7)**:075407, February 2011.
57. S. Y. Huang, X. Fan, D. Qu, Y. P. Chen, W. G. Wang, J. Wu, T. Y. Chen, J. Q. Xiao, and C. L. Chien. Transport magnetic proximity effects in platinum. *Physical Review Letters*, **109(10)**:107204, September 2012.
58. Stephan Geprägs, Sibylle Meyer, Stephan Altmannshofer, Matthias Opel, Fabrice Wilhelm, Andrei Rogalev, Rudolf Gross, and Sebastian T. B. Goennenwein. Investigation of induced pt magnetic polarization in Pt/$Y_3Fe_5O_{12}$ bilayers. *Applied Physics Letters*, **101(26)**:262407, December 2012.
59. Y. M. Lu, Y. Choi, C. M. Ortega, X. M. Cheng, J. W. Cai, S. Y. Huang, L. Sun, and C. L. Chien. Pt magnetic polarization on $Y_3Fe_5O_{12}$ and magnetotransport characteristics. *Physical Review Letters*, **110(14)**:147207, April 2013.
60. T. Kikkawa, K. Uchida, Y. Shiomi, Z. Qiu, D. Hou, D. Tian, H. Nakayama, X.-F. Jin, and E. Saitoh. Longitudinal spin seebeck effect free from the proximity nernst effect. *Physical Review Letters*, **110**:067207, Feb 2013.
61. K. Uchida, J. Xiao, H. Adachi, J. Ohe, S. Takahashi, J. Ieda, T. Ota, Y. Kajiwara, H. Umezawa, H. Kawai, G. E. W. Bauer, S. Maekawa, and E. Saitoh. Spin seebeck insulator. *Nature Materials*, **9(11)**:894–897, 11 2010.
62. Jiang Xiao, Gerrit E. W. Bauer, Kenchi Uchida, Eiji Saitoh, and Sadamichi Maekawa. Theory of magnon-driven spin seebeck effect. *Physical Review B*, **81(21)**:214418, Jun 2010.
63. Silas Hoffman, Koji Sato, and Yaroslav Tserkovnyak. Landau–lifshitz theory of the longitudinal spin seebeck effect. *Physical Review B*, **88**:064408, Aug 2013.

64. S. M. Rezende, R. L. Rodríguez-Suárez, R. O. Cunha, A. R. Rodrigues, F. L. A. Machado, G. A. Fonseca Guerra, J. C. Lopez Ortiz, and A. Azevedo. Magnon spin-current theory for the longitudinal spin-seebeck effect. *Physical Review B*, **89**:014416, Jan 2014.
65. K. Uchida, H. Adachi, T. An, T. Ota, M. Toda, B. Hillebrands, S. Maekawa, and E. Saitoh. Long-range spin seebeck effect and acoustic spinâpumping. *Nature Materials*, **10**(**10**):737–741, 10 2011.
66. M. Weiler, H. Huebl, F. S. Goerg, F. D. Czeschka, R. Gross, and S. T. B. Goennenwein. Spin pumping with coherent elastic waves. *Physical Review Letters*, **108**:176601, Apr 2012.

7

Spin Seebeck Effect

Jiang Xiao
Department of Physics, Fudan University, Shanghai, China

7.1 Introduction

The Seebeck effect, discovered almost two centuries ago by Thomas Johann Seebeck, is a thermoelectric effect that directly converts a temperature difference into an electric voltage [1]. The efficiency of such conversion, characterized by the thermopower (Seebeck coefficient), is proportional to the energy derivative of the conductivity because the conversion relies on the difference in the conductivity of conduction electrons at different energies. Since the Seebeck effect is due to the conduction electron transport, it only exists in conductors. According to the Onsager reciprocal relations [2], there also exists the inverse effect of the Seebeck effect – the Peltier effect, which converts an electric voltage into a temperature difference.

As the spin analogy of the Seebeck effect, the spin Seebeck effect converts a temperature difference into a spin voltage (or spin current). This effect was first experimentally observed by Uchida *et al.* [3] in a ferromagnetic metal Py using Pt as a detector for the spin current. This observation was initially attributed to the thermally induced spin accumulation in Py, similar to the conventional Seebeck in normal metals, but with two spin channels having different thermopower. However, the spin Seebeck effect was found to be a nonlocal effect, which differs from the conventional Seebeck effect, which is local and depends only on the local temperature gradient. Another intriguing result is the length scale over which the spin Seebeck effect sustains. The observed spin voltage signal sustain in millimeter size sample, while a spin accumulation can only sustain up to the spin relaxation length, which is typically much shorter than micrometer.

Ever since the first experiment reported in 2008, the spin Seebeck effect lacked a plausible theoretical explanation. Until 2010, Xiao *et al.* [4] proposed a mechanism based on

Spintronics for Next Generation Innovative Devices, First Edition. Edited by Katsuaki Sato and Eiji Saitoh.
© 2015 John Wiley & Sons, Ltd. Published 2015 by John Wiley & Sons, Ltd.

the thermal spin pumping, i.e., a spin current is generated at the interface between a ferromagnet (F) and a nonmagnet (N) when the magnons in F and the electrons in N have different temperatures, whereas the temperature difference can be achieved by applying a temperature bias. The length scale of the measured spin current then is directly related to the length scale of the temperature differences, which is governed by, instead of the spin relaxation length, a much longer relaxtion length between magnons and phonons. Different from the spin accumulation picture, this picture does not rely on conduction electrons at all. This means that the spin Seebeck effect should exist in ferromagnetic insulator or semiconductor as well. This was later confirmed in a ferromagnetic insulator, yttrium iron garnet (YIG), by Uchida *et al.* [5] and in a semiconductor (Ga,Mn)As, by Jaworski *et al.* [6].

With the realization in all types of ferromagnetic materials: ferromagnetic metals [3] insulators [5], and semiconductors [6] the spin Seebeck effect seems to be a universal effect in ferromagnets. Because of its universality and potential applications in energy harvesting and heat management in microstructures [7], the spin Seebeck effect attracts a lot of attention and initiated a new research direction called spin caloritronics [8], which focuses on the interplay between heat and spin.

After several years of intensive experimental and theoretical efforts trying to understand the underlying physics of the spin Seebeck effect, we have already achieved a lot, but there are still many unsolved puzzles. In this chapter, we try to present the most fundamental principles of the spin Seebeck effect, and also point out what are yet to be understood.

In Section 7.1, we start with some basic concepts that are crucial in understanding the spin Seebeck effect, including the Landau–Lifshitz–Gilbert (LLG) phenomenology for describing the magnetization dynamics, the spin-transfer torque, and spin-pumping effects for exchanging angular momentum between conduction electron spins and the magnetization, the fluctuation–dissipation theorem (FDT) that is important when thermal fluctuation involved, as well as the spin Hall effect (SHE) as an efficient detecting method for the spin current. In Section 7.2, we briefly present the major experimental results for the spin Seebeck effect in all major materials and two different geometric configurations used in the experiments. Section 7.3 presents the theoretical mechanism that governs the spin Seebeck effect. Some general discussions and potential applications are included in Section 7.4.

7.1.1 Landau–Lifshitz–Gilbert Phenomenology

At temperatures well below the Curie temperature, the equilibrium magnetization $\mathbf{M}$ of a bulk ferromagnet saturates to a maximum value $|\mathbf{M}| = M_s$, and the magnetic state can be described by a unit vector order parameter $\mathbf{m} = \mathbf{M}/M_s$. The evolution of this spatially variable order parameter is phenomenologically described by the Landau–Lifshitz (LL) equation [9]

$$\dot{\mathbf{m}}(\mathbf{r}, t) = -\gamma \mathbf{m}(\mathbf{r}, t) \times \mathbf{H}_{\text{eff}}(\mathbf{r}), \tag{7.1}$$

where γ is the gyromagnetic ratio and $\mathbf{H}_{\text{eff}} = -\partial_{\mathbf{M}} F[\mathbf{M}]$ is the effective magnetic field determined by the free energy functional $F[\mathbf{M}]$. Equation (7.1) shows that the magnetization precesses around the effective magnetic field and the free energy is preserved. In order to take into account the energy dissipation, a damping term is added into the LL equation [10],

$$\dot{\mathbf{m}}(\mathbf{r}, t) = -\gamma \mathbf{m}(\mathbf{r}, t) \times [\mathbf{H}_{\text{eff}}(\mathbf{r}) + \mathbf{h}(\mathbf{r}, t)] + \alpha \mathbf{m}(\mathbf{r}, t) \times \dot{\mathbf{m}}(\mathbf{r}, t), \tag{7.2}$$

where α is the Gilbert damping constant. Equation (7.2) is called the LLG equation. According to FDT (see Section 7.1.3) [11, 12], dissipation is always accompanied with fluctuation, therefore a thermal random magnetic field $\mathbf{h}(\mathbf{r}, t)$ associated with the Gilbert damping in Equation (7.2) is included accounting for the fluctuations [13].

By including the anisotropy and exchange energies, the effective field can be written as $\mathbf{H}_{\text{eff}} = \mathbf{H}_0 + (D/\gamma\hbar)\nabla^2\mathbf{m}(\mathbf{r}, t)$, where (i) $\mathbf{H}_0 = (\omega_0/\gamma)\hat{\mathbf{z}}$ includes the external field and the uniaxial anisotropy field (both in the $\hat{\mathbf{z}}$ direction), (ii) the exchange field $(D/\gamma\hbar)\nabla^2\mathbf{m}$ is due to spatial variation of magnetization, and D is the exchange constant. Because of the anisotropy field, $\mathbf{m}\|\hat{\mathbf{z}}$ at thermal equilibrium. Near thermal equilibrium, $m_{x,y} \ll 1$ and $m_z \simeq 1$, we may linearize and Fourier transform the LLG equation

$$\begin{pmatrix} m_x(\mathbf{k}, \omega) \\ m_y(\mathbf{k}, \omega) \end{pmatrix} = \chi(\mathbf{k}, \omega) \begin{pmatrix} h_x(\mathbf{k}, \omega) \\ h_y(\mathbf{k}, \omega) \end{pmatrix}, \tag{7.3}$$

where $\mathbf{k}, \omega$ are the magnon wave vector and frequency, respectively, and χ is the transverse dynamic magnetic susceptibility:

$$\chi(\mathbf{k}, \omega) = -\frac{1/(1+\alpha^2)}{(\omega - \omega_{\mathbf{k}}^{+})(\omega - \omega_{\mathbf{k}}^{0})} \begin{pmatrix} \omega_{\mathbf{k}} - i\alpha\omega & -i\omega \\ i\omega & \omega_{\mathbf{k}} - i\alpha\omega \end{pmatrix}, \tag{7.4}$$

with $\hbar\omega_{\mathbf{k}} = \hbar\omega_0 + D|\mathbf{k}|^2$ and $\omega_{\mathbf{k}}^{\pm} = \pm\omega_{\mathbf{k}}/(1 \pm i\alpha)$. The macrospin approximation corresponds to $\mathbf{k} = 0$.

7.1.2 Spin-Transfer Torque and Spin Pumping

Spin-transfer torque (STT) [15, 16] and spin-pumping effects [17] are a pair of reciprocal effects that are related through the Onsager relation. They describe how conduction electron spins and localized spins (magnetic moments) are coupled to each other. Quantum mechanically, these two effects can be understood as the processes shown in Figure 7.1. The diagram at the left interface describes the microscopic process of the STT: an electron–hole pair annihilates at the interface and creates a magnon in F, i.e., spin current absorption excites magnetization. The diagram at the right interface describes the reciprocal effect – spin pumping, where a magnon in F is annihilated and an electron–hole pair is created in N, i.e., magnetization dynamics emits spin current from F to N.

Using the scattering theory [18], the spin-transfer torque acting on the magnetization in F by a spin accumulation $\boldsymbol{\mu}_s$ in N can be expressed as

$$\mathbf{N}_{\text{stt}} = \frac{1}{4\pi}[g_r \mathbf{m} \times (\mathbf{m} \times \boldsymbol{\mu}_s) + g_i \mathbf{m} \times \boldsymbol{\mu}_s], \tag{7.5}$$

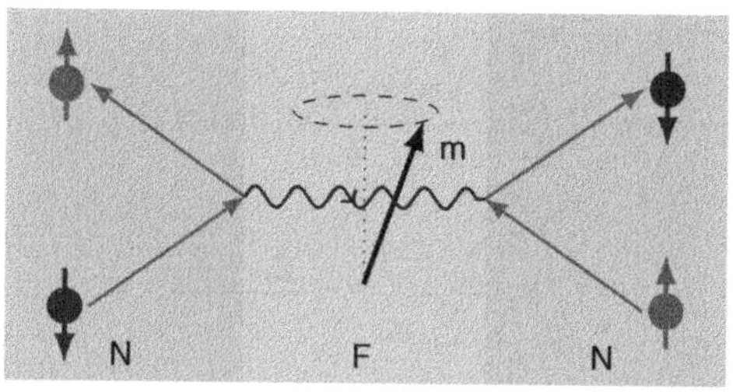

Figure 7.1 Spin-transfer torque and its reciprocal effect – spin pumping. Adapted with permission from Ref. [14]. Copyrighted by the American Physical Society

where g_r and g_i are, respectively, the real and imaginary parts of the spin-mixing conductance of the F|N interface. The reciprocal spin-pumping current emitted by a time-dependent magnetization $\mathbf{m}(t)$ is [17]

$$\mathbf{I}_{\mathrm{sp}} = \frac{\hbar}{4\pi}[g_r \mathbf{m}(t) \times \dot{\mathbf{m}}(t) + g_i \dot{\mathbf{m}}(t)]. \tag{7.6}$$

The first term in Equation (7.6) has the same form as the Gilbert damping for the magnetization in Equation (7.2), which means that the spin-pumping effect enhances the magnetization damping and dissipates angular momentum of the magnetization. Because the imaginary part of the mixing conductance is usually much smaller than the real part and the time average of the g_i terms in Equations (7.5) and (7.6) vanishes for thermally activated magnetization dynamics, we disregard the g_i terms throughout the chapter.

7.1.3 Fluctuation–Dissipation Theorem

The FDT, established by Callen and Welton [11], relates the fluctuations of physical quantities of a system to its dissipative properties when subjected to an external interaction. The close relationship between the two lies in their common origin of the fluctuation at thermal equilibrium and the dissipative force one must work against. The most famous examples of FDT include the Brownian motion [19] and the Johnson–Nyquist noise [20, 21]. An important implication of FDT is that any dissipative entity is also a source of fluctuations.

The manifest of FDT for the magnetic system is the following. Due to the interactions with the lattice and/or electrons, the magnetization, when undergoing a deterministic precessional motion, experiences dissipative magnetic Gilbert damping [10]. Reciprocally, the thermal fluctuation of the lattice and/or electrons also exerts random torque on the magnetization and causes the magnetization fluctuation. This random torque can be modeled by a random magnetic field $\mathbf{h}(t)$ in the LLG equation (Equation (7.2)) [13]. In principle, the magnetic system can couple to multiple dissipative/noise sources, such as the lattice, electrons, or contacts etc. Assuming that the nth noise source (at temperature T_n) contributes α_n to the magnetic damping such that the total damping $\alpha = \sum_n \alpha_n$, according to FDT and assuming different noise sources are uncorrelated, the random field $\mathbf{h}^{(n)}$ originated from the nth source satisfies the following white noise correlation [13]:

$$\langle \gamma h_i^{(n)}(t) \gamma h_j^{(n)}(0) \rangle = \frac{2\alpha_n \gamma k_B T_n}{M_s V} \delta_{ij} \delta(t), \tag{7.7}$$

where M_s and V are the saturation magnetization and the total volume, respectively. The total random field $\mathbf{h} = \sum_n \mathbf{h}^{(n)}$ has the correlation

$$\langle \gamma h_i(t) \gamma h_j(0) \rangle = \frac{2\alpha \gamma k_B T_F^m}{M_s V} \delta_{ij} \delta(t) \equiv \sigma^2 \delta_{ij} \delta(t), \tag{7.8}$$

where

$$T_F^m = \frac{\sum_n \alpha_n T_n}{\sum_n \alpha_n} = \frac{\sum_n \alpha_n T_n}{\alpha}, \tag{7.9}$$

can be regarded as the magnetization temperature, which is simply a weighted average of the temperatures for all thermal source with which the magnetization is coupled. The weight is determined by how strongly the magnetization is coupled to the noise source, which is proportional to its contribution to the dissipation: α_n.

In particular, for an F|N junction, which is of our main interests, the N contact is a dissipative/noise source for the magnetization. Because the first term in Equation (7.6) has the same form as the Gilbert damping, the spin pumping enhances the magnetic damping by

$$\alpha' = \frac{\gamma\hbar}{M_s V} g_r. \tag{7.10}$$

According to FDT and Equation (7.7), this additional dissipation should be accompanied with an increased magnetization fluctuation, which is caused by the STT noise on $\mathbf{m}$ and can be captured by a random magnetic field $\mathbf{h}'(t)$ with the correlation as Equation (7.7) [22]:

$$\langle \gamma h'_i(t) \gamma h'_j(0) \rangle = \frac{2\alpha' \gamma k_B T_N^e}{M_s V} \delta_{ij}\delta(t) \equiv \sigma'^2 \delta_{ij}\delta(t), \tag{7.11}$$

which is determined by the coupling strength α' and the electron temperature T_N^e in N.

7.1.4 Spin Hall Effect

The SHE, originated from the relativistic spin–orbit interaction, converts charge current into transverse spin current in metals or semiconductors (see Figure 7.2(a)): [23–31]

$$\mathbf{j}_s = \theta_{SH} \boldsymbol{\sigma} \times \mathbf{j}_c, \tag{7.12}$$

where $\mathbf{j}_c$ is the injecting charge current, $\boldsymbol{\sigma}$ is the spin polarization, and $\mathbf{j}_s$ is the transverse spin current induced by the SHE. The efficiency of such conversion is characterized by the spin Hall angle θ_{SH}. The reverse effect of the SHE (see Figure 7.2(b)), i.e., the inverse spin Hall effect (ISHE), converts a spin current into a charge current [32]:

$$\mathbf{j}_c = \theta_{SH} \boldsymbol{\sigma} \times \mathbf{j}_s. \tag{7.13}$$

The SHE and ISHE provide a simple method of generating and detecting spin current electrically, and are often used in experiments for these two purposes.

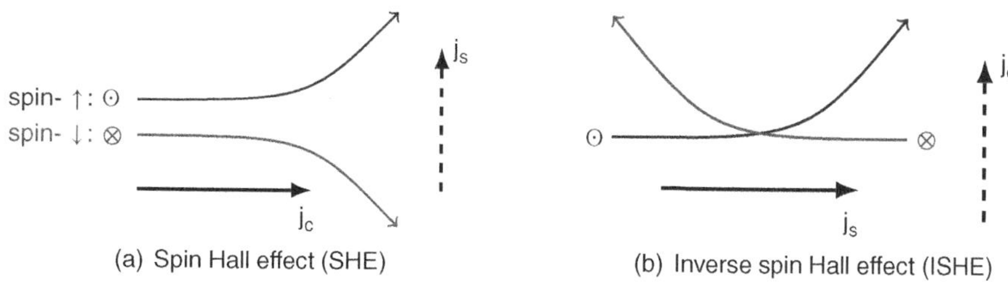

Figure 7.2 *The SHE (left) and inverse SHE (right). The symbol ⊙/⊗ denotes the spins pointing out of/into the paper*

7.2 Experiments

Experimentally, there are two different configurations for measuring the spin Seebeck effect, (i) the transverse configuration (see Figure 7.3(a)) with temperature gradient perpendicular (transverse) to the spin current flow, and (ii) the longitudinal configuration (see Figure 7.4(a)) with temperature gradient parallel (longitudinal) to the spin current flow. The transverse configuration can be applied to all ferromagnetic materials, metal, semiconductor, or insulators. While the longitudinal configuration only works for the magnetic insulators, because the spin Seebeck effect is indistinguishable from the artifacts such as the anomalous Nernst effect, which is present in metals and has the same symmetry as the spin Seebeck effect. Initially, all experiments used the transverse configuration in Figure 7.3(a) [3, 5, 6, 33, 34]. However, the transverse configuration involves complicated

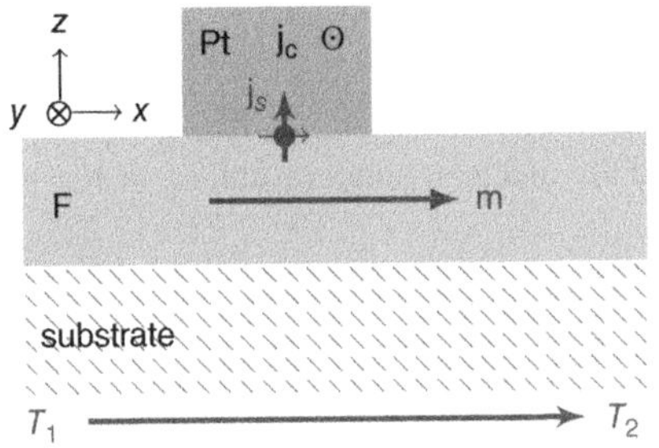

(a) The transverse configuration for measuring the spin Seebeck effect, where the F is placed on top of a substract (thicknesses are not to scale) and the temperature gradient (in $\hat{\mathbf{x}}$-direction) is transverse to the induced spin current $\mathbf{j}_s$ (in $\hat{\mathbf{z}}$-direction). An electric current (or voltage) is measured in the $\hat{\mathbf{y}}$-direction via ISHE in Pt.

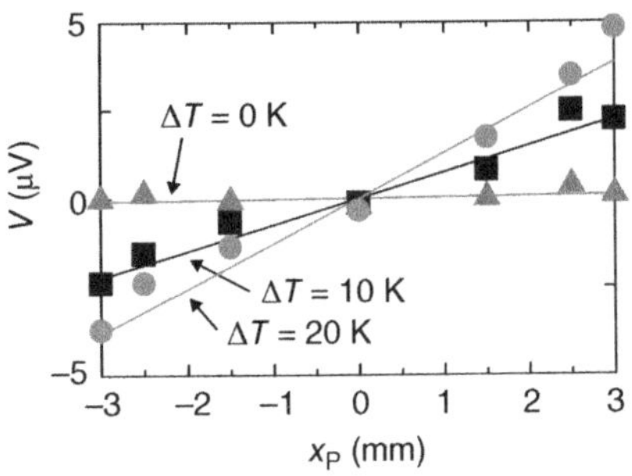

(b) Py result : Inverse spin Hall voltage vs. Pt bar position along the temperature gradient direction with various temperature differences. (3)

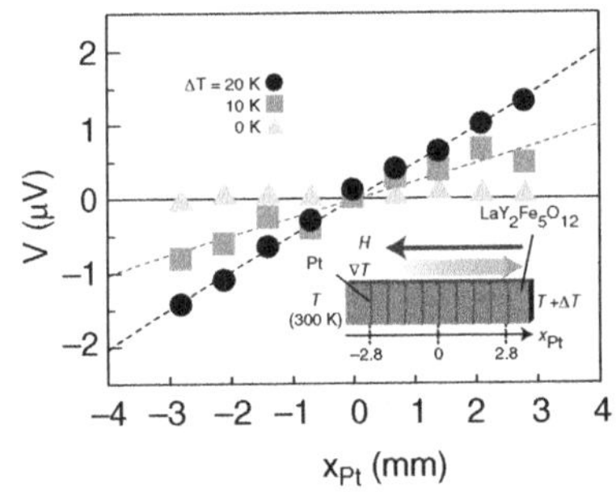

(c) YIG result. Inverse spin Hall voltage vs. Pt bar position along the temperature gradient direction at various temperature *bias*. (5)

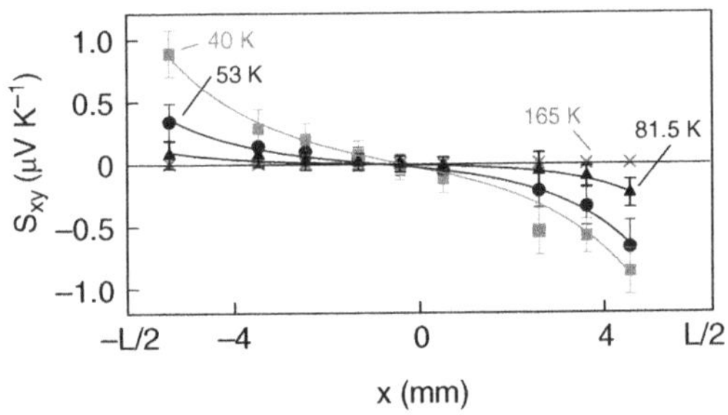

(d) GaMnAs result. The spin Seebeck coefficient (inverse spin Hall voltage renormalized by temperature bais and geometric sizes) vs. Pt bar position along the temperature gradient direction at different *average* temperatures. (6)

Figure 7.3 *(a) The transverse configuration geometry, and the Spin Seebeck effect for (b) magnetic metal, (c) magnetic insulator YIG, and (d) semiconductor GaMnAs* temperature distribution manipulation and depends crucially on the choice of substrate. Therefore, most recent experiments use the longitudinal configuration [35–42], which is simpler for the study and application of the spin Seebeck effect. In this section, we use a few representative results to show the qualitative characters of the spin Seebeck experiments for these two different configurations.

7.2.1 Transverse Configuration

Figure 7.3(a) shows a schematic diagram of the transverse configuration for measuring the spin Seebeck effect. In this configuration, a ferromagnetic film of millimeter size is placed on top of a substrate and a Pt bar is put over the ferromagnetic film at a location where the spin current is intended to measure. The temperature gradient (in the $\hat{\mathbf{x}}$-direction) is applied perpendicular to the spin injection (in the $\hat{\mathbf{z}}$-direction) into the Pt. This configuration can be applied to all types of magnetic materials, including metallic, semiconducting, and insulating. These experiments helped to establish the idea that the spin Seebeck effect is a universal phenomenon in magnetic materials, and is an effect resulting from the collective excitations of magnons instead of conduction electrons. The latter point is qualitatively different from the conventional thermoelectrics, for which all physics are governed by the behavior of conduction electrons.

The very first experiment demonstrating the spin Seebeck effect was conducted on ferromagnetic metal Py using the transverse configuration by the Saitoh group in Tohoku University in 2008 [3]. As the temperature gradient is applied along the film in the $\hat{\mathbf{x}}$-direction, the spin Hall voltage V is measured across the Pt bar in the $\hat{\mathbf{y}}$-direction. The position dependence of V is obtained by changing the location of Pt along $\hat{\mathbf{x}}$, and is shown in Figure 7.3(b) for the different temperature bias. It is checked experimentally that the observed voltage shows the cross-product behavior as in Equation (7.13) when rotating the direction of $\mathbf{m}$. This experiment demonstrates, for the first time, that a thermal gradient can induce a spin current, or a spin Seebeck effect, the spin analogy of the conventional charge Seebeck effect. It can be seen from Figure 7.3(b) that the magnitude of the spin Hall voltage from the spin Seebeck effect is proportional to the temperature bias, which is expected, and the same as conventional Seebeck effect. However, the spin Hall voltage magnitude vanishes at the center of the F film and has an opposite sign at the hot and cold ends. This suggests that the spin Seebeck effect (in the transverse configuration) is a nonlocal effect; thus Pt senses its relative location on F. On the contrary, the conventional Seebeck effect is a local effect and only depends on the local temperature gradient. The spin Seebeck effect was originally understood as a thermally induced spin accumulation in the ferromagnetic metal, which diffuses into the Pt and gives rise to the inverse spin Hall voltage across the Pt. However, this does not explain the nonlocal character. Another puzzle is the length scale over which the spin Seebeck effect sustains. Experimentally, this length scale (see Figure 7.3) is comparable to the length of the F film, or millimeter here. However, if any spin accumulation is induced thermally in F, such spin accumulation should not survive in a length scale much longer than the spin relaxation length, which is much shorter than micrometer.

In 2010, Xiao *et al.* proposed a new physical mechanism based on thermal spin pumping driven by thermal magnons [4], which suggests that the spin Seebeck effect should also appear in ferromagnetic insulators or semiconductors. It was soon confirmed experimentally in magnetic insulator YIG [5] and magnetic semiconductors GaMnAs [6], whose results are shown in Figures 7.3(c) and (d), respectively. Qualitatively, the spin Seebeck effect is almost identical for ferromagnetic metal Py, magnetic insulator YIG, and semiconductor GaMnAs. Even the quantitative magnitudes are about the same. The strikingly similar results for very different materials suggest that indeed the conduction electrons play no important role.

7.2.2 Longitudinal Configuration

The theory of the spin Seebeck effect (see Section 7.3) points out that the spin current flowing into the Pt detector is driven by the temperature difference between the magnons in F and electrons in N. In the transverse configuration, such a temperature difference along the spin current injection direction is created by a transverse temperature gradient. This nontrivial conversion involves a complicated interplay between magnon, phonon, and the substrate. To eliminate such complications, a longitudinal configuration as shown in Figure 7.4(a), where the temperature gradient is parallel to the spin injection direction, is proposed to measure the spin Seebeck effect directly [35]. The inverse spin Hall voltage across the Pt is again found to be proportional to the temperature gradient. The longitudinal configuration provides a much more clean way of creating spin current from a thermal gradient. The longitudinal configuration has the constraint that it can only be used for magnetic insulators, because the spin Seebeck signal is inseparable from the anomalous Nernst effect.

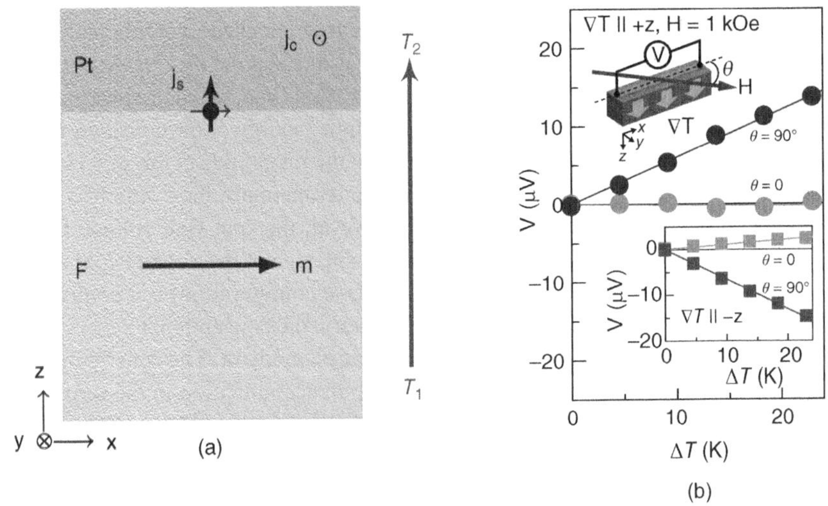

Figure 7.4 Spin Seebeck effect for YIG using longitudinal configuration. (a) The longitudinal configuration for measuring the spin Seebeck effect, where the temperature gradient (in $\hat{z}$-direction) is parallel to the induced spin current $\mathbf{j}_s$ (in $\hat{z}$-direction), which is measured as a charge current (voltage) in the $\hat{y}$-direction via ISHE in Pt. (b) The spin Seebeck coefficient (inverse spin Hall voltage renormalized by temperature bais and geometric sizes) vs. Pt bar position along the temperature gradient direction at different average temperatures. *(35)*

7.3 Theory

In this section, we explain the basic physical mechanism of the spin Seebeck effect. Almost all existing theories are based on the thermal spin pumping proposed by Xiao *et al.* [4] With this idea, a typical spin Seebeck effect can usually be explained by a three-step procedure: (i) a thermal gradient creates nonequilibrium magnons at the F|N interface, (ii) the nonequilibrium magnons pump a thermal spin current across the F|N interface, (iii) the ISHE in the heavy metal converts the pumped spin current into a spin Seebeck voltage.

7.3.1 Thermal Spin Pumping

To illustrate the basic idea, let us first consider a simple F|N structure as shown in Figure 7.5, in which the magnetization is a single domain and can be regarded as a macrospin $\mathbf{M} = M_s V\mathbf{m}$, where $\mathbf{m}$ is the unit vector parallel to the magnetization and $\mathbf{m}||\hat{\mathbf{z}}$ at equilibrium. The magnetization in F and electron spins in N can exchange angular momentum through the spin-transfer torque and spin-pumping effects as discussed in Section 7.1.2. Since we are interested in the thermally induced effects, we need to distinguish the temperatures of different quasiparticles, which include the electron, phonon, and magnon in F, as well as the phonon and electron in N. Considering the relaxation times within the magnon, phonon, and electron subsystems are much shorter than the magnon–phonon and magnon–electron relaxation times, we may assume that these quasiparticles are thermally equilibrate internally; hence, temperature for each subsystem can be defined: $T_F^{\mathrm{p/e/m}}$ for phonon/electron/magnon temperature in F and $T_N^{\mathrm{p/e}}$ for phonon/electron temperature in N. We further assume that the electron–phonon interaction is strong enough such that $T_F^e = T_F^p \equiv T_F$ and $T_N^e = T_N^p \equiv T_N$. Because magnon is weakly coupled to the other subsystems, the magnon temperature may deviate: $T_F^m \neq T_F$. In a steady state, the magnon temperature, as in Equation (7.9), is determined by its coupling with these subsystems, which serve as dissipation and noise sources for magnon.

As shown in Equation (7.6), the magnetization dynamics can pump spin current into adjacent N contact. At finite temperatures, magnetization dynamics is thermally activated

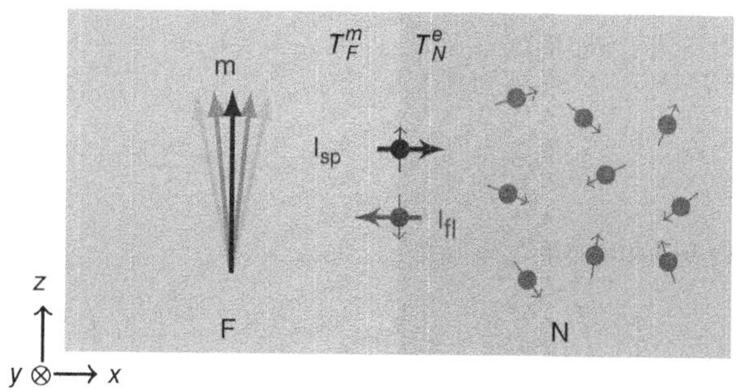

Figure 7.5 Thermal spin pumping: the thermally activated magnetization **m** *(at temperature T_F^m) pumps spin current* $\mathbf{I}_{sp}$ *from F to N, and the thermally activated electron spins (at temperature T_N^e) exert a fluctuating torque* $\mathbf{I}_{fl}$ *on* **m**. *The total spin current crossing the F|N interface is* $\mathbf{I}_s = \mathbf{I}_{sp} + \mathbf{I}_{fl}$, *whose DC component vanishes at thermal equilibrium*

and pumps a spin current (noise) into N:

$$\mathbf{I}_{\text{sp}}(t) = \frac{M_s V}{\gamma}\alpha' \mathbf{m}(t) \times \dot{\mathbf{m}}(t). \tag{7.14}$$

On the other hand, electron spins in N also fluctuate at finite temperatures, and exert a random torque on the magnetization in F. This torque noise is described by the random field $\mathbf{h}'$ given by Equation (7.11):

$$\mathbf{I}_{\text{fl}}(t) = -\frac{M_s V}{\gamma}\mathbf{m}(t) \times \gamma \mathbf{h}'(t), \tag{7.15}$$

therefore the total spin current flowing across F|N interface is [22, 43]

$$\mathbf{I}_{\text{s}}(t) = \mathbf{I}_{\text{sp}}(t) + \mathbf{I}_{\text{fl}}(t) = \frac{M_s V}{\gamma}[\alpha' \mathbf{m}(t) \times \dot{\mathbf{m}}(t) - \mathbf{m}(t) \times \gamma \mathbf{h}'(t)], \tag{7.16}$$

It can be understood that $\mathbf{I}_{\text{sp}}$ depends on how strongly $\mathbf{m}$ fluctuates or the magnon temperature T_F^m, while $\mathbf{I}_{\text{fl}}$ is determined by the electron temperature T_N^e in N. Since both $\mathbf{I}_{\text{sp}}$ and $\mathbf{I}_{\text{fl}}$ are fluctuating, only the DC component $\langle \mathbf{I}_{\text{s}}(t) \rangle$ is of our interest. At thermal equilibrium, $T_F^m = T_F = T_N = T_N^e$, there should be no DC (spin) current flowing across the interface, i.e., $\langle \mathbf{I}_{\text{s}}(t) \rangle = 0$. At nonequilibrium situation, assuming the equilibrium magnetization pointing in the $\hat{\mathbf{z}}$ direction, the $\hat{\mathbf{x}}$ and $\hat{\mathbf{y}}$ component of $\langle \mathbf{I}_{\text{s}}(t) \rangle$ vanish due to the symmetry, and only the $\hat{\mathbf{z}}$ component remains

$$\langle I_z \rangle = \frac{M_s V}{\gamma}[\alpha' \langle m_x \dot{m}_y - m_y \dot{m}_x \rangle - \gamma \langle m_x h'_y - m_y h'_x \rangle]. \tag{7.17}$$

The equal time correlators such as $\langle m_x(0) m_y(0) \rangle$ in Equation (7.17) can be calculated using the correlation of random fields in Equations (7.8) and (7.11) and the transverse dynamic magnetic susceptibility (for macrospin):

$$\chi(\omega) = \frac{1}{(\omega_0 - i\alpha\omega)^2 - \omega^2}\begin{pmatrix} \omega_0 - i\alpha\omega & -i\omega \\ i\omega & \omega_0 - i\alpha\omega \end{pmatrix}. \tag{7.18}$$

For instance,

$$\langle m_i(t) m_j(0) \rangle = \frac{\sigma^2}{2\alpha}\int \frac{\chi_{ij}(\omega) - \chi_{ji}^*(\omega)}{i\omega} e^{-i\omega t}\frac{d\omega}{2\pi}, \tag{7.19a}$$

$$\langle m_i(t) h'_j(0) \rangle = \frac{\sigma'^2}{\gamma}\int \chi_{ij}(\omega) e^{-i\omega t}\frac{d\omega}{2\pi}. \tag{7.19b}$$

The time derivative of Equation (7.19a) is

$$\langle \dot{m}_i(t) m_j(0) \rangle = -\frac{\sigma^2}{2\alpha}\int [\chi_{ij}(\omega) - \chi_{ji}^*(\omega)] e^{-i\omega t}\frac{d\omega}{2\pi} = \frac{\gamma k_B T_F^m}{M_s V}\begin{pmatrix} \alpha & 1 \\ -1 & \alpha \end{pmatrix}. \tag{7.20}$$

By inserting Equations (7.19b) and (7.20) with $t \to 0$ into Equation (7.17), we find the thermally driven DC spin current across the F|N interface:

$$\langle I_z \rangle = L_{\text{sT}}(T_F^m - T_N^e) \text{ with } L_{\text{sT}} = \frac{\gamma \hbar g_r k_B}{2\pi M_s V}, \tag{7.21}$$

where L_{sT} is the interfacial spin Seebeck coefficient. From Equation (7.21), we conclude that the DC spin current is proportional to the temperature difference between the magnons and electrons temperatures and polarized along the average magnetization.

In extended ferromagnetic films, the macrospin model breaks down and we have to consider magnon excitations at all wave vectors. The space–time magnetization autocorrelation function can be derived from the LLG equation:

$$\langle \dot{m}_i(0,0)m_j(0,0)\rangle = \frac{\gamma k_B T_F^m}{M_s V_a}\begin{pmatrix} \alpha & 1 \\ -1 & \alpha \end{pmatrix} \text{ with } V_a = \frac{2}{3\zeta_{5/2}}\lambda^3, \tag{7.22}$$

where $\lambda = \sqrt{4\pi D/k_B T_F^m}$ is the thermal de Broglie wavelength. Assuming that the magnon temperature does not change appreciably in the volume $V_a \ll V$, the results in Equation (7.21) remain the same by replacing $V \to V_a$, and the spin Seebeck current and coefficient are

$$\langle I_z\rangle = L_{\mathrm{sT}}(T_F^m - T_N^e) \text{ with } L_{\mathrm{sT}} = \frac{\gamma \hbar g_r k_B}{2\pi M_s V_a}. \tag{7.23}$$

A quantitative study of this expression is given in Ref. [4].

Based on similar ideas of thermal spin pumping, other theories are also developed, which include a linear response theory by Adachi *et al.* for both transverse and longitudinal configurations, and a theory by Hoffman *et al.* for the longitudinal configuration.

7.3.2 Nonequilibrium Magnon

Equation (7.23) tells us that as long as there is a temperature difference between the magnon and electron across the F|N interface, a spin Seebeck current is induced. In practice, there are many ways of realizing this temperature imbalance, or of creating the nonequilibrium magnon. The most straightforward way is to apply a temperature gradient directly across the F|N interface as in the longitudinal configuration. A not so trivial way is to apply a temperature gradient parallel to the F|N interface as in the transverse configuration. Historically, the later nontrivial way was first used because of the original incorrect understanding of the spin Seebeck effect. There are mainly two different mechanisms to create the nonequilibrium magnon: (i) the magnon-driven mechanism and (ii) the phonon drag mechanism.

Magnon-driven mechanism: For a finite size magnet, when subject to a (phonon) temperature gradient, Sanders and Walton [44] showed that the magnon temperature deviates from the local phonon temperature because the phonon subsystem can exchange heat with the external thermal reservoirs, while the magnon subsystem cannot. Therefore, the magnon subsystem has to receive/release heat from/to the phonon subsystem. Consequently, the magnon subsystem is cooler/hotter than phonon subsystem at the hot/cool end. A schematic plot of the phonon and magnon temperatures is shown in Figure 7.6 for a magnetic film of length L with its two ends anchored to thermal reservoirs of different temperatures. The curvature of the magnon temperature profile vanishes at the ends because the magnon cannot exchange heat with the outside. The magnon and phonon temperature difference $\Delta T_{pm} = T_p - T_m$ has the following hyperbolic position dependence [4, 44]:

$$\Delta T_{\mathrm{pm}}(x) = \eta \frac{\sinh(x/\lambda)}{\sinh(L/2\lambda)}\Delta T, \tag{7.24}$$

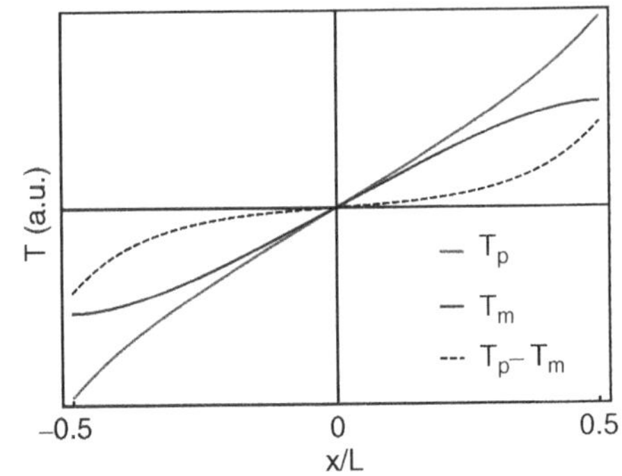

Figure 7.6 The schematic temperature profiles for phonon (red) and magnon (blue), and their difference (dashed) for a magnetic film of length L and a temperature difference is applied at the ends at $x = \pm L/2$

where ΔT is the temperature bias applied over the ends of the film, η determines the overall amplitude, and λ is the magnon–phonon relaxation length.

For the transverse configuration (see Figure 7.3(a)), the induced spin Seebeck current for a Pt placed at x is directly proportional to the local temperature difference $\Delta T_{mp}(x) = T_m(x) - T_p(x) = T_F^m - T_N^e$, assuming that the electron in N is equilibrated with the phonon in N and F: $T_N^e = T_p(x)$. This temperature difference (dashed curve in Figure 7.6) is hyperbolic as shown in Equation (7.24), which resembles the line shape observed in the experiments (see Figure 7.3) with the features of vanishing at the center and having opposite signs at the two ends. The length scale of this hyperbolic curve λ is determined by the magnon–phonon relaxation time, which tells how fast their temperatures equilibrate with each other. For the longitudinal configuration (see Figure 7.4(a)), the Pt is placed at the end of the magnet, where the magnon temperature deviates from the phonon temperature the most. In this sense, the longitudinal configuration is more efficient in extracting the spin current from the thermal imbalance. In combination with the thermal spin pumping and the ISHE, the nonequilibrium magnon temperature deviation from the phonon temperature qualitatively explain the main features of the spin Seebeck experiment for both the transverse and longitudinal configurations.

Phonon-driven mechanism: There are some strange behaviors of the spin Seebeck effect that cannot be explained by the magnon-driven mechanism above. For example, in the experiment on GaMnAs using the transverse configuration, when the magnetic film is sliced into two pieces as shown in Figure 7.7(a), the spin Seebeck signal remains almost unchanged. In the magnon-driven mechanism, with the scratch, there is no way for magnon to sense both edges simultaneously, therefore the vanishing spin Seebeck signal at the center is not possible. A more extreme experiment is to remove the part of the F film that is not right beneath the Pt as shown in Figure 7.7(b), in which the spin Seebeck effect is still present and has the same qualitative features as those in Figure 7.3. Both of these two experiments suggest that the magnon has to interact with the substrate, mostly likely through phonons in the substrate. Indeed, the spin Seebeck effect in the transverse configuration strongly depends on the choice of the substrate material. Typically, the substrate uses material with long phonon lifetime, such as sapphire. If the substrate is replaced by silica glass, which has short phonon lifetime, the spin Seebeck effect disappears. Another behavior

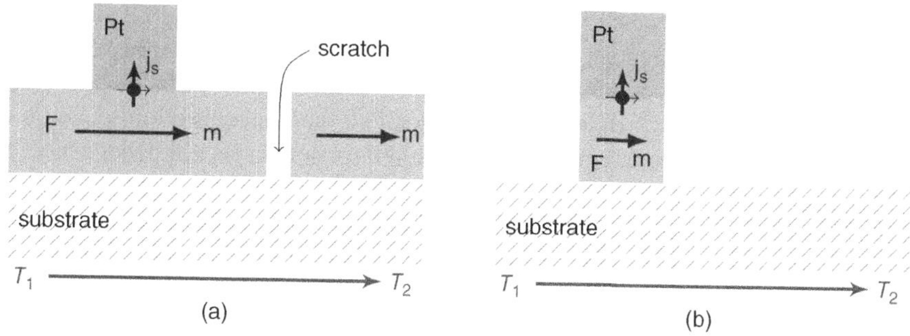

Figure 7.7 Spin Seebeck effect remains for the two variant transverse configurations that break the continuum F film. (a) The transverse configuration with a scratch. (b) The transverse configuration with Pt-F wire extending in the out-of-paper direction

that cannot be explained by the magnon-driven mechanism is the enhancement of the spin Seebeck effect at low temperatures. This enhancement is attributed to the phonon drag, similar to the phonon drag-induced thermopower enhancement in the conventional thermoelectrics. All of these strange behaviors in the transverse configuration are due to the substrate phonons. The role of phonons from the substrate is discussed in detail in the review by Adachi *et al.* [45].

7.3.3 Alternative Theory

Most existing theories are based on the thermal spin-pumping picture described in Section 7.3.1, which requires the magnon subsystem deviates from the underlying phonon/electorn subsystem. Tikhonov *et al.*, however, proposed a nonmagnon mechanism for ferromagnetic metal-based spin Seebeck effect [46]. This theory assumes that the phonon in the substrate can be separated into thermal magnons at high frequencies and subthermal magnons at low frequencies. The former stores the energy and the latter transport energy due to their long relaxation length. The subthermal magnons can deviate from the local thermal equilibrium and drag along the electron spins in F into the Pt, which is converted into an electric voltage by ISHE. This theory can only explain the spin Seebeck effect in ferromagnetic metals, but not in insulators.

7.4 Summary

In this chapter, we briefly reviewed the main experimental results of the spin Seebeck effect and provided a simple physical picture based on the thermal spin-pumping effect utilizing the fluctuation–dissipation theorem. To summarize, the spin Seebeck effect is a thermally driven spin current induced by the imbalance between the thermal fluctuation of magnetization of F and that of electrons spins in N. Such imbalance can be realized by applying a thermal gradient in the phonon subsystem, which couples to the magnetization and the phonon/electron subsystems differently.

Although we have achieved so much in understanding the spin Seebeck effect, there are still many open questions in this field. First, recent experiment using the Brillouin light scattering (BLS) spectroscopy did not measure the temperature difference between the magnon and phonon subsystems within the resolution of a few kelvin [47]. A numerical simulation suggests the magnon–phonon temperature difference in a transverse configuration with a 100 K temperature bias is less than 20 μK, which is far below than the experimental resolution [48]. Both studies suggest that the nonequilibrium magnon may not be described by a single temperature, but could have some spectral dependence. Second, the role of phonon or the substrate is still far from clear, including the length scale problem in the transverse configuration and the phonon drag. Part of the reason is that it is extremely difficult to calculate the nonequilibrium spatial and spectral distribution of electron or phonon and magnetization, not to mention the various relaxation processes among these quasiparticles. Third, the Pt on ferromagnetic metal is known to have the proximity effect so that the Pt becomes magnetic at the interface, which makes the spin Seebeck effect is contaminated with the anomalous Nernst effect [49]. Recent experiment on shows the spin Seebeck effect is present for Au|YIG structure [40, 41], in which it is known there is no proximity effect in Au. Another experiment, which includes a comparison study combining the spin pumping, spin Seebeck effect, and spin Hall magnetoresistance [50], suggests all three phenomena have a common origin, and thus rule out the possibility of the proximity effect of Pt [42]. Nevertheless, the Pt|F structure has complications for many studies including the spin Seebeck effect, as well as the spin Hall magnetoresistance etc.

The spin Seebeck effect can be used in many applications, for example, thermospin generators or thermoelectric generators for harvesting waste heat and solar energy. Because the spin Hall voltage is proportional to the length of the Pt bar, the spin Seebeck output voltage can be scaled linearly by simply increasing the size of the Pt and do not require any complicated thermopile struct uring as in the conventional thermoelectrics [7]. Other applications like position sensing device are also proposed [51]. The spin Seebeck effect is a fascinating topic, which started out as an analog of the conventional Seebeck effect, and eventually turned out to have completely different origin. The studies about the spin Seebeck effect are broadened to much larger scope and initiated a new field called spin caloritronics, which is aimed to utilizing spin degrees of freedom to control heat and vice versa.

References

1. Ashcroft NW, Mermin ND. *Solid State Physics*. Brooks/Cole; 1985.
2. Onsager L. Reciprocal relations in irreversible processes. *I. Physical Review.* 1931;**37(4)**:405.
3. Uchida K, Takahashi S, Harii K, *et al.* Observation of the spin Seebeck effect. *Nature.* 2008;**455(7214)**:778–781.
4. Xiao J, Bauer GEW, Uchida Kc, *et al.* Theory of magnon-driven spin Seebeck effect. *Physical Review B.* 2010;**81(21)**:214418.
5. Uchida K, Xiao J, Adachi H, *et al.* Spin Seebeck insulator. *Nat Mater.* 2010;**9(11)**: 894–897.
6. Jaworski CM, Yang J, Mack S, *et al.* Observation of the spin-Seebeck effect in a ferromagnetic semiconductor. *Nature Materials.* 2010;**9(11)**:898–903.
7. Kirihara A, Uchida Ki, Kajiwara Y, *et al.* Spin-current-driven thermoelectric coating. *Nature Materials.* 2012;**11(8)**:686–689.

8. Bauer GEW, Saitoh E, Wees BJv. Spin caloritronics. *Nature Materials.* 2012;**11(5)**: 391–399.
9. Landau LD, Lifsic EM, Pitaevskii LP, *et al. Statistical physics* Vol. 9. Butterworth-Heinemann; 1980.
10. Gilbert T. A phenomenological theory of damping in ferromagnetic materials. *Magnetics, IEEE Transactions on.* 2004;**40**:3443–3449.
11. Callen HB, Welton TA. Irreversibility and generalized noise. *Physical Review.* 1951;**83(1)**:34.
12. Landau LD, Lifshits EM, Pitaevskii LP. *Statistical Physics* Vol. 5. Butterworth-Heinemann; 1980.
13. Brown WF. Thermal fluctuations of a single-domain particle. *Physical Review.* 1963;**130(5)**:1677.
14. Bauer GE, Tserkovnyak Y. Spin-magnon transmutation. *Physics.* 2011;**4**:40.
15. Berger L. Emission of spin waves by a magnetic multilayer traversed by a current. *Physical Review B.* 1996;**54(13)**:9353.
16. Slonczewski JC. Current-driven excitation of magnetic multilayers. *Journal of Magnetism and Magnetic Materials.* 1996;**159(1–2)**:L1.
17. Tserkovnyak Y, Brataas A, Bauer GEW. Enhanced Gilbert damping in thin ferromagnetic films. *Physical Review Letters.* 2002;**88(11)**:117601.
18. Brataas A, Nazarov YV, Bauer GEW. Finite-element theory of transport in ferromagnet-normal metal systems. *Physical Review Letters.* 2000;**84(11)**:2481.
19. Einstein A. *Investigations on the Theory of the Brownian Movement.* Courier Dover Publications; 1956.
20. Johnson JB. Thermal agitation of electricity in conductors. *Physical Review.* 1928;**32(1)**:97.
21. Nyquist H. Thermal agitation of electric charge in conductors. *Physical Review.* 1928;**32(1)**:110.
22. Foros J, Brataas A, Tserkovnyak Y, Bauer GEW. Magnetization noise in magnetoelectronic nanostructures. *Physical Review Letters.* 2005;**95(1)**:016601–016604.
23. Dyakonov M, Perel V. Current-induced spin orientation of electrons in semiconductors. *Physics Letters A.* 1971;**35(6)**:459–460.
24. Hirsch JE. Spin Hall effect. *Physical Review Letters.* 1999;**83(9)**:1834.
25. Zhang S. Spin Hall effect in the presence of spin diffusion. *Physical Review Letters.* 2000;**85(2)**:393.
26. Takahashi S, Maekawa S. Hall effect induced by a spin-polarized current in superconductors. *Physical Review Letters.* 2002;**88(11)**:116601.
27. Murakami S, Nagaosa N, Zhang SC. Dissipationless quantum spin current at room temperature. *Science.* 2003;**301(5638)**:1348–1351.
28. Sinova J, Culcer D, Niu Q, *et al.* Universal intrinsic spin Hall effect. *Physical Review Letters.* 2004;**92(12)**:126603.
29. Kato YK, Myers RC, Gossard AC, Awschalom DD. Observation of the spin Hall effect in semiconductors. *Science.* 2004;**306(5703)**:1910–1913.
30. Wunderlich J, Kaestner B, Sinova J, Jungwirth T. Experimental observation of the spin-Hall effect in a two-dimensional spin-orbit coupled semiconductor system. *Physical Review Letters.* 2005;**94(4)**:047204.

31. Takahashi S, Maekawa S. Spin current, spin accumulation and spin Hall effect. *Science and Technology of Advanced Materials.* 2008;**9**(**1**):014105.
32. Saitoh E, Ueda M, Miyajima H, Tatara G. Conversion of spin current into charge current at room temperature: Inverse spin-Hall effect. *Applied Physics Letters.* 2006;**88**(**18**):182509–182513.
33. Bosu S, Sakuraba Y, Uchida K, *et al.* Spin Seebeck effect in thin films of the Heusler compound Co_{2}MnSi. *Physical Review B.* 2011;**83**(**22**):224401.
34. Jaworski CM, Yang J, Mack S, *et al.* Spin-Seebeck effect: A phonon driven spin distribution. *Physical Review Letters.* 2011;**106**(**18**):186601.
35. Uchida Ki, Adachi H, Ota T, *et al.* Observation of longitudinal spin-Seebeck effect in magnetic insulators. *Applied Physics Letters.* 2010;**97**(**17**):172505.
36. Uchida Ki, Nonaka T, Ota T, Saitoh E. Longitudinal spin-Seebeck effect in sintered polycrystalline (Mn,Zn)Fe_2O_4. *Applied Physics Letters.* 2010;**97**(**26**):262504.
37. Weiler M, Althammer M, Czeschka FD, *et al.* Local charge and spin currents in magnetothermal landscapes. *Physical Review Letters.* 2012;**108**(**10**):106602.
38. Meier D, Kuschel T, Shen L, *et al.* Thermally driven spin and charge currents in thin NiFe_{2}O_{4}/Pt films. *Physical Review B.* 2013;**87**(**5**):054421.
39. Ramos R, Kikkawa T, Uchida K, *et al.* Observation of the spin Seebeck effect in epitaxial Fe_3O_4 thin films. *Applied Physics Letters.* 2013;**102**(**7**):072413.
40. Qu D, Huang SY, Hu J, Wu R, Chien CL. Intrinsic spin Seebeck effect in Au/YIG. *Physical Review Letters.* 2013;**110**(**6**):067206.
41. Kikkawa T, Uchida K, Shiomi Y, *et al.* Longitudinal spin Seebeck effect free from the proximity Nernst effect. *Physical Review Letters.* 2013;**110**(**6**):067207.
42. Weiler M, Althammer M, Schreier M, *et al.* Experimental test of the spin mixing interface conductivity concept. *Physical Review Letters.* 2013;**111**(**17**):176601.
43. Xiao J, Bauer GEW, Maekawa S, Brataas A. Charge pumping and the colored thermal voltage noise in spin valves. *Physical Review B (Condensed Matter and Materials Physics).* 2009;**79**(**17**):174415–174419.
44. Sanders DJ, Walton D. Effect of magnon–phonon thermal relaxation on heat transport by magnons. *Physical Review B.* 1977;**15**(**3**):1489.
45. Adachi H, Uchida Ki, Saitoh E, Maekawa S. Theory of the spin Seebeck effect. *Reports on Progress in Physics.* 2013;**76**(**3**):036501.
46. Tikhonov KS, Sinova J, Finkelstein AM. Spectral non-uniform temperature and nonlocal heat transfer in the spin Seebeck effect. *Nature Communications.* 2013;4.
47. Agrawal M, Vasyuchka VI, Serga AA, *et al.* Direct measurement of magnon temperature: New insight into magnon–phonon coupling in magnetic insulators. *Physical Review Letters.* 2013;**111**(**10**):107204.
48. Schreier M, Kamra A, Weiler M, *et al.* Magnon, phonon, and electron temperature profiles and the spin Seebeck effect in magnetic insulator/normal metal hybrid structures. *Physical Review B.* 2013;**88**(**9**):094410.
49. Huang SY, Fan X, Qu D, *et al.* Transport magnetic proximity effects in platinum. *Physical Review Letters.* 2012;**109**(**10**):107204.
50. Chen YT, Takahashi S, Nakayama H, *et al.* Theory of spin Hall magnetoresistance. *Physical Review B.* 2013;**87**(**14**):144411.
51. Uchida Ki, Kirihara A, Ishida M, *et al.* Local spin-Seebeck effect enabling two-dimensional position sensing. *Japanese Journal of Applied Physics.* 2011;**50**:120211.

8

Spin Conversion at Magnetic Interfaces

Tomoyasu Taniyama
Materials and Structures Laboratory, Tokyo Institute of Technology, Nagatsuta, Yokohama, Japan

8.1 Introduction

Spintronic devices, in which the electron spin angular momentum plays an equally critical role to the electron charge, have led to a further major technological innovation in traditional magnetics, electronics, and photonics, and are now at the important stage of developing entirely new concept [14]. Since the device functionality in a spintronic device is based on the conversion of the electron spin angular momentum into the other degree of freedom or material properties such as the electron charge, the photon helicity, the heat, the magnetic state, etc., the principle of the spin conversion is being widely discussed [1, 2, 18, 28, 39, 40, 43, 48]. Spin light-emitting diodes (spin-LED) [26, 55] and spin transfer torque memories (spin-RAM) [8, 50] are typical examples based on the spin conversion technology, thereby the spin angular momentum injected from a spin polarized electron source electrode is converted into the corresponding photo helicity in a spin-LED [15] and the magnetoresistance in a spin-RAM. Consequently, the information about the photon helicity, i.e., the optical polarization, of light emission from the LED is utilized in a device. In order to realize such spintronic devices, spin-injection technology is obviously important. Half metallic materials which have either spin-up or spin-down channels at the Fermi energy are in development as a spin source material [9, 11, 41]. The use of spin-filtering effect using a ferromagnetic insulator, etc., is also another approach to achieving efficient spin injection with a definite advantage over other ferromagnetic metal/semiconductor interfaces

Spintronics for Next Generation Innovative Devices, First Edition. Edited by Katsuaki Sato and Eiji Saitoh.
© 2015 John Wiley & Sons, Ltd. Published 2015 by John Wiley & Sons, Ltd.

[25, 33, 34, 36, 37, 46, 57]. Despite such success in spin-injection technology, external control of the spin polarization of electrons in a source electrode cannot be achieved once such spin-injection source materials are integrated in a device. The device functionality is therefore limited depending on the spin polarization of the materials integrated. However, if the spin polarization of the source electrode can be tuned by an external means, the controllability of the spin polarization could open the new pathway to a step forward to the development of entirely new multifunctional spintronic devices; this remains a challenging issue in current research. The focus of this chapter is on reviewing the current status of this subject with special emphasis on the spin-filtering effect at the magnetic Fe and Fe_3O_4/semiconductor interfaces [59, 60] and spin polarized electron-induced magnetic phase transition in a chemically ordered FeRh alloy to be used as an externally tunable spin source magnet [39].

This chapter is organized as follows. In the next section, the mechanism of the conversion of electron spin into the photon helicity in GaAs is briefly described according to the optical transition selection rules in the radiative electron–hole recombination process. This section is followed by recent experimental results of spin injection into GaAs across a Fe_3O_4 spin filter. Optical spin orientation method, which coverts the photo helicity into the electron spin, is also described to examine the spin-filtering effect at a Fe/AlGaAs interface, the efficiency of which can be optimized by electric voltage across the interface. In the last section, the effect of injecting spin-polarized electrons into a chemically ordered FeRh alloy which shows the antiferromagnetic–ferromagnetic phase transition is presented in order to see how spin-polarized electrons influence the magnetic phase transition. The review is concluded with the future prospect of spintronics technology.

8.2 Optical Detection of Electron Spins

To evaluate the spin polarization of electrons injected into the direct band gap semiconductor GaAs, the optical transition selection rules in the electron–hole recombination process are widely used [15]. As spin-polarized electrons are injected into GaAs as shown in Figure 8.1(a), the electrons of the conduction band minimum $S_{1/2}$ at Γ_6 radiatively recombine with holes in the valence band which splits into fourfold $P_{3/2}$ at Γ_8 consisting of the twofold degenerate heavy- and light-hole subbands. According to the selection rules, the

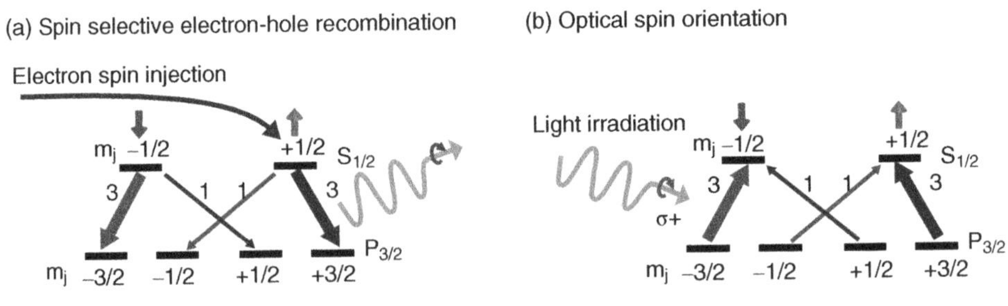

Figure 8.1 (a) Spin selective electron–hole recombination processes according to the optical transition selection rules. (b) Optical spin orientation processes, which are the reverse to the recombination processes

transition probabilities from the conduction band to the heavy- and the light-hole subbands are different by a factor of 3, so that the intensity of the right-handed circularly polarized (RCP) light emitted in the recombination process is three times larger than that of the left-handed circularly polarized (LCP) light; the transition from $m_{1/2}$ of $S_{1/2}$ to $m_{3/2}$ of $P_{3/2}$ causes the emission of RCP light. Therefore, the spin polarization of electrons injected and the circular polarization of the light emission have the following relationship:

$$2P_{\text{circ}} = P_S, \tag{8.1}$$

where P_S and P_{circ} denote the spin polarization of electrons when the electrons recombine with holes and the circular polarization of the emitted light, respectively. The electrically injected electron spins, however, depolarize or relax until the electron–hole recombination occurs in GaAs, and the spin-relaxation process, e.g., D'yakonov–Perel' mechanism [12], Bir–Aronov–Pikus mechanism [5], Elliott–Yafet mechanism [16, 61], hyperfine interaction mechanism [13], leads to a smaller circular polarization of light emission than that expected for the spin polarization of electrons just after injected. Taking such spin-relaxation effects into account, the circular polarization can be written as follows [27]:

$$2P_{\text{circ}} = P_S = \frac{P_S^0 + (1-\nu)\tau_R/2\tau_s}{1 + (1+\nu)\tau_R/2\tau_s}, \tag{8.2}$$

where P_S^0 is the spin polarization just after the electrons are injected into GaAs, τ_R is the recombination time, τ_s is the spin-relaxation time, and ν is the ratio of the electron population between the upper and lower Zeeman levels given by the Fermi–Dirac distribution. If the Zeeman polarization is vanishing, i.e., $\nu = 1$, then the circular polarization results in the following equation:

$$2P_{\text{circ}} = P_S = \frac{P_S^0}{1 + \tau_R/\tau_s}. \tag{8.3}$$

It should therefore be noted that the spin polarization estimated from the P_{circ} provides a lower limit of the spin polarization of electrons injected into GaAs.

Optical spin orientation, on the other hand, provides a very powerful approach to generating spin-polarized electrons in GaAs, allowing us to study the spin-filtering effect across ferromagnetic material/semiconductor interfaces [3, 21–24, 30–32, 42–44, 49, 51, 56, 58]. The optical spin-orientation process is fundamentally the reverse to the electron–hole recombination process as illustrated in Figure 8.1(b). As GaAs is irradiated by RCP light with the photon energy of the band gap of GaAs, the electron excitation from the heavy-hole valence subband and the light-hole subband to the conduction band minimum occurs, thereby spin-down and spin-up electrons are generated, respectively. Since the transition probability from the heavy-hole subband is a factor of 3 larger than that from the light-hole subband, 50% spin-polarized electrons are generated in the GaAs. Using this method, the spin-filtering effect at the Fe/GaAs interface will be discussed in the following section.

8.3 Spin Filtering Effect of Fe_3O_4 Thin Layers

Ferrimagnetic Fe_3O_4 has been investigated for a very long time due to its intriguing magnetic and electrical properties [35]. One of the most interesting features is the Verwey transition at which Fe^{3+} and Fe^{2+} ions order in the crystal lattice, accompanying the metal–insulator phase transition. When electrons tunnel across a thin Fe_3O_4 layer below the Verwey transitio temperature (T_V), the tunneling probabilities for spin-up and spin-down channels should be different due to the distinct exchange-split tunneling barrier height, providing spin-dependent electron transmission across the Fe_3O_4 barrier. Such a spin-dependent feature of the barrier is termed spin filtering, hence the thin Fe_3O_4 layer works as a spin filter as depicted in Figure 8.2(a). Above the T_V, on the other hand, Fe_3O_4 is a metallic ferrimagnet and the spin-filtering effect would be less significant. In the following, the spin-filtering effect of Fe_3O_4 is discussed using the optical spin detection technique described in the previous section.

Fe_3O_4/AlGaAs–GaAs quantum-well heterostructures shown in Figure 8.2(c) were used to study the spin-filtering effect of Fe_3O_4. Details of the sample fabrication process can be found in Ref. [60]. The electroluminescence (EL) from the Fe_3O_4/AlGaAs–GaAs quantum-well heterostructure was measured under spin-injection conditions in magnetic fields applied normal to the plane surface. The RCP and LCP light components of the emission were obtained separately as I^R and I^L and the circular polarization of the emitted light was estimated using the equation $P_{\mathrm{circ}} = (I^L - I^R)/(I^L + I^R)$, from which a measure of spin polarization injected into the GaAs quantum well was obtained using Equation 8.1.

Figure 8.3(a) shows the circular polarization-resolved EL spectra obtained this way. A clear difference between the RCP component and LCP component reveals that spin-polarized electrons are injected into the GaAs quantum well across the thin Fe_3O_4 layer. The temperature dependence of the spin polarization, P_S, shown in Figure 8.3(b) demonstrates more interesting electron transmission properties of the Fe_3O_4 barrier layer, i.e., P_S drastically increases with decreasing temperature below the Verwey transition. The feature is compatible with the magnetic and electrical properties of a ferrimagnetic insulating barrier layer described above, that is, the spin-injection efficiency is enhanced below T_V since the ferrimagnetic insulating Fe_3O_4 thin layer works as a spin filter. Results for a control Fe spin injector shown in Figure 8.3(b) also ensure that spin-filtering effect for Fe_3O_4 is very efficient and viable to inject spin-polarized electrons into GaAs, although the high efficiency of spin injection is obtained at low temperatures.

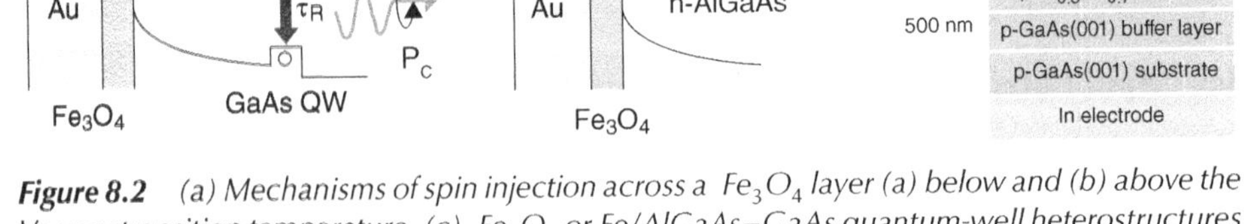

Figure 8.2 *(a) Mechanisms of spin injection across a Fe_3O_4 layer (a) below and (b) above the Verwey transition temperature. (c) Fe_3O_4 or Fe/AlGaAs–GaAs quantum-well heterostructures used*

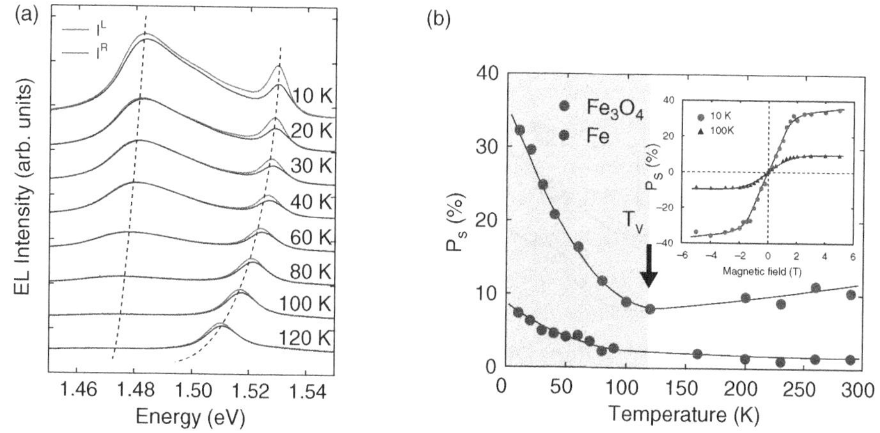

Figure 8.3 *(a) Electroluminescence spectra measured at temperatures from 10 to 120 K under the spin-injection condition. (b) Temperature-dependent P_S of electrons injected into a GaAs quantum well. The inset shows the field dependence of spin polarization for the Fe_3O_4 spin filter. Data for a Fe spin injector are also shown for reference. Modified with permission from Ref. (Wada* et al. *(2010c))*

To date, the ability of spin filtering has been demonstrated for ferromagnetic insulators such as EuS, EuO [19, 34, 36, 38, 47], and $BiMnO_3$ [20] at low temperatures. In order to implement spin-filtering effect in practical spintronic devices, however, materials with insulating ferro- or ferrimagnetic properties even at room temperature, are much preferred. The ferrite materials such as $CoFe_2O_4$ and $NiFe_2O_4$ [7] could be promising candidates for spin-filtering materials used for room temperature operation. This issue is still one of the key challenges for spin-injection-based electronic devices [54].

8.4 Electric Tunable Spin Resonant Tunneling Effect

In the previous section, efficient spin injection into a GaAs quantum well across a Fe_3O_4 spin filter is considered analyzing the circular polarization of light emission in the electron–hole recombination process. Another approach to discussing the spin-filtering effect is to use the optical spin orientation method. Once an AlGaAs–GaAs quantum-well structure, shown in Figure 8.4(a), is irradiated with circularly polarized light, spin-polarized electrons are excited in the quantum well, where the spin polarization depends on the optical polarization of the light irradiation. The electrons generated in the quantum well then drift toward the AlGaAs surface, reaching the ferromagnetic thin layer/AlGaAs interface. Since the transmission probabilities for spin-up and spin-down electrons should be different due to the spin-filtering effect of the Fe/AlGaAs interface, different photocurrents under RCP and LCP light irradiation would be obtained, thereby spin-filtering effect can be explored measuring the circular polarization-dependent photocurrent (spin-dependent photocurrent).

A number of studies have reported on spin-filtering effect using the optical spin orientation method followed by a pioneering work by Prins *et al.* [3, 21–24, 30–32, 42–44, 49,

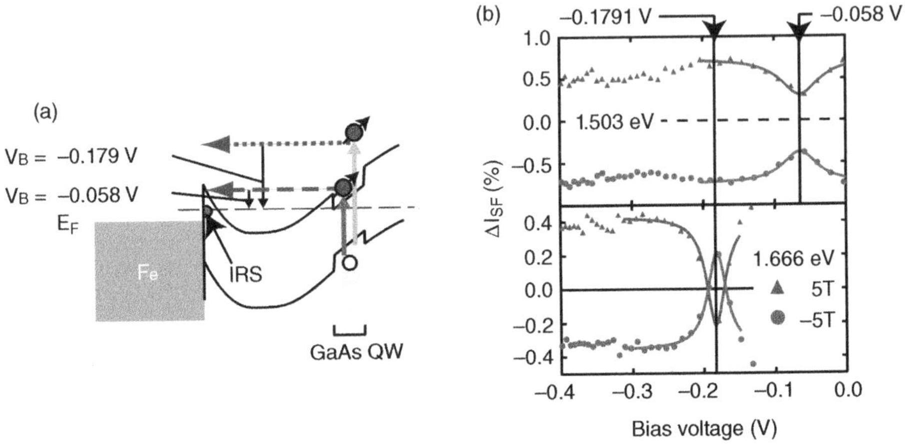

Figure 8.4 *(a) Schematic diagram of spin-dependent tunneling processes of excited electrons at the Fe/GaAs QW interface. The blue and green arrows represent the electron excitation at 1.503 and 1.666 eV, respectively.* $V_B = -0.179$ *V,* -0.058 *V denote the applied bias voltage for resonance. The dashed line is the energy level of localized states. (b) Bias voltage dependence of* ΔI_{SF} *under the excitation of 1.503 and 1.666 eV. Modified with permission from Ref. [58]. (©2010 APS)*

51, 56, 58, 59]. Recently, we proposed a unique approach to investigating the spin-filtering effect across a Fe/AlGaAs–GaAs quantum-well structure using a combined method of circular polarization resolved EL and optical spin orientation [58]. In our approach, the GaAs quantum well was irradiated with circularly polarized light with the band gap energy which was determined by EL spectra, whereby electrons with well-defined spin polarization could be generated according to the optical transition processes in Figure 8.1(b). Figure 8.4(b) shows the effective spin-filtering photocurrent ΔI_{SF} obtained this way, i.e., the circular polarization-dependent photocurrent normalized by the corresponding polarization-independent photocurrent, as a function of the bias voltage applied across the Fe/AlGaAs–GaAs heterostructure junction. The most prominent feature in Figure 8.4(b) is the dip at the bias voltage of −0.058 V and −0.179 V for the excitation at 1.503 eV and 1.666 eV, respectively. Even the inversion of the sign of ΔI_{SF} is observed. The dip feature can be understood in terms of a Breit–Wigner resonant tunneling process via localized states at the interface. According to the spin-dependent Breit–Wigner resonant tunneling, the transmission probability of spin-up (down) electrons for the RCP (LCP) excitations $T_{R(L)}^{\uparrow(\downarrow)}$ can be written as

$$T_{R(L)}^{\uparrow(\downarrow)}(E) \propto \frac{\Gamma_F^{\uparrow(\downarrow)}\Gamma_A^{\uparrow(\downarrow)}}{(E-E_i)^2 + (\Gamma_F^{\uparrow(\downarrow)} + \Gamma_A^{\uparrow(\downarrow)})^2/4}, \tag{8.4}$$

where E_i, $\Gamma_F^{\uparrow(\downarrow)}/\hbar$ ($\Gamma_A^{\uparrow(\downarrow)}/\hbar$) are the energy level of a localized state and the spin-dependent coupling between the localized state and Fe (AlGaAs) electrodes, respectively [10]. If the energy of electrons is located at the resonance and the condition $\Gamma_A^{\uparrow(\downarrow)} \ll \Gamma_F^{\uparrow(\downarrow)}$ is fulfilled, $T_{R(L)}^{\uparrow(\downarrow)}$ is proportional to $\Gamma_A^{\uparrow(\downarrow)}/\Gamma_F^{\uparrow(\downarrow)}$. Therefore, the relationships $T_R^{\uparrow(\downarrow)} \propto (1 \pm P_A)/(1 \pm P_F)$

and $T_L^{\uparrow(\downarrow)} \propto (1 \mp P_A)/(1 \pm P_F)$ are obtained, where $P_{F(A)}$ is the spin polarization of electrons in Fe (AlGaAs), and accordingly $\Gamma_{F(A)}^{\uparrow(\downarrow)} \propto 1 \pm P_{F(A)}$ [45], since the RCP (LCP) excitation provides the parallel (antiparallel) spin orientation between electrons in the Fe and AlGaAs. Using these equations, ΔI_{SF} at resonance can be expressed as

$$\Delta I_{\mathrm{SF}} \propto (T_R^{\uparrow} + T_R^{\downarrow}) - (T_L^{\uparrow} + T_L^{\downarrow}) = -\frac{4P_A P_F}{1 - P_F^2} < 0. \tag{8.5}$$

This yields a negative value of ΔI_{SF}. At off resonance ($E - E_i \gg \Gamma_F^{\uparrow(\downarrow)} + \Gamma_A^{\uparrow(\downarrow)}$), on the other hand, $T_{L(R)}^{\uparrow(\downarrow)}$ is proportional to $\Gamma_A^{\uparrow(\downarrow)}\Gamma_F^{\uparrow(\downarrow)}$ and a positive $\Delta I_{\mathrm{SF}} \propto 4P_A P_F > 0$ can be obtained. Therefore, we consider that the dip feature in Figure 8.4(b) is associated with the resonant tunneling via the localized states at the Fe/AlGaAs interface due to the asymmetry of the coupling between the localized states and Fe and AlGaAs, $\Gamma_A^{\uparrow(\downarrow)} \ll \Gamma_F^{\uparrow(\downarrow)}$. From the results, the bias-voltage-dependent inversion of the spin-dependent photocurrent clearly indicates that the electron spin transmission probabilities across the interface for spin-up and spin-down are different and the interface works as a spin filter. In other words, the spin-filtering efficiency can be tuned by electric bias voltage and the tunable capability of the spin-filtering effect could be incorporated in spintronic devices.

8.5 Spin-Injection-induced Magnetic Phase Transition in FeRh

Another key challenge to tailoring multifunctional spintronic devices is to control the magnetic phase transition of materials by an external means. A suggested solution could be found in the spin-injection-induced magnetic phase transition in chemically ordered FeRh alloys. The FeRh alloy with the CsCl-type chemical ordering has been studied extensively [4, 6, 17, 29, 52, 53, 62] and the most interesting property is the antiferromagnetic–ferromagnetic phase transition at around 370 K (Figure 8.5(a)).

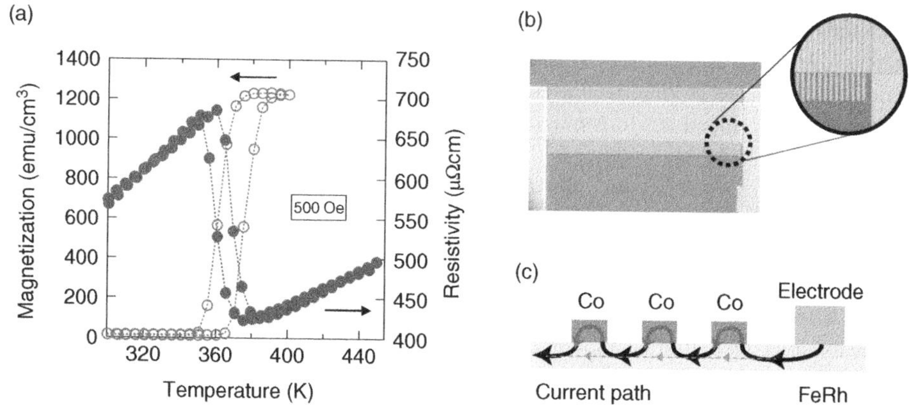

Figure 8.5 (a) Temperature dependence of magnetization and resistivity of a FeRh film. (b) Optical microscope image of a FeRh/Co wires junction structure. (c) An illustration of current path at the interfaces. Modified with permission from Ref. (Naito et al. (2011)). (©2011 AIP)

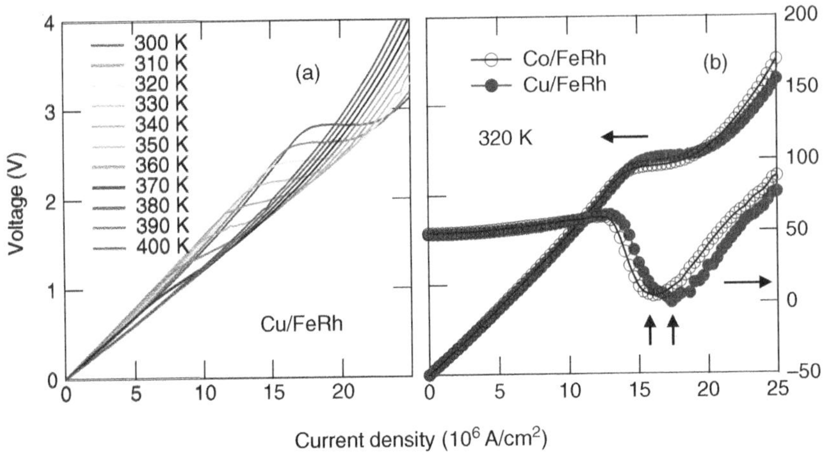

Figure 8.6 *(a) I–V characteristics of the FeRh/Co wires junction structure at various temperatures and (b) those at 320 K for FeRh/Co and FeRh/Cu wires junction structures. dV/dI are also shown. Modified with permission from Ref. (Naito* et al. *(2011)). (©2011 AIP)*

Since the magnetic phase transition is associated with a combined effect of both s–d exchange interaction between Fe atoms and spin fluctuation, the magnetic states should be modulated given that the spin accumulation and/or spin transfer torque effect occurs in FeRh under spin-injection conditions. In order to explore the spin-injection effect, the device structure consisting of a FeRh wire covered with an array of Co wires is used as shown in Figure 8.5(b). Since the resistivity of Co is three orders of magnitude smaller than that of FeRh, the current flows in a series of Co and FeRh segments as depicted in Figure 8.5(c) and spin accumulation and/or spin transfer torque effect should occur at the Co/FeRh interfaces. This allows for observation of the effect of spin injection on the magnetic phase transition of FeRh. Figure 8.6(a) is the current–voltage (I–V) characteristics measured at different temperatures. A clear feature appears in the I–V characteristics measured at temperatures below 370 K, where the slope of the curves changes distinctly at a particular current density, corresponding to the current-induced antiferromagnetic–ferromagnetic phase transition. We consider that there are two causes for the current-induced antiferromagnetic–ferromagnetic phase transition, one of which is the heating effect and the other is the spin-modulation effect due to spin accumulation or spin transfer torque at the interface of FeRh/Co. In order to check the origin of the current-induced effect, I–V curves for a nonmagnetic Cu/FeRh junction reference sample were also collected as shown in Figure 8.6(b). While both samples show the current-induced magnetic phase transition in FeRh, it should be noted that the current density where the change in the slope occurs is different from each other; the Co/FeRh sample exhibits the feature at a lower current density than the Cu/FeRh sample. The results indicate that spin-injection effect assists the magnetic phase transition for the Co/FeRh sample although heating effect also contributes to the phase transition. Therefore, provided a more significant spin-injection effect in FeRh, the spin-injection-induced magnetic phase transition would open a new pathway to create different magnetic states, enabling a spin-polarization tunable source magnet in this approach.

8.6 Summary and Future Prospects

In this chapter, some recent developments in electron spin conversion technology have been reviewed. The approaches to developing spin conversion technology could offer the promise of a new type of spin sources with potential advantage in the external controllability of spin polarization, i.e., spin-filtering efficiency tuned by electric bias voltage and magnetic phase transition induced by spin injection. However, both technologies are still at a very pioneering stage and the fundamental mechanisms related to this phenomena, such as spin accumulation effect, spin transfer torque effect, etc., need to be elucidated. The author envisages that further work will overcome the difficulty in the challenging issues and these technologies together offer the promise of surpassing the limits of conventional electric charge base semiconductor devices, leading to entirely new functional spintronic devices.

Acknowledgements

The authors would like to acknowledge the financial support of the Precursory Research for Embryonic Science and Technology (PRESTO) program of the Japan Science and Technology Agency, grants-in-aid for Scientific Research (nos. 18686050 and 22360265) from the Ministry of Education, Culture, Sports, Science and Technology, Japan, the Mitsubishi Foundation, the Shimadzu Science Foundation, the Kato Science Foundation, the Nippon Sheet Glass Foundation for Materials Science and Engineering, the Iketani Science and Technology Foundation and the Toray Science Foundation.

References

1. An T, Vasyuchka VI, Uchida K, Chumak AV, Yamaguchi K, Harii K, Ohe J, Jungfleisch MB, Kajiwara Y, Adachi H, Hillebrands B, Maekawa S and Saitoh E 2013 Unidirectional spin-wave heat conveyer. *Nature Mater.* **12**, 549–553.
2. Ando K, Morikawa M, Trypiniotis T, Fujikawa Y, Barnes CHW and Saitoh E 2010 Photoinduced inverse spin-hall effect: Conversion of light-polarization information into electric voltage. *Appl. Phys. Lett.* **96**, 082502–1–082502–3.
3. Andresen SE, Steinmuller SJ, Ionescu A, Wastlbauer G, Guertler CM and Bland JAC 2003 Role of electron tunneling in spin filtering at ferromagnet/semiconductor interfaces. *Phys. Rev. B* **68**, 073303–1–073303–4.
4. Baldasseroni C, Bordel C, Gray AX, Kaiser AM, Kronast F, Herrero-Albillos J, Schneider CM, Fadley CS and Hellman F 2012 Temperature-driven nucleation of ferromagnetic domains in ferh thin films. *Appl. Phys. Lett.* **100**, 262401–1–262401–4.
5. Bir GL, Aronov AG and Pikus GE 1975 Spin relaxation of electrons due to scattering by holes. *Sov. Phys. JETP* **33**, 705–712.
6. Bordel C, Juraszek J, Cooke DW, Baldasseroni C, Mankovsky S, Minár J, Ebert H, Moyerman S, Fullerton EE and Hellman F 2012 Fe spin reorientation across the metamagnetic transition in strained ferh thin films. *Phys. Rev. Lett.* **109**, 117201–1–117201–5.

7. Caffrey NM, Fritsch D, Archer T, Sanvito S and Ederer C 2013 Spin-filtering efficiency of ferrimagnetic spinels cofe$_2$o$_4$ and nife$_2$0$_4$. *Phys. Rev. B* **87**, 024419–1–016602–4.
8. Chen E, Apalkov D, Diao Z, Driskill-Smith A, Druist D, Lottis D, Nikitin V, Tang X, Watts S, Wang S, Wolf SA, Ghosh AW, Lu J, Poon SJ, Stan M, Butler WH, Gupta S, Mewes CKA, Mewes T and Visscher PB 2010 Advances and future prospects of spin-transfer torque random access memory. *IEEE Trans. Magn.* **46**, 1873–1878.
9. Coey JMD and Chien CL 2003 Half-metallic ferromagnetic oxides. *MRS Bulletin* **28**, 720–724.
10. Datta S 1995 *Electronic Transport in Mesoscopic Systems*. Cambridge University press, Cambridge.
11. de Groot RA, Mueller FM, van Engen PG and Buschow KHJ 1983 New class of materials: Half-metallic ferromagnets. *Phys. Rev. Lett.* **50**, 2024–2027.
12. D'yakonov MI and Perel' VI 1971 Spin orientation of electrons associated with interband absorption of light in semiconductors. *Sov. Phys. JETP* **33**, 1053–1059.
13. D'yakonov MI and Perel' VI 1974 Optical orientation in a system of electrons and lattice nuclei in semiconductors. theory. *Sov. Phys. JETP* **96**, 177–183.
14. (ed. Heinrich B and Bland JAC) 2005 *Ultrathin Magnetic Structures IV: Applications o Nanomagnetism*. Springer-Verlag, Berlin Heidelberg.
15. (ed. Meier F and Zakharchnya BP) 1984 *Optical Orientation*. North-Holland, Amsterdam.
16. Elliott RJ 1954 Theory of the effect of spin-orbit coupling on magnetic resonance in some semiconductors. *Phys. Rev.* **96**, 266–279.
17. Fallot M 1938 Les alliages du fer avec les métaux de la famille du platine. *Ann. Phys.* **10**, 291–332.
18. Fiederling R, Kelm K, Reuscher G, Ossau W, Schmidt G, Waag A and Molenkamp LW 1999 Injection and detection of a spin-polarized current in a light-emitting diode. *Nature* **402**, 787–790.
19. Filip AT, LeClair P, Smits CJP, Kohlhepp JT, Swagten HJM, Koopmans B and de Jonge WJM 2002 Spin-injection device based on eus magnetic tunnel barriers. *Phys. Rev. Lett.* **81**, 1815–1817.
20. Gajek M, Bibes M, Marthélémy A, Bouzehouane K, Fusil S, Varela M, Fontcuberta J and Fert A 2005 Spin filtering through ferromagnetic bimno$_3$ tunnel barriers. *Phys. Rev. B* **72**, 020406–1–016602–4.
21. Hirohata A, Steinmüller SJ, Cho WS, Xu YB, Gürtler CM, Wastlbauer G, Bland JAC and Holmes SN 2002 Ballistic spin filtering across ferromagnet/semiconductor interfaces at room temperature. *Phys. Rev. B* **66**, 035330–1–242102–11.
22. Hövel S, Gerhardt NC, Hofmann MR, Lo FY, Reuter D, Wieck AD, Schuster E, Keune W, Wende H, Petracic O and Westerholt K 2008 Electrical detection of photoinduced spins both at room temperature and in remanence. *Appl. Phys. Lett.* **92**, 242102–1–242102–3.
23. Isakovic AF, Carr DM, Strand J, Schultz BD, Palmstrøm CJ and Crowell PA 2001 Optical pumping in ferromagnet-semiconductor heterostructures: Magneto-optics and spin transport. *Phys. Rev. B* **64**, 161304–1–161304–4.
24. Isber S, Park YJ, Moodera JS and Heiman D 2008 Spin injection into ferromagnetic co$_2$mnal by optical absorption in gaas. *J. Appl. Phys.* **103**, 07D713–1–07D713–3.

25. Jiang X, Wang R, Shelby RM, Macfarlane RM, Bank SR, Harris JS and Parkin SSP 2005 Highly spin-polarized room-temperature tunnel injector for semiconductor spintronics using mgo(100). *Phys. Rev. Lett.* **94**, 056601–1–056601–4.
26. Jonker BT, Erwin SC, Petrou A and Petukhov AG 2003 Electrical spin injection and transport in semiconductor spintronic devices. *MRS Bulletin* **28**, 740–748.
27. Kawaharazuka A, Ramsteiner M, Herfort J, Scönherr HP, Kostial H and Ploog KH 2004 Spin injection from fe_3si into gaas. *Appl. Phys. Lett.* **85**, 3492–3494.
28. Kosaka H, Inagaki T, Rikitake Y, Imamura H, Mitsumori Y and Edamatsu K 2009 Spin state tomography of optically injected electrons in a semiconductor. *Nature* **457**, 702–705.
29. Kouvel JS and Hartelius CC 1962 Anomalous magnetic moments and transformations in the ordered alloy ferh. *J. Appl. Phys.* **33**, 1343–1344.
30. Kurebayashi H, Steinmuller SJ, Laloë JB, Trypiniotis T, Easton S, Ionescu A, Yates JR and Bland JAC 2007 Initial/final state selection of the spin polarization in electron tunneling across an epitaxial fe/gaas (001) interface. *Appl. Phys. Lett.* **91**, 102114–1–102114–3.
31. Kurebayashi H, Trypiniotis T, Lee K, Easton S, Ionescu A, Farrer I, Ritchie DA, Bland JAC and Barnes CHW 2010 Electrical determination of the spin relaxation time of photoexcited electrons in gaas. *Appl. Phys. Lett.* **96**, 022505–1–045301–3.
32. Lampel G 1968 Nuclear dynamic polarization by optical electronic saturation and optical pumping in semiconductors. *Phys. Rev. Lett.* **20**, 491–493.
33. Manago T and Akinaga H 2002 Spin-polarized light-emitting diode using metal/insulator/semiconductor structures. *Appl. Phys. Lett.* **81**, 694–696.
34. Miao GX, Münzenberg M and Moodera JS 2011 Tunneling path toward spintronics. *Rep. Prog. Phys.* **74**, 036501–1–036501–19.
35. Miles PA, Westphal WB and von Hippel A 1957 Dielectric spectroscopy of ferromagnetic semiconductors. *Rev. Mod. Phys.* **29**, 279–307.
36. Moodera JS, Miao GX and Santos TS 2010 Frontiers in spin-polarized tunneling. *Phys. Today* **63**, 46–51.
37. Moodera JS, Santos TS and Nagahama T 2007 The phenomena of spin-filter tunnelling. *J. Phys.: Condens. Matter* **19**, 165202–1–165202–24.
38. Nagahama T, Santos TS and Moodera JS 2007 Enhanced magnetotransport at high bias in quasimagnetic tunnel junctions with eus spin-filter barriers. *Phys. Rev. Lett.* **99**, 016602–1–016602–4.
39. Naito T, Suzuki I, Itoh M and Taniyama T 2011 Effect of spin polarized current on magnetic phase transition of ordered ferh wires. *J. Appl. Phys.* **109**, 07C911–1–07C911–3.
40. Ohno Y, Young DK, Beschoten B, Matsukura F, Ohno H and Awschalom DD 1999 Electrical spin injection in a ferromagnetic semiconductor heterostructure. *Nature* **402**, 790–792.
41. Palmstrøm C 2003 Epitaxial heusler alloys: new materials for semiconductor spintronics. *MRS Bulletin* **28**, 725–2027.
42. Park YJ, Hickey MC, Veenhuizen MJV, Chang J, Heiman D, Perry CH and Moodera JS 2011 Efficient spin transfer phenomena in fe/mgo/gaas structure. *J. Phys.:Condens. Matter* **23**, 116002–1–162104–5.

43. Prins MWJ, van Kempen H, van Leuken H, de Groot RA, van Roy W and de Boeck J 1995 Spin-dependent transport in metal/semiconductor tunnel junctions. *J. Phys.: Condens. Matter* **7**, 9447–9464.
44. Renucci P, Truong VG, Jaffrès H, Lombez L, Binh PH, Amand T, George JM and Marie X 2010 Spin-polarized electroluminescence and spin-dependent photocurrent in hybrid semiconductor/ferromagnetic heterostructures: An asymmetric problem. *Phys. Rev. B* **82**, 195317–1–195317–10.
45. Sahoo S, Kontos T, Furer J, Hoffmann C, Gräber M, Cottet A and Schönenberger C 2005 Electric field control of spin transport. *Nature Phys.* **1**, 99–105.
46. Saito H, Breton JCL, Zayets V, Mineno Y, Yuasa S and Ando K 2010 Efficient spin injection into semiconductor from an fe/gao$_x$ tunnel injector. *Appl. Phys. Lett.* **96**, 012501–1–012501–3.
47. Santos TS and Moodera JS 2004 Observation of spin filtering with a ferromagnetic euo tunnel barrier. *Phys. Rev. B* **69**, 241203–1–241203–4.
48. Satoh T, Terui Y, Moriya R, Ivanov BA, Ando K, Saitoh E, Shimura T and Kuroda K 2012 Directional control of spin wave emission by spatially shaped light. *Nature Photonics* **6**, 662–666.
49. Shen C, Trypiniotis T, Lee KY, Holmes SN, Mansell R, Husain M, Shah V, Li XV, Kurebayashi H, Farrer I, de Groot CH, Leadley DR, Bell G, Parker EHC, Whall T, Ritchie DA and Barnes CHW 2010 Spin transport in germanium at room temperature. *Appl. Phys. Lett.* **97**, 162104–1–162104–3.
50. Slonczewski JC 1996 Current driven excitation of magnetic multilayers. *J. Magn. Magn. Mater.* **159**, L1–L7.
51. Steinmüller SJ, Gürtler CM, Wastlbauer G and Bland JAC 2005 Separation of electron spin filtering and magnetic circular dichroism effects in photoexcitation studies of hybrid ferromagnet/gaas schottky barrier structures. *Phys. Rev. B* **72**, 045301–1–045301–5.
52. Suzuki I, Koike T, Itoh M, Taniyama T and Sato T 2009 Stability of ferromagnetic state of epitaxially grown ordered ferh thin films. *J. Appl. Phys.* **105**, 07E501–1–07E501–3.
53. Suzuki I, Naito T, Itoh M, Sato T and Taniyama T 2011 Clear correspondence between magnetoresistance and magnetization of epitaxially grown ordered ferh thin films. *J. Appl. Phys.* **109**, 07C717–1–07C717–3.
54. Takahashi YK, Kasai S, Furubayashi T, Mitani S, Inomata K and Hono K 2010 High spin-filter efficiency in a co ferrite fabricated by a thermal oxidation. *Appl. Phys. Lett.* **96**, 072512–1–072512–3.
55. Taniyama T, Wada E, Itoh M and Yamaguchi M 2011 Electrical spin injection and transport in semiconductor spintronic devices. *NPG Asia Mater.* **3**, 65–73.
56. Taniyama T, Wastlbauer G, Ionescu A, Tselepi M and Bland JAC 2003 Spin-selective transport through fe/alo$_x$/gaas (100) interfaces under optical spin orientation. *Phys. Rev. B* **68**, 134430–1–134430–5.
57. van't Erve OMJ, Kioseoglou G, Hanbicki AT, Li CH, Jonker BT, Mallory R, Yasar M and Petrou A 2004 Comparison of fe/schottky and fe/al$_2$o$_3$ tunnel barrier contacts electrical spin injection into gaas. *Appl. Phys. Lett.* **84**, 4334–4336.

58. Wada E, Shirahata Y, Naito T, Itoh M, Yamaguchi M and Taniyama T 2010a Inversion of spin photocurrent due to resonant transmission. *Phys. Rev. Lett.* **105**, 156601–1–156601–4.
59. Wada E, Shirahata Y, Naito T, Itoh M, Yamaguchi M and Taniyama T 2010b Spin polarized electron transmission into gaas quantum well across fe_3o_4: Optical spin orientation analysis. *Appl. Phys. Lett.* **97**, 172509–1–172509–3.
60. Wada E, Watanabe K, Shirahata Y, Itoh M, Yamaguchi M and Taniyama T 2010c Efficient spin injection into gaas quantum well across fe_3o_4 spin filter. *Appl. Phys. Lett.* **96**, 102510–1–102510–3.
61. Yafet Y 1963 g factors and spin-lattice relaxation of conduction electrons In *Solid State Physics* (ed. Seitz F and Turnbull D) vol. 14 Academic Press New York pp. 1–98.
62. Yuasa S, Akiyama T, Miyajima H and Otani Y 1995 Change in the resistivity of bcc and bct ferh alloys at first-order magnetic transitions. *J. Phys. Soc. Jpn.* **64**, 3978–3985.

9

Carbon-based Spintronics

Masashi Shiraishi
Graduate School of Engineering, Department of Electronic Science and Engineering, Kyoto University, Kyoto, Japan

9.1 Introduction

An electron possesses two degrees of freedom; charge and spin. Both degrees of freedom have been individually utilized in semiconductor physics and magnetism, yielding a tremendous amount of practical applications and expansion of horizons in basic science. Spintronics is thought to be a fusion domain of semiconductor physics and magnetism because both degrees of freedom in electrons are utilized simultaneously. In 1986, Gruenberg *et al.* reported the antiferromagnetic coupling of Fe in Fe/Cr/Fe layers, where magnetoresistance ratio was observed to be several percentages [1]. In 1988, Fert and co-workers reported the first giant magnetoresistance (GMR) of $\sim 40\%$ by using Fe/Cr multilayers, which opened a door for this new research field, spintronics [2]. After their discovery, the similar GMR effects were observed in Co/Cr multilayers [3] and so on. In 1992, a new system named granular was found where magnetic particles were dispersed in a nonmagnetic matrix [4]. Co/Cu, Co/Ag, and Co/Au are typical examples in the granular systems, and the magnetization direction of the ferromagnetic particles governs the spin transport phenomenon and the GMR ratio.

The GMR was a novel physical feature and the MR ratio was comparatively high, so several practical applications, such as a magnetic reproducing head, were realized. In GMR, the Cr layers played roles of spin channel, namely, conducting spins were injected and transferred in the Cr. In contrast, insulating layers can be inserted between

Spintronics for Next Generation Innovative Devices, First Edition. Edited by Katsuaki Sato and Eiji Saitoh.
© 2015 John Wiley & Sons, Ltd. Published 2015 by John Wiley & Sons, Ltd.

two ferromagnetic layers instead of conductive layers, which allows us to achieve spin-dependent tunneling transport. This is a tunneling magnetoresistance (TMR) effect. Miyazaki [5] and Moodera [6] individually observed the TMR effect in 1995, where the TMR ratio was up to 20% at room temperature (RT). They introduced an Al–O insulating barrier layer that was sandwiched between two ferromagnetic layers. Granular TMR was also reported [7] and unique temperature dependence of the tunnel resistance was revealed which was due to a charging effect [8]. In 2004, MgO single crystal was newly introduced in TMR devices as a tunneling barrier, which made coherent spin tunneling possible in a Fe/MgO/Fe system [9], and currently the TMR ratio is increased up to ~ 600% at RT [10].

Whereas the above history of spintronics is that of metallic spintronics, spintronics using inorganic semiconductors (GaAs, Si and, etc.) has been vigorously investigated. Among these inorganic semiconductors, GaAs could be used to realize spin transistors as proposed by Das and Datta [11], where injected spins into GaAs are rotated by an applied gate voltage because of the existence of a strong spin–orbit interaction. On the contrary, Si is a comparatively light element and has lattice inversion symmetry, and thus spin MOS field effect transistors (FETs) are expected to be realized. Currently, several research groups are investigating spin injection and spin transport intensively [12]. This field is known as the second pillar of spintronics, namely, semiconductor spintronics. Since 1999, a third pillar of spintronics attracts much attention from the people in spintronics and in molecular electronics, that is, molecular spintronics. Although detailed theoretical backgrounds will be described in detail in the next section, where a molecule comparatively exhibits a smaller spin–orbit interaction. A spin–orbit interaction is known as the interaction, which induced loss of spin coherence, thus, a material with a smaller spin–orbit interaction is needed to realize quantum computation systems and a so-called Sugahara–Tanaka-type spin MOSFETs [13]. Currently, nanocarbonaceous molecules (graphene, carbon nanotube, and fullerene) and organic molecules are intensively investigated for further progress in this research field, and a number of important and attractive result have been obtained. Furthermore, spin-dependent tunneling transport via molecules also exhibits novel physical features that were not observed in metallic and inorganic semiconductor spintronics.

The purpose of this chapter is to overview the history of molecular spintronics and to introduce important findings in the field. Section 9.1 is an introductory part, and Section 9.2 is mainly for introducing theories and important concepts, which should be understood to study molecular spintronics. Section 9.3 is for describing spin-dependent transport (spin injection and spin-dependent tunneling, namely, GMR and TMR effects) via molecules, and it consists of two parts; spintronics using nanocarbonaceous molecules and that using organic molecules. Section 4 is the summary of this chapter.

9.2 Theories and Importance Concepts in Spin-Dependent Transport and Spin Relaxation

In this chapter, several important concepts are introduced from a theoretical point of view. The first concenpt is a spin–orbit interaction, which induces spin relaxation. Then, the conductance mismatch is discussed, and pure spin current is introduced.

9.2.1 Spin–Orbit Interaction

The spin–orbit interaction is a purely relativistic effect, which is derived from the Dirac equation in the natural unit as

$$i\frac{\partial}{\partial t}\psi(x,t) = \left[-i\left(\sum_{i=1}^{3}\alpha_i\frac{\partial}{\partial x_i}\right) + \beta m\right]\psi(x,t), \tag{9.1}$$

$$\alpha_i = \begin{pmatrix} 0 & \sigma_i \\ \sigma_i & 0 \end{pmatrix}, \sigma_1 = \begin{pmatrix} 0 & 1 \\ 1 & 0 \end{pmatrix}, \sigma_2 = \begin{pmatrix} 0 & -i \\ i & 0 \end{pmatrix}, \sigma_3 = \begin{pmatrix} 1 & 0 \\ 0 & -1 \end{pmatrix},$$

$$\beta = \begin{pmatrix} 1 & 0 \\ 0 & -1 \end{pmatrix},$$

where t is the time, $\psi(x,t)$ is a wave function, x is a coordinate in a three-dimensional space, m is mass, α and β are 4×4 matrices, σ_i is a 2×2 Pauli's spin matrix and i is an imaginary unit. Space and time coordinates are equivalent in relativistic quantum mechanics, and then the wave function possesses four components (3 + 1), where the upper two components of the wave function are for a positive energy solution, and by contrast, the lower two components are for a negative energy solution. In other words, the positive and the negative energy solutions are solutions for an electron and a positron, respectively. It should be noted that we can describe a spin degree of freedom by the Dirac equation as a spinor. Here, the wave functions for an electron and a positron are mixed in the Dirac equation because of the existence of an off-diagonal σ_i matrix in α_i. In order to investigate interactions of the Dirac electrons in a nonrelativistic limit, an appropriate diagonalization of the Dirac Hamiltonian is needed. When we look back Equation (9.1), the Dirac Hamiltonian is described as follows:

$$H = \begin{pmatrix} m \cdot I & \sigma \cdot p \\ \sigma \cdot p & -m \cdot I \end{pmatrix} = \alpha \cdot p + \beta m, \tag{9.2}$$

where I is a 2×2 unit matrix and p is momentum. Writing $U_F = e^{+iS}$ with S as Hermitian and not explicitly time dependent, the unitary transformation is

$$\psi' = e^{+iS}\psi, \tag{9.3}$$

$$i\frac{\partial}{\partial t}\psi = e^{+iS}H\psi = e^{+iS}He^{-iS}\psi' = H'\psi', \tag{9.4}$$

and where H' contains no off-diagonal components by construction. This transformation is called as the Foldy–Wouthuysen transformation. When we choose $U_F = e^{+iS}$ as

$$e^{+iS} = \exp(\beta\alpha \cdot p\theta(p)) = \cos|p|\theta + \frac{\beta\alpha \cdot p}{|p|}\sin|p|\theta. \tag{9.5}$$

The Dirac Hamiltonian under the unitary transformation becomes

$$H' = \left(\cos|p|\theta + \tfrac{\beta\alpha\cdot p}{|p|}\sin|p|\theta\right)(\alpha \cdot p + \beta m)\left(\cos|p|\theta - \tfrac{\beta\alpha\cdot p}{|p|}\sin|p|\theta\right). \tag{9.6}$$

Because α and β are anticommutable ($\{\alpha, \beta\} = 0$)

$$\begin{aligned}H' &= \left(\cos|p|\,\theta + \frac{\beta\alpha\cdot p}{|p|}\sin|p|\theta\right)(\alpha\cdot p + \beta m)\left(\cos|p|\,\theta - \frac{\beta\alpha\cdot p}{|p|}\sin|p|\theta\right)\\ &= (\alpha\cdot p + \beta m)\left(\cos|p|\,\theta - \frac{\beta\alpha\cdot p}{|p|}\sin|p|\theta\right)^2\\ &= (\alpha\cdot p + \beta m)\exp(-2\beta\alpha\cdot p\theta)\\ &= \alpha\cdot p\left(\cos 2\,|p|\,\theta - \frac{m}{|p|}\sin 2|p|\theta\right) + \beta(m\cos 2|p|\theta + |p|\sin 2|p|\theta).\end{aligned} \tag{9.7}$$

When we choose θ as

$$\tan 2|p|\theta = \frac{|p|}{m}, \tag{9.8}$$

off-diagonal element, α, can be eliminated, and

$$\begin{aligned}H' &= \alpha\cdot p\left(1 - \frac{m}{|p|}\tan 2|p|\theta\right)\cos 2|p|\theta + \beta(m - |p|\tan 2|p|\theta)\cos 2|p|\theta\\ &= \beta\left(m + \frac{|p|^2}{m}\right)\frac{m}{\sqrt{p^2+m^2}}\\ &= \beta\sqrt{p^2+m^2}.\end{aligned} \tag{9.9}$$

For further generalization, we introduce an electromagnetic field

$$\begin{aligned}H &= \sigma(p - eA) + \beta m + e\phi\\ &= \beta m + \theta + \varepsilon,\\ \theta &= \sigma(p - eA),\\ \varepsilon &= e\phi,\end{aligned} \tag{9.10}$$

where e is an electric charge, A is the vector potential, and φ is a scalar potential. Here, θ has off-diagonal elements. Remember that β and θ are anticommutable, whereas β and ε are commutable. We introduce the unitary transformation here again as

$$i\frac{\partial}{\partial t}e^{-iS}\psi = H\psi = He^{-iS}\psi' = e^{-iS}\left(i\frac{\partial}{\partial t}\psi\right) + \left(i\frac{\partial}{\partial t}e^{-iS}\right)\psi',$$

and hence,

$$i\frac{\partial}{\partial t}\psi' = \left[e^{iS}\left(H - i\frac{\partial}{\partial t}\right)e^{-iS}\right]\psi' = H'\psi'. \tag{9.11}$$

The Foldy–Wouthuysen transformation is implemented three times by using the Baker–Hausdorff formula (see Appendix for the detail), and then the Hamiltonian can be written as

$$\begin{aligned}H' &= \beta\left\{m + \frac{(p-eA)^2}{2m} - \frac{p^4}{8m^3}\right\} + e\phi - e\frac{1}{2m}\beta(\sigma\cdot B)\\ &-\frac{ie}{8m^2}\sigma\cdot\mathrm{rot}(E) - \frac{e}{4m^2}\sigma\cdot(E\times p) - \frac{e}{8m^2}\mathrm{div}(E).\end{aligned} \tag{9.12}$$

This is the Hamiltonian of the Dirac equation for an electron with an electromagnetic field in a nonrelativistic limit, and

$$-\frac{ie}{8m^2}\sigma \cdot \mathrm{rot}(E) - \frac{e}{4m^2}\sigma \cdot (E \times p), \tag{9.13}$$

is the spin-orbit Hamiltonian. When we assume spherically symmetric potential, *V*(*r*), for simplicity, the first term of Equation (9.13) goes to zero. Finally, we obtain the spin-orbit Hamiltonian as

$$\begin{aligned} H_{\text{spin-orbit}} &= \frac{e^2}{4m}\frac{1}{r}\frac{\partial V}{\partial r}\sigma \cdot L, \\ L &= r \times p. \end{aligned} \tag{9.14}$$

The above discussion is valid even when the atomic number, Z, is not equal to one (hydrogen), and it is notable that this Hamiltonian is proportional to Z^4 (Z: atomic number) under this assumption. Because the potential, *V*(*r*), is a Coulomb interaction from a nucleus to a rotating electron around the nucleus, the spin-orbit Hamiltonian with a relativistic effect is described as

$$H_{\text{spin-}orbit} = \frac{Ze^2}{4mr^3}\sigma \cdot L. \tag{9.15}$$

When we calculate a classical spin-orbit Hamiltonian, the value is twice as large as the value in Equation (9.15), which is attributed to Thomas precession. Here, *r* is inversely proportional to the atomic mass, Z, because a heavier atom possesses larger radii for outer-shell electrons, hence it can be understood that the spin–orbit interaction is proportional to Z^4 and light elements induce the smaller spin–orbit interaction [14].

9.2.2 Conductance Mismatch

Next, we discuss the conductance mismatch [15] and a spin drift/diffusion model is introduced for explaining the essence of the feature. An electric current, J^{charge}, can be written as follows:

$$\begin{aligned} J^{\text{charge}} &= J_\uparrow + J_\downarrow \\ &= \sum_{s=\uparrow\downarrow}(J_s^{\text{drift}} + J_s^{diffusion}) \\ &= \sum_{s=\uparrow\downarrow}(\sigma_s E + eD_s \mathrm{grad}(n_s)) \\ &= \sum_{s=\uparrow\downarrow}(\sigma_s \mathrm{grad}(\phi) + eD_s \mathrm{grad}(n_s)), \end{aligned} \tag{9.16}$$

where $\uparrow, \downarrow$ indicates spin directions (up and down), J^{drift} and $J^{\text{diffusion}}$ are a drift and diffusion current of spins, respectively, n_s is a spin carrier density, σ_s is the conductivity ($= e^2 n_s \tau_s / m^*$, τ_s is the momentum relaxation time, and m^* is an effective mass), *E* is an electric field and D_s is a diffusion constant. Here the Einstein's relationship is introduced and Equation (9.16) can be rewritten as

$$J_s = \sigma_s \mathrm{grad}(\bar{\mu})/e, \quad \bar{\mu} = \mu_s - e\phi, \tag{9.17}$$

where $\bar{\mu}$ is an electrochemical potential and μ is a chemical potential. In addition, differential equations for spin and charge are

$$\frac{\partial(\mathrm{spin})}{\partial t} + \mathrm{div}J_s = \frac{\sigma_s}{e}\frac{\bar{\mu}_\uparrow - \bar{\mu}_\downarrow}{l_{sf}^2}, \tag{9.18a}$$

$$l_{sf} = (D_s \tau_{sf})^{1/2},$$

$$\frac{\partial(\mathrm{charge})}{\partial t} + \mathrm{div}(J_\uparrow + J_\downarrow) = 0, \tag{9.18b}$$

where l_{sf} and τ_{sf} are spin diffusion length and time, respectively. Equation (9.18b) means charge current conservation, whereas Equation (9.18b) means that spins cannot be conserved because of spin relaxation. From the above equations, spin diffusion equations become

$$\Delta(\bar{\mu}_\uparrow - \bar{\mu}_\downarrow) = (\bar{\mu}_\uparrow - \bar{\mu}_\downarrow)/l_{sf}, \tag{9.19a}$$

$$\Delta(\sigma_\uparrow \bar{\mu}_\uparrow - \sigma_\downarrow \bar{\mu}_\downarrow) = 0, \tag{9.19b}$$

where Equation (9.19a) shows spin diffusion, and Equation (9.19b) shows an electric neutrality condition.

We apply the above equations for a practical case, namely, a heterojunction of nonmagnet (NM) and ferromagnet (FM) without a tunneling barrier at $z = 0$ (see Figure 9.1). The

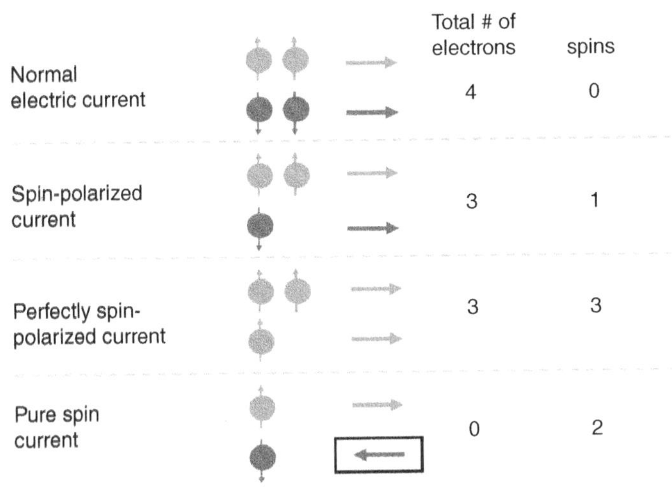

Figure 9.1 *Concept of pure spin current*

electrochemical potential for up- and down-spins in the NM and the FM can be written as for the NM side

$$\bar{\mu}_{\uparrow} = \frac{A}{\sigma_{\uparrow}} \exp\left(-\frac{z}{l_{sf,\,F}}\right) + Bz + C, \quad \bar{\mu}_{\downarrow} = \frac{A}{\sigma_{\downarrow}} \exp\left(-\frac{z}{l_{sf,\,F}}\right) + Bz + C, \tag{9.20}$$

for the FM side

$$\bar{\mu}_{\uparrow} = \frac{a}{\sigma} \exp\left(-\frac{z}{l_{sf,\,N}}\right) + bz, \quad \bar{\mu}_{\downarrow} = \frac{a}{\sigma} \exp\left(-\frac{z}{l_{sf,\,N}}\right) + bz, \tag{9.21}$$

where A,B,C,a, and b are constant, $l_{sf,\,N}$ and $l_{sf,\,F}$ are spin-relaxation length in the NM and the FM, respectively. Then, the spin polarized current for each side can be calculated by using Equation (9.17), and the up-spin current in the FM side, for example, becomes

$$eJ_{\uparrow} = \sigma_{\uparrow} \left\{ \frac{A}{\sigma_{\uparrow} l_{sf,\,F}} \exp\left(\frac{z}{l_{sf,\,N}}\right) + B \right\}. \tag{9.22}$$

Because the electrochemical potential and the up- and down-spin currents are continuous at $z = 0$, we can determine the constants (A,B,C,a and b) with ease and obtain the final form of the spin polarized current at the interface between the NM and the FM as

$$\begin{aligned} \frac{J_{\uparrow} - J_{\downarrow}}{J_{\uparrow} + J_{\downarrow}} &= \frac{\beta}{1 + \frac{r_N}{r_F}}, \\ \beta &= \frac{\sigma_{\uparrow} - \sigma_{\downarrow}}{\sigma_{\uparrow} + \sigma_{\downarrow}} = \frac{\sigma_{\uparrow} - \sigma_{\downarrow}}{\sigma}, \\ r_N &= \frac{l_{sf,\,N}}{2\sigma}, \\ r_F &= \frac{l_{sf,\,F}}{4}(\sigma_{\uparrow}^{-1} + \sigma_{\downarrow}^{-1}), \end{aligned} \tag{9.23}$$

where β is the spin polarization of the conductivity, r_N and r_F is spin resistance in the NM and the FM, respectively. As one can see, the spin resistance is inversely proportional to conductivity, and if the spin resistance in NM (namely, the resistivity) is much larger than that in FM, Equation (9.23) becomes zero. This directly indicates that we cannot expect spin injection into NM if the NM is highly resistive. Obviously, a metallic heterojunction is not the case. In the case of inorganic semiconductor and molecules, this conductivity mismatch becomes a serious problem for spin injection. It should be noted that the conductivity mismatch cannot be avoided even if one chooses an FM electrode with the same Fermi level as the levels of conduction or valence bands of a semiconductor. In such as case, we may avoid a Schottky barrier formation but we still face with the conductance mismatch because the conductivity of the FM and the NM is largely different.

There are several ways to avoid or solve this problem; one is insertion of a tunneling barrier (interfacial spin-dependent resistance) and the other is utilization of FM with lower conductivity, such as LaSrMnO (LSMO). Both ways are commonly used in spintronics, and here we show the effect of inserting a tunneling barrier from a theoretical point of view. Even when the tunneling barrier is inserted, the electrochemical potential for up- and down-spin is same as Equations (9.20) and (9.21), and the spin polarized current can be written as same as in the previous case. The difference is manifested in the continuity condition of the electrochemical potential at $z = 0$, namely, the electrochemical potential for up- and down-spin at $z = 0$ is continuous when the spin-dependent interfacial resistance is included

$$\bar{\mu}_\uparrow = \frac{A}{\sigma_\uparrow} + C + r_{i\uparrow} eJ_\uparrow = \frac{a}{\sigma}, \tag{9.24}$$

$$\bar{\mu}_\downarrow = \frac{A}{\sigma_\downarrow} + C + r_{i\downarrow} eJ_\downarrow = -\frac{a}{\sigma}, \tag{9.25}$$

where r_i is the spin-dependent interfacial resistance. The spin polarized current at FM/tunneling barrier/NM is described as

$$\begin{aligned} \frac{J_\uparrow - J_\downarrow}{J_\uparrow + J_\downarrow} &= \frac{r_F\beta + r_i\beta''}{r_F + r_N + r_i}, \\ \beta'' &= \frac{r_{i\uparrow}^{-1} - r_{i\downarrow}^{-1}}{r_{i\uparrow}^{-1} + r_{i\downarrow}^{-1}}, \\ r_{i\uparrow} &= 2(1-\beta'')r_i, \quad r_{i\downarrow} = 2(1+\beta'')r_i. \end{aligned} \tag{9.26}$$

It should be noted that the spin polarization does not go to zero even when r_N is much larger than r_F if the interfacial spin-dependent resistance r_i is compared to r_N,

$$\begin{aligned} \frac{J_\uparrow - J_\downarrow}{J_\uparrow + J_\downarrow} &= \frac{r_F\beta + r_i\beta''}{r_F + r_N + r_i} = \frac{r_F\beta/r_i + \beta''}{(r_F/r_i) + (r_N/r_i) + 1} \\ &\approx \frac{\beta''}{(r_N/r_i) + 1} \ (\text{if } r_N, r_i >> r_F). \end{aligned} \tag{9.27}$$

Equation (9.27) clearly shows that the tunnel barrier plays a dominant role in realizing the injection of spin-polarized current into NM.

In the final part of this subsection, we briefly show the concept of a magnetoresistance effect by using TMR as an example. The resistance in parallel and antiparallel magnetization alignments is written as R_P and R_{AP}, and then the MR ratio is defined as

$$\begin{aligned} \frac{R_{AP} - R_P}{R_P} &= \frac{2P_1P_2}{1 - P_1P_2}, \\ P &= \frac{D_{\text{Majority}} - D_{\text{minority}}}{D_{\text{Majority}} + D_{\text{minority}}}. \end{aligned} \tag{9.28}$$

where $P_{1,2}$ is spin polarization of each ferromagnet and D is a density of state (DOS) of majority and minority spins. The magnetization alignments of the FM electrodes (parallel or antiparallel) are controlled by applying an external magnetic field parallel to a long axis of the electrodes. The FM electrodes should have different coercive forces by changing their geometries, and we can control the alignments in this manner. The spin polarization is determined by the difference of the DOS for up- and down-spin at the Fermi level. Because the polarization is often positive for ferromagnets, the MR ratio is often positive. However, when we change a bias voltage for a TMR device, the DOS at the pseudo-Fermi level can also be changed and the spin polarization can change its sign. This can induce the decrease/increase of the MR ratio or inverse of the sign of the MR ratio. Experimentally, the negative MR ratio was sometimes observed especially when LaSrMnO was used as one ferromagnetic electrode for molecular spin devices. This is thought to be ascribable to the inverse of the DOS at a finite bias voltage, and the bias voltage dependence of the DOS, namely, the sign of the MR ratio should be observed as the band structure changes. To certify this, a detailed analysis of the bias voltage dependence is demanded.

9.2.3 Pure Spin Current

Pure spin current is now playing one of the most important roles in spintronics, and simultaneously provides steadfast evidence for successful spin injection in condensed mattes including molecules. So, a study on spin transport without showing generation of pure spin current does not fully support the claim of successful spin injection. Figure 9.1 shows what pure spin current is. An electric current is a charge flow, where the same numbers of up- and down-spin flow in the same direction. A spin-polarized current is a type of charge current, but the numbers of up- and down-spin are not the same. Only one kind of spins flows in one direction in a perfectly polarized current. A pure spin current is a completely different current from the ones mentioned above. The same number of up- and down-spins flows in *opposite directions*, and thus there is no charge flow in a pure spin current. Notably, the down-spin current to the left is equivalent to the up-spin flow to the right, because of time reversal symmetry (keep in mind that the time reversal operation induces a reversal of motion and spin direction simultaneously). Hence, only the spin angular momentum flows in one direction without charge flow. Moreover, a pure spin current ideally possesses a time-reversal symmetry resulting in a dissipationless current (experimentally, a diffusive pure spin current is often generated, which is a dissipative current). An example of a dissipationless motion is a harmonic oscillator without friction; the equation describing this motion is

$$m\frac{d^2x}{dt^2} = -kx, \tag{9.29}$$

$$m\frac{d^2x}{dt^2} = -kx - \kappa\frac{dx}{dt}, \tag{9.30}$$

where m is the mass, x is the position, t is the time, k is the spring constant, and κ is the friction coefficient. Equation (9.29) shows the motion of a harmonic oscillator without friction, whereas Equation (9.30) shows the oscillator with friction. As one can see, Equation (9.29) has time reversal symmetry under the $t \rightarrow -t$ operation, while Equation (9.30) does not. Therefore, it can be said that a motion with time reversal symmetry is dissipationless.

Several methods have been reported so far for generating pure spin current, i.e., the electrical, dynamical and thermal methods. Electrical generation of a pure spin current is achieved as follows (see Figure 9.2(a)): one metallic wire, Al, with two ferromagnetic Co electrodes is fabricated, and electric current is injected by using one circuit (the left side). Spin accumulation in the Al wire beneath Co1 takes place, but it does not occur in the nonmagnetic junctions. The accumulated spins flow in the left circuit because of the electric field, and also diffuse in the right circuit because spins diffuse isotropically (when we assume that the majority spin is up-spin, the up-spin diffuses to the right, see Figure 9.2(c)).

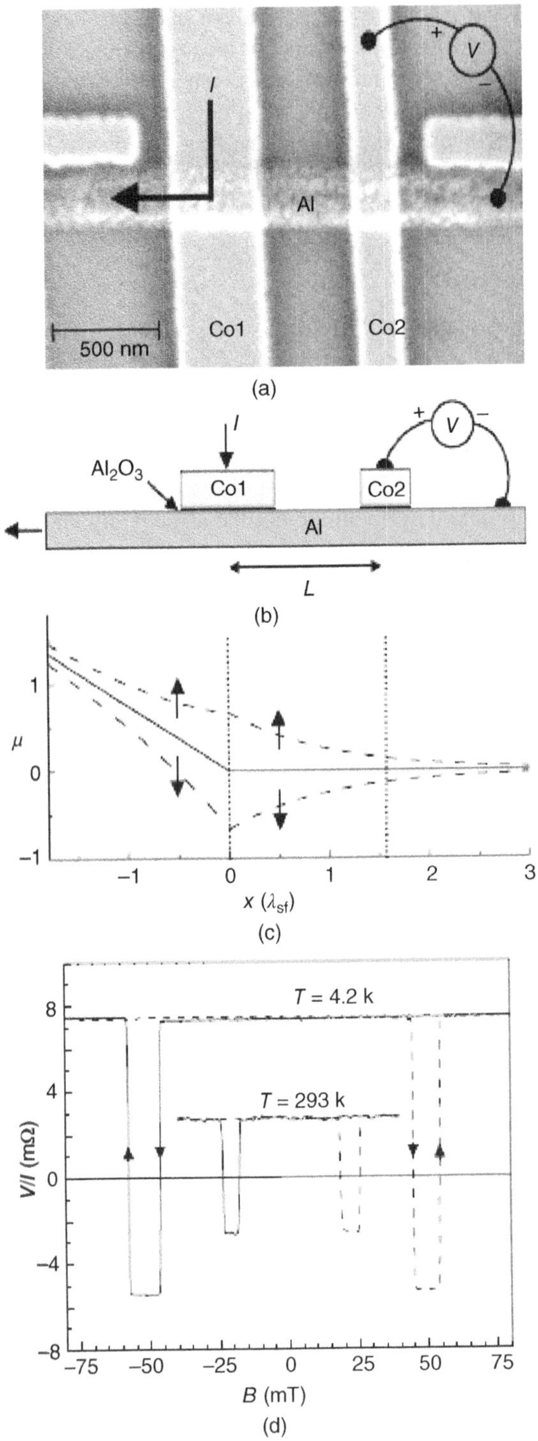

Figure 9.2 *(a) A scanning electron microscopic image of a Co/AlO/Al spin valve. (b) A cross–sectional schematic of the metallic spin valve. (c) Position dependence of electrochemical potential for up- and down-spin. (d) Observed spin signals at 4.2 and 293 K*

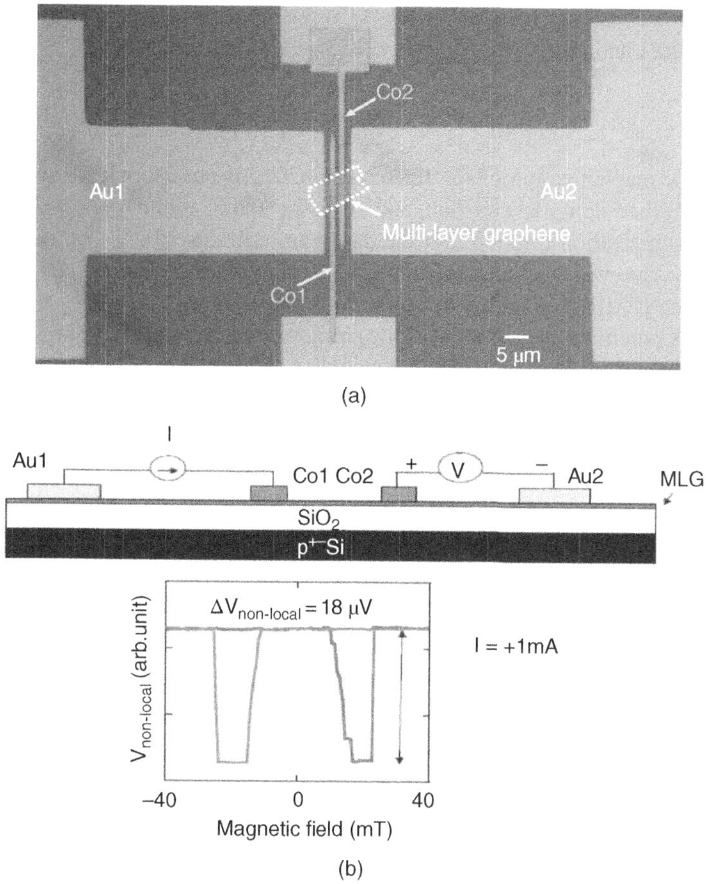

Figure 9.3 *An example of generating and detecting spin current in graphene*

Here, Al is a nonmagnet and no electric current passes through Co2. Thus, the down-spin of which number is the same as that of up-spin flows to the left in order to maintain carrier neutrality and spin balance. Since the up-spin flows to the right and the down-spin flows to the left, a pure spin current can be generated electrically. When the spin direction of Co_2 is changed by an external magnetic field, Co2 can detect the electrochemical potential of the up-spins propagating to the right or the down-spins propagating to the left in the Al wire, resulting in the reversal of output voltages, as shown in Figure 9.2(d). An example of generating and detecting of the pure spin current in graphene is shown in Figure 9.3. The NL-4T method can eliminate spurious effects, for example, anisotropic MR, the local Hall effect and so on, because no electric current is used in the detection circuit. This is an obvious advantage of this method for proving definite spin injection in solids. The electrical method for generating pure spin current is widely introduced in metallic and semiconductor spin valves, such as those using Al [16], Cu [17], Si [18], GaAs [19], multi- and single-layer graphene [20, 21] up to room temperature and single-walled carbon nanotubes [22] at 4 K, and allowed us to obtain reproducible results and discuss the underlying physics.

The second method is a dynamical one, namely, spin-pumping. Mizukami and co-workers proposed the dynamical method [23, 24], and reported the modulation of the

Gilbert damping constant of Cu in NiFe(Py)/Cu junction. The magnetization dynamics is described by the Landau–Lifsitz–Gilbert (LLG) equation as follows:

$$\frac{dM}{dt} = \gamma H_{eff} \times M + \alpha M \times \frac{dM}{dt}, \tag{9.31}$$

where M is the magnetization of the ferromagnet, H_{eff} is an external magnetic field, t is time, α is the Gilbert damping constant, and the first and the second terms are the field term describing magnetization rotation and the damping term describing torque, respectively. When a microwave with the suitable frequency is applied to the ferromagnet, ferromagnetic resonance (FMR) can occur. The damping torque is suppressed by the microwave under the FMR condition, which induces the pumping of spins into the nonmagnet attached to the ferromagnet, due to spin angular momentum conservation (see Figure 9.4). Saitoh

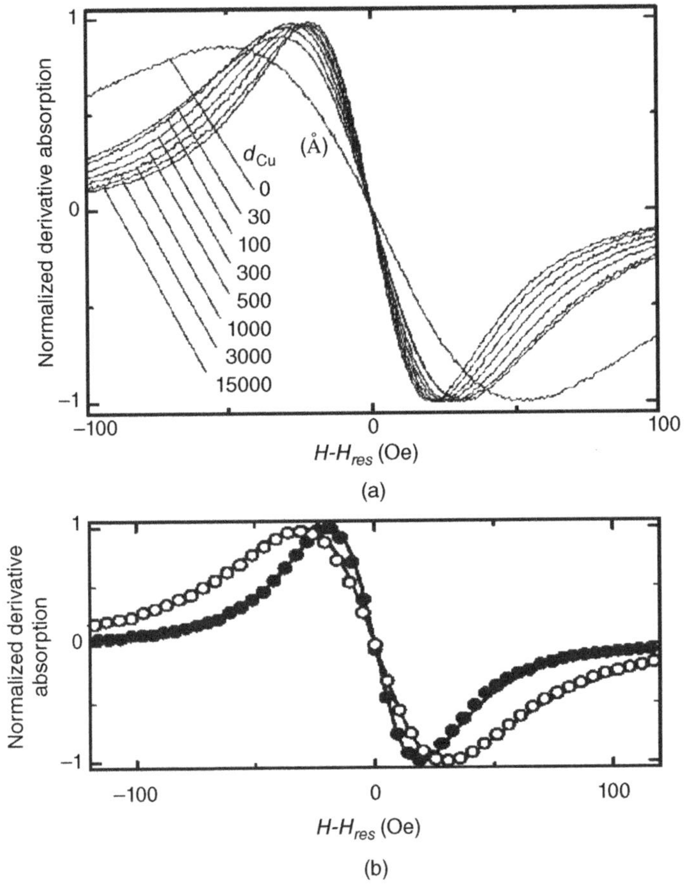

Figure 9.4 *(a) Normalized ferromagnetic resonance (FMR) spectra for Cu/Py (3 nm)/Cu (d_{Cu})/Pt films with various d_{Cu}. (b) Normalized FMR spectra for Cu/Py (3 nm)/Cu (10 nm)/Pt (open circles) and Cu/Py (3 nm)/Cu (10 nm) (solid circles) films, respectively. The line width is obviously larger by attaching Pt, which directly indicates spin pumping into Pt, resulting in the enhancement of spin relaxation (Gilbert damping constant) in the system*

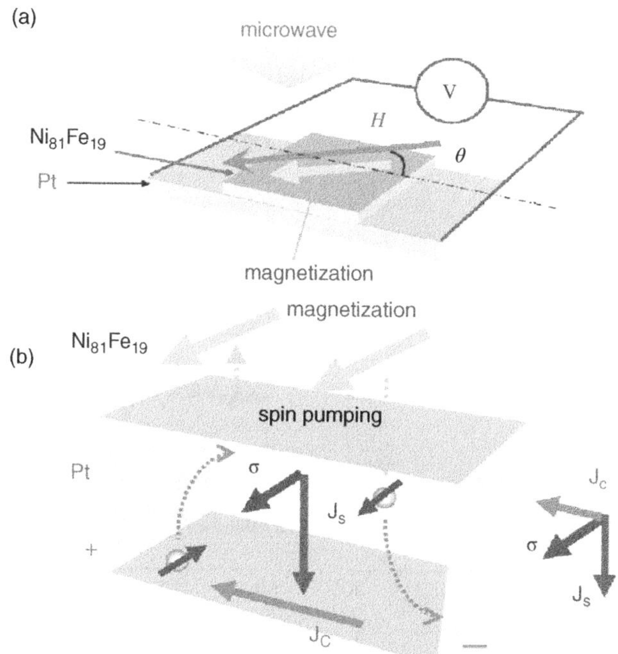

Figure 9.5 *(a) Schematic illustration of the NiFe/Pt sample. (b) Schematic illustration of dynamical spin injection into Pt, resulting in a charged current (conversion of spin current to charged current)*

et al. used this method in order to inject spins in Pt, and observed the inverse spin Hall effect (Figures 9.5 and 9.6), which is a reciprocal effect of the spin Hall effect [25]. Spin injection by spin pumping was also used for GaAs [26] and p-Si [27] in addition to various nonmagnetic metals; it is noted that spin injection into a magnetic insulator can be realized by spin pumping. Kajiwara *et al.*, generated a spin-wave spin current in yttrium iron garnet (YIG; $Y_3Fe_5O_{12}$) by using spin pumping, which propagated an order of cm [28] (see Figure 9.7). This study is quite significant, since spin propagation was successfully achieved whereas YIG is electrically insulative. Recently, Shikoh *et al.* have experimentally achieved the generation and propagation of a normal pure spin current (not spin-wave spin current) in p-Si at room temperature [29]. This study enables us to generate a normal pure spin current in molecules, and Tang *et al.* firstly demonstrated to generate and transport the dynamically generated pure spin current in single-layer graphene [30]. Here, note that a number of control experiments are indispensable for concluding successful spin transport in the dynamical method. Tsukahara *et al.* reported that ferromagnets can generate self-induced electromotive forces under the FMR due to its spin–orbit interaction [31]. This self-induced effect induces the modulation of FMR signals, which is quite similar with the signals when dynamical spin injection is succeeded. Hence, previous studies [29, 30] carefully carried out necessary control experiments to rule out this effect. Notable is that a study exhibiting only the modulation of the FMR linewidth [32] does not completely support the claim of spin transport and should be reinvestigated.

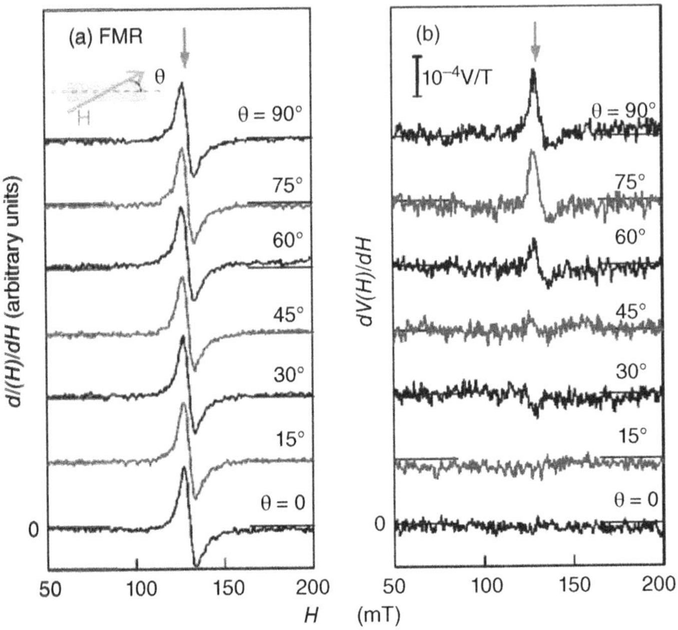

Figure 9.6 (a) FMR signals as a function of the angles of the external magnetic field. (b) Output voltages due to the generated charged current as a function of the angles of the external magnetic field

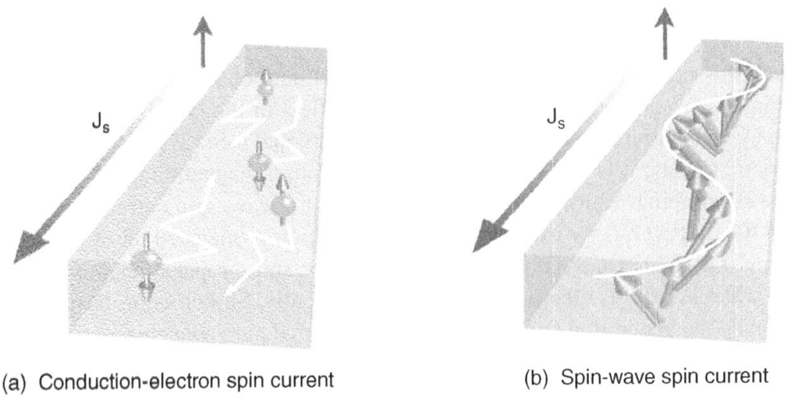

Figure 9.7 (a) Schematic illustration of a normal pure spin current. (b) Schematic illustration of a spin-wave spin current

9.3 Spin-Dependent Transport via Molecules

9.3.1 Various Origins of Magnetoresistance

Before introducing experimental results, we summarize various origins of MR except for spin-dependent transport, which are known to induce spurious signals. The most important origin is anisotropic MR (AMR), which is an effect that resistance of FM changes as a relative angle between an electric current direction and a magnetization direction changes. The AMR effect becomes obvious in 3d transition ferromagnetic metals and their alloys, for instance, NiFe, CoNi, and FeCo [33], and the value of the MR ratio is typically 1%–10%. AMR is the phenomenon that the electric current is modulated by the magnetization, and the electric current is induced by a motion of electrons. Because an electron has a spin degree of freedom, it is thought that the origin of the AMR is correlation between the spin and the motion of the charge, namely, the spin–orbit interaction. Experimentally, the AMR ratio exhibited clear dependence of a magnetic moment per atom [33], which is thought to be a direct proof. It is noteworthy that the AMR is found in every ferromagnet and we cannot ignore this effect when we investigate spin-dependent transport. Before utilizing a new FM as an electrode, the measurement of the AMR should be carried out for a precise spin transport measurement.

As shown in Figure 9.8, we assume that the magnetization and the electric current are not parallel. Then,

$$\begin{pmatrix} E_x \\ E_y \\ E_z \end{pmatrix} = \begin{pmatrix} \rho_\perp & \rho_H & 0 \\ \rho_H & \rho_\perp & 0 \\ 0 & 0 & \rho_\parallel \end{pmatrix} \begin{pmatrix} J_x \\ J_y \\ J_z \end{pmatrix}, \tag{9.32}$$

is a generalized relationship between the electric field and the current. Here, $\rho_\perp$, $\rho_\parallel$ and ρ_H are conductivity perpendicular to the z-axis, that parallel to the z-axis, and the Hall conductivity, respectively. Alternating the relationship to the equation which describes resistance, $\rho = \mathbf{EJ}/J^2$,

$$\begin{aligned} \rho &= (E_x J_x + E_y J_y + E_z J_z)/J^2 \\ &= \rho_\perp + (\rho_\parallel - \rho_\perp) J_z^2/J^2 \\ &\equiv \rho_\perp + (\rho_\parallel - \rho_\perp)\cos^2\theta, \end{aligned} \tag{9.33}$$

is obtained. Here, because no external magnetic field is applied and the magnetization of the FM is random, Equation (9.33) has to be averaged. Because $\langle \cos^2\theta \rangle = 1/3$, then

$$\begin{aligned} \rho &= \rho_\perp + (\rho_\parallel - \rho_\perp)/3 \\ &= \frac{2}{3}\rho_\perp + \frac{1}{3}\rho_\parallel \equiv \bar{\rho}. \end{aligned} \tag{9.34}$$

When $\boldsymbol{J} \parallel \boldsymbol{M}$, θ in Equation (9.33) is equal to 0, by contrast, when $\boldsymbol{J} \perp \boldsymbol{M}$, θ in Equation (9.33) is equal to $\pi/2$. From the above discussion, the AMR ratio is described as

$$\frac{\Delta\rho}{\bar{\rho}} = \frac{\rho_\parallel - \rho_\perp}{\bar{\rho}}. \tag{9.35}$$

In a practical case for spin devices, each FM electrode can exhibit AMR. When resistance hysteresis is tried to be measured, the magnetization direction of both FM electrodes is set to be parallel in the beginning. This means that the magnetization of the FM is not random, but uniform. Next, an external magnetic field is applied in order to induce magnetization reversal in the FM electrodes. As described above, the resistance of the electrodes changes in the course of the application of the external magnetic field. When the magnetic

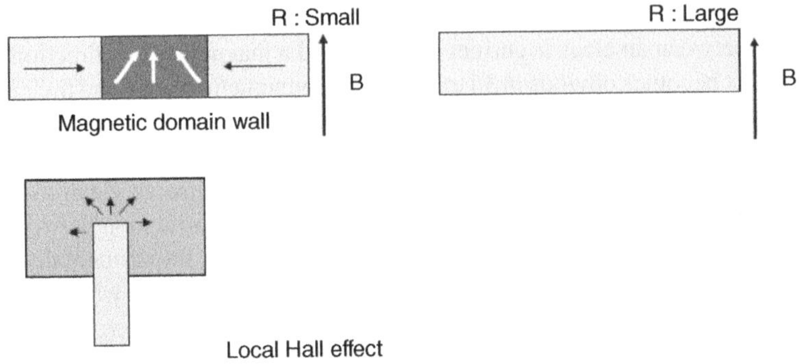

Figure 9.8 *Various origins of spurious signals. The top panel shows anisotropic MR effect, which is attributed to the generation of a magnetic domain wall in a ferromagnetic electrode during magnetization reversal. The external magnetic field (B) is applied perpendicular to the ferromagnetic electrodes. The bottom panel shows a schematic of local Hall effect, which occurs when the spin transport channel is not completely covered by ferromagnet*

field becomes large enough, the resistance comes back to the original value, because the uniform magnetization of the electrodes is recovered. In this process, resistance hysteresis can be observed in a so-called local measurement scheme [34], which is the reason why the contribution of the AMR should be eliminated by careful characterizations or by introducing the other experimental techniques. A nonlocal four-terminal method is a potential method for excluding this spurious effect, and this is widely introduced in spintronics [34]. The detail will be described later.

The other origins of spurious signals are (1) local Hall effect, and (2) too large contact resistance between FM and molecules. The former occurs when the spin transport channel is not completely covered by an FM electrode (see Figure 9.8) because the leakage magnetic field from the top-edge of the electrode affects the motion of spin carriers. The latter can induce large noise in spin signals and such a noise may be misunderstood as resistance hysteresis. Furthermore, reliability of the conventional "local" measurement scheme has not been fully clarified in molecular spintronics, which will be discussed in the following chapter, and the "nonlocal" four-terminal measurement scheme is strongly needed when we certify the spin injection into new materials.

9.3.2 Molecular Spintronics using Nanocarbonaceous Molecules

The door for molecular spintronics was opened by the first report on spin-dependent transport via multiwalled carbon nanotubes (MWNT) in 1999 [35]. Tsukagoshi and co-workers fabricated spin valve devices, where an MWNT acted as a spin channel. Figure 9.9 shows the data which they obtained, and the resistance hysteresis was observed up to 20 K. The splitting of the resistance hysteresis was not clear enough in some samples, because the difference of the coercive force in the Co electrode was not designed. Nonetheless, this is recognized as the first report on the spin transport via molecule. Because a field of spintronics and that of carbon nanotubes started to garner much attention at that time, this report provides large impact as a pioneering study. After this study, many people started to try observing spin transport via MWNT and also via single-walled carbon nanotubes

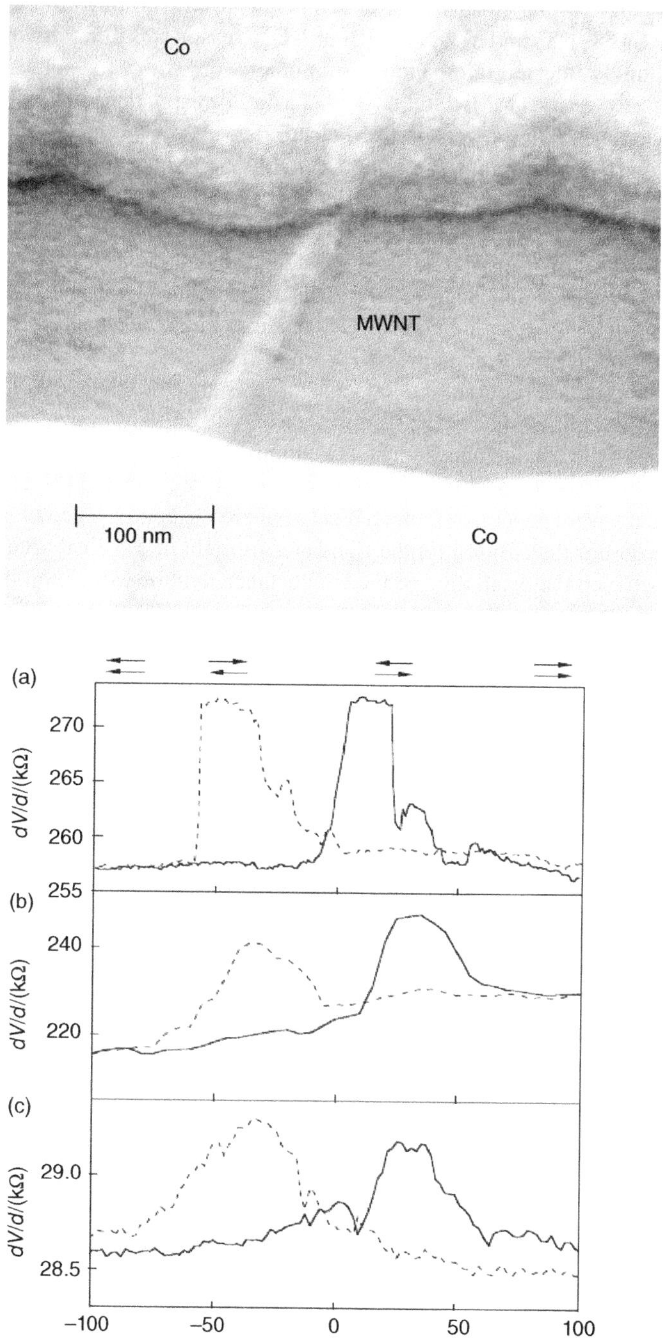

Figure 9.9 *The spin valve geometry (top) and resistance hysteresis observed at 4.2 K (bottom)*

(SWNTs). However, not so much progress was not achieved, although there were several reports on MR via SWNTs and MWNTs [36–38]. One reason was difficulty in proving that spins were definitely injected into nanotubes, in other words, how to exclude possibilities of spurious signals such as AMR and how to clarify correspondence between resistance hysteresis and magnetization reversal of FM electrodes.

In order to solve this important issue, Sahoo *et al.* [39] and Man *et al.* [40] proposed a new approach. They reported gate-tunable spin transport via a SWNT and an MWNT by using PdNi ferromagnetic electrodes. There are two advantages in their study; (1) realization of low contact resistance by using the PdNi alloy, and (2) gate modulation of the channel conductance. The role of the PdNi electrode was to decrease contact resistance. Pd was famous for realizing an Ohmic contact to a SWNT [41] and the contact resistance between the FM electrode and the nanotubes was suppressed down to several kΩ when Pd was incorporated into Ni. First Sahoo *et al.* reported the low contact resistance in PdNi/MWNT [42], and it help us knowing the importance of the significance of realizing low contact resistance. The gate modulation of the conductance induced Coulomb blockade in the nanotube, and then a clear Coulomb oscillation was observed, as was observed previously [43]. Hence, on- and off-resonant states were generated in the SWNTs, where the conductance of the nanotubes was oscillatory controlled. Based on the Landauer formula in a ballistic regime, the correspondence between the oscillation of the conductance and the MR ratio observed in the nanotube was theoretically explained, which was recognized as a proof of spin injection and transport in the nanotube (Figure 9.10).

In 2006, an important milestone in molecular spintronics was realized, which is the generation of a pure spin current in a molecule [22]. Because spurious effects are eliminated in this method, the nonlocal method allows us to obtain reliable results and to implement valuable discussions. The study by Tombros provided clear proof of the spin injection into a SWNT at 4 K (Figure 9.11), and also called people's attention to the reliability of the conventional local two-terminal method. The spin signal intensity in the nonlocal method should be a half of that in the local method because difference of electrochemical potential that is experimentally observed in the nonlocal method is only for up- or down-spin while the difference in the local is for both (Figure 9.12) [44]. Whereas this was experimentally observed in metallic spin valves [45], the intensity in the local method observed in their study was 35 times larger than that in the nonlocal, and the authors pointed out the importance of clarification of the reliability of the local method. In other words, their study taught us that we cannot conclude spin accumulation and spin injection as far as we use the local method only. In this sense, it should be emphasized that (1) MR-induced resistance hysteresis, (2) generation of nonlocal pure spin current and (3) Hanle-type spin precession [46] (described later) should be observed when we try to prove spin injection.

The other important milestone in nanotube spintronics is an observation of very large MR ratio by using a LSMO/MWNT/LSMO spin valve [47]. The authors introduced two LSMO electrodes, which were expected as half-metallic electrodes at low temperature, and observed the MR ratio of ~60% at 5 K and estimated the spin relaxation length to be 50μm assuming the MWNT mean free path of ~100 nm.

Based on the progress of experimental techniques and accumulation of understandings how to obtain a reliable result in nanotube spintronics, graphene spintronics was brilliantly established and has been dramatically developed since 2007. Graphene is currently one of the hottest materials in solid-state physics, because a number of interesting and unique

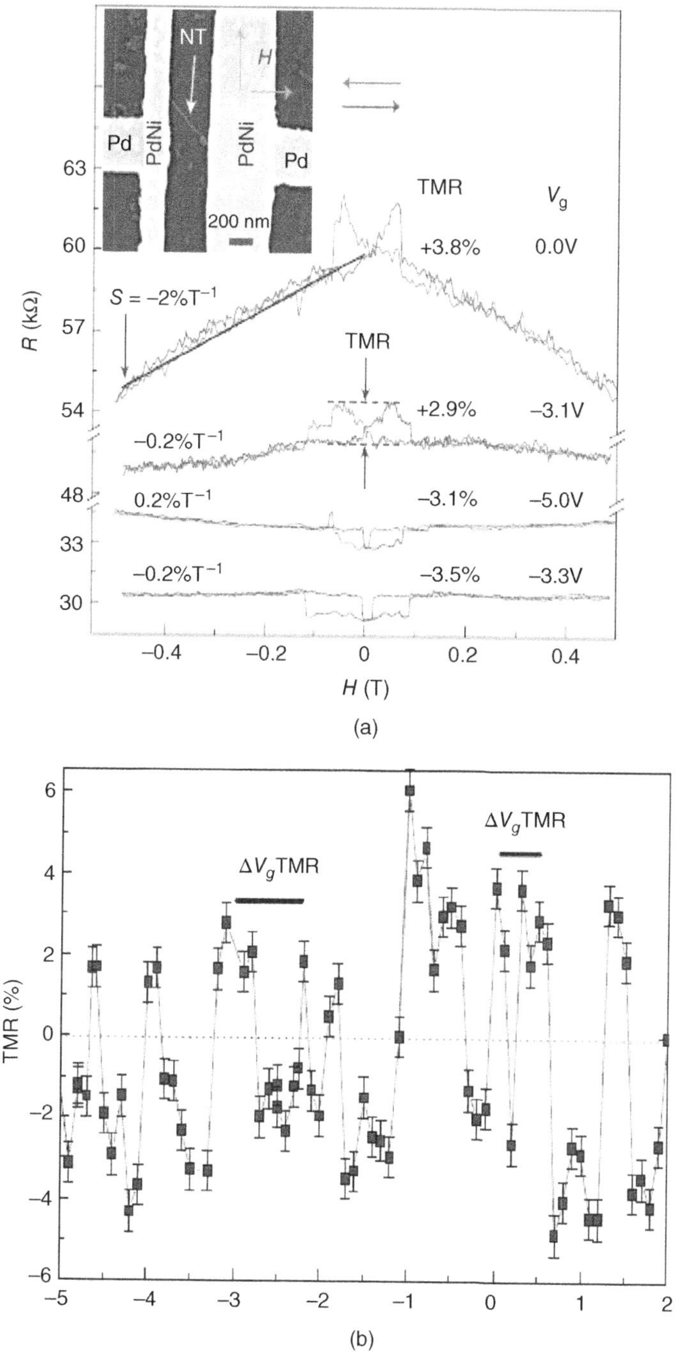

Figure 9.10 *(a) The observed MR effects in the PdNi-contacted nanotubes. The sign and the values of the MR ratio oscillated as the gate voltage was changed. (b) The oscillation of the MR ratio. The black closed squares are experimental data and the red solid line shows the theoretical fitting*

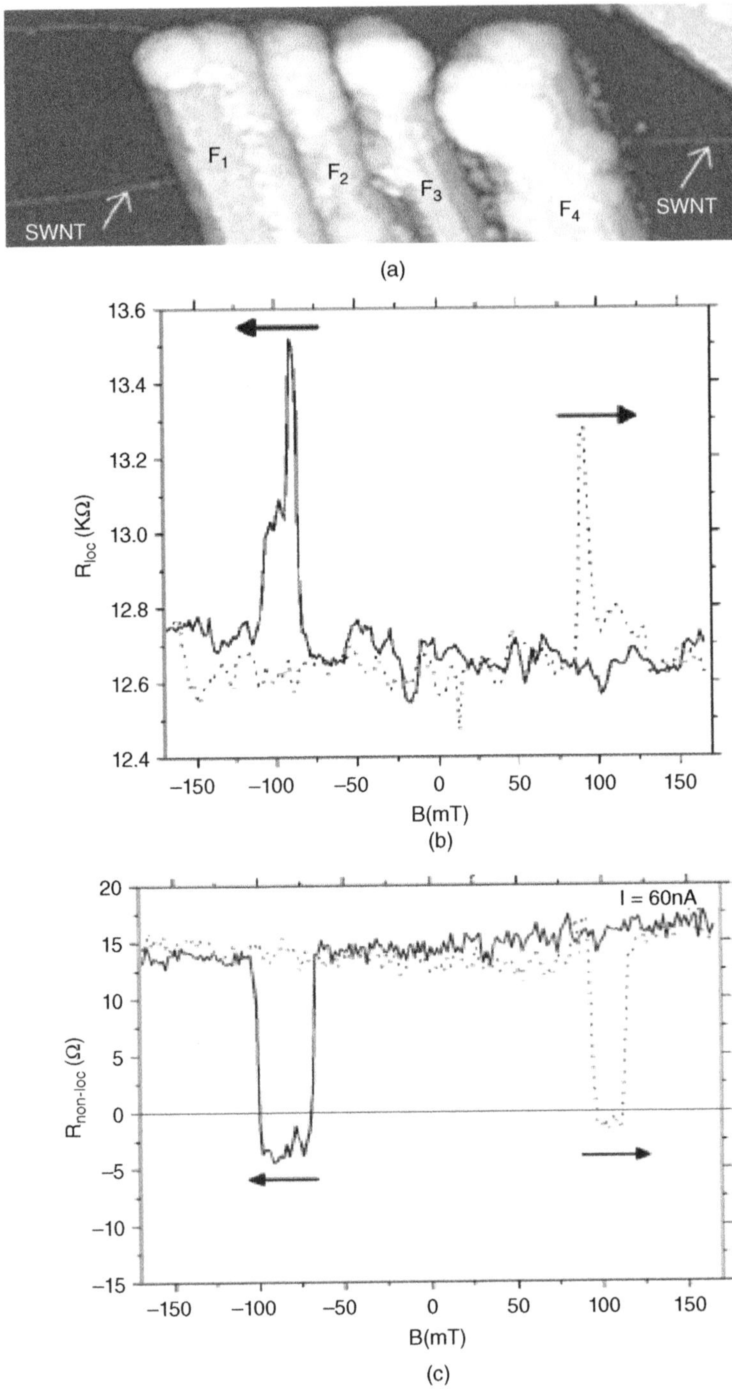

Figure 9.11 *(a) An scanning electron microscopic image of a SWNT spin valve with four ferromagnetic electrodes. (b) A spin signal in the local geometry at 4.2 K (Injection current was 10 nA). (c) A spin signal in the nonlocal geometry at 4.2 K (Injection current was 30 nA)*

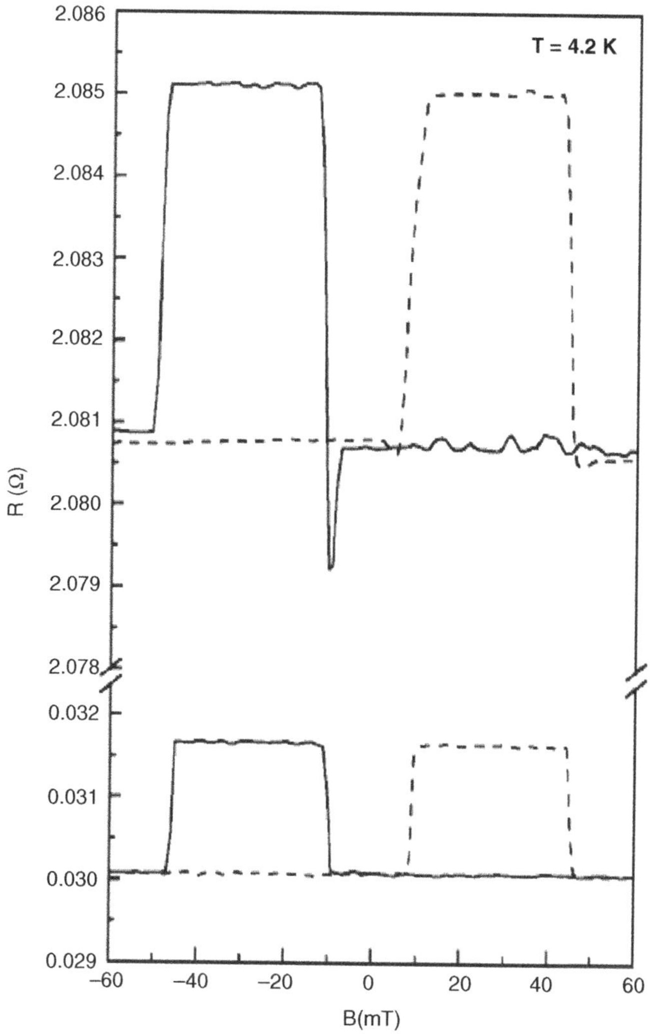

Figure 9.12 *Spin signals in the conventional "local" method (top curve) and "nonlocal" method (bottom curve) observed in Py spin valves at 4.2 K. The ratio of the signal intensities is equal to 2*

physics is realized in it. Integer quantum Hall effect at room temperature [48], fractional Hall effect at low temperature [49], a proximity effect [50] and field effect transistors with an extraordinarily large mobility [51] have been realized, because the zero-mass Dirac fermion system can be realized in graphene. Introduction of a spin function in graphene was another attractive challenge, and reliable and reproducible results on spin injection and generation of a spin current in single- and multilayer graphene (SLG and MLG) at room temperature were firstly reported by several groups individually [20–22], which was

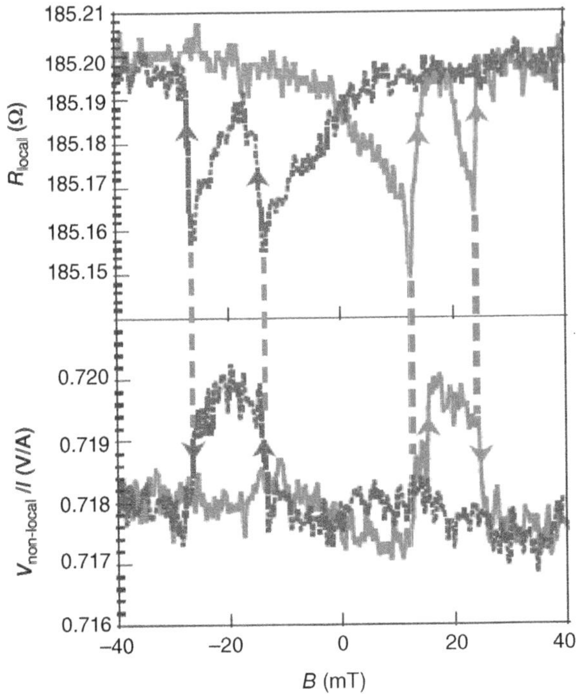

Figure 9.13 *(Top panel) MR signal due to AMR effect observed in a MLG spin valve at RT. (Bottom panel) Spin injection signal in a MLG spin valve at RT*

achieved by introducing the nonlocal method. The advantage of the nonlocal method was shown in the literature [20], because AMR effect observed in the local method was fully eliminated in the nonlocal method (Figure 9.13). Furthermore, the Hanle-type spin precession was observed in SLG [21] and MLG [53], which is the direct proof for the spin injection (Figure 9.14(a) and (b)). In order to generate the precession, an external magnetic field perpendicular to the spin transport channel (in this case, graphene) is applied (see Figure 9.15). The in-plane spins, which are injected into graphene, starts to precess along the magnetic field. The spins diffuse to the detector electrode (Co2) with precessing. The output signal depends on the relative angle between the transferred spins and the detector spins, namely, the signal oscillates as the spins precess. The oscillation and a crossing of the spin signals in parallel and antiparallel magnetization alignments allow us to conclude the precession of the injected spins. The nonlocal resistance with the precession can be described as

$$\frac{V_{\text{nonlocal}}}{I_{\text{inject}}} = \frac{P^2}{\sigma A/D} \int_0^{\infty} \frac{1}{\sqrt{4\pi Dt}} \exp\left(-\frac{L^2}{4Dt}\right) \cos(\omega t) \exp\left(-\frac{t}{\tau_{sf}}\right) dt, \tag{9.36}$$

where P is spin polarization, σ and A are conductivity and a cross-section of the channel, D is a diffusion constant, L is a gap length between two FM electrodes, ω is the Lamour frequency ($= g\mu_B B/\hbar, t$ is time and τ_{sf} is the spin coherent time. The analytical solution of

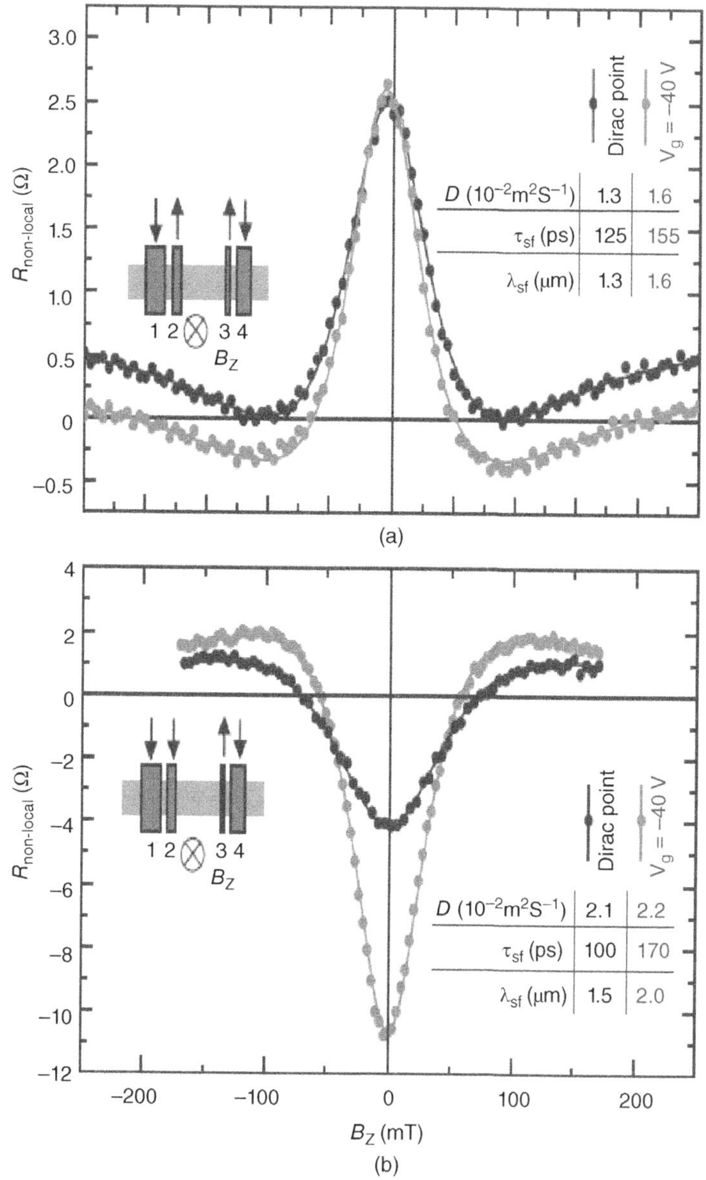

Figure 9.14 *Hanle-type spin precession observed in SLG. (a) The precession in the parallel alignment. (b) The precession in the anti-parallel alignment. The red line is for $V_g = -40$ V, and the black line is for the Dirac point*

Equation (9.36) was described in the literature [54]

$$\frac{V_{\text{nonlocal}}}{I_{\text{inject}}} = \frac{P^2 \lambda_N}{2\sigma A} \exp\left(-\frac{L}{\lambda_N}\right)(1 - \omega^2 \tau_{sf}^2)^{-1/4}$$

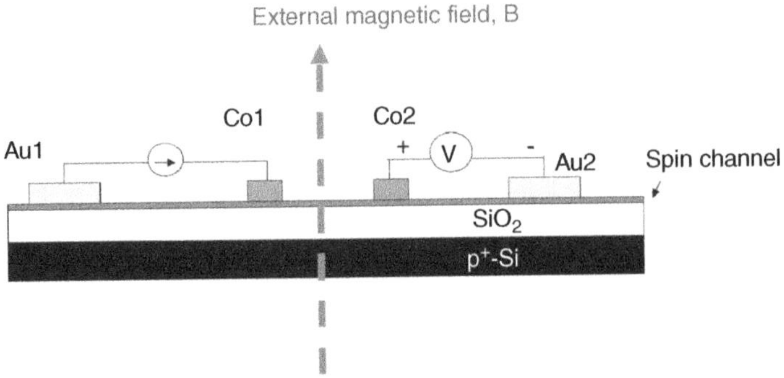

Figure 9.15 *A schematic of observing the Hanle-type spin precession. The external magnetic field is applied perpendicular to the spin channel*

$$\exp\left(-\frac{L}{\lambda_N}\right)\left\{\sqrt{\frac{1}{2}\left(\sqrt{1-\omega^2\tau_{sf}}+1\right)}-1\right\}$$
$$\times\cos\left\{\frac{\tan^{-1}(\omega\tau)}{2}+\frac{L}{\lambda_N}\sqrt{\frac{1}{2}(\sqrt{1-\omega^2\tau_{sf}}-1)}\right\}. \tag{9.37}$$

By using the equation, a couple of important spin transport parameters (diffusion constant, spin coherent time and spin polarization) are estimated, and spin coherent length, $l_{sf,}$ is also calculated from the following relationship:

$$l_{sf} = \sqrt{D\tau_{sf}}. \tag{9.38}$$

An example is shown in Figure 9.14(a) and (b). It is notable that one can also estimate spin relaxation length by using results on length dependence of spin signals, but the estimation is verified only in the nonlocal regime because it was pointed out that all of spin signals in the local regime is not due to spin accumulation as mentioned above. As far as investigated, the spin coherent length and time in graphene is typically 1–2 μ m and 100–200 ps, which is currently apart from the values in a theoretical prediction and should be improved in the future.

A number of studies have been implemented in graphene spintronics, so far, and anisotropic spin relaxation [55], spin drift in graphene [56, 57], correspondence between carrier densities and spin coherent time [58], unprecedented robustness of spin polarization in graphene [53], electron–hole asymmetry in spin transport [59], modification of spin transport by a chemical doping [60], investigation of tunneling barrier [61, 62], demonstration of gate-induced modulation of spin signals [51, 63–65] and so on. Here, the anisotropic spin relaxation is introduced as one example of an attractive spin transport character in graphene. They injected spins perpendicular to the graphene plain by applying the external perpendicular magnetic field of above 1.5 T, and compared spin relaxation time under $B = 0$ T and 2 T. Then, they found that spin relaxation of the spins perpendicular to the graphene plain is 20% faster than that parallel to the plain, which was

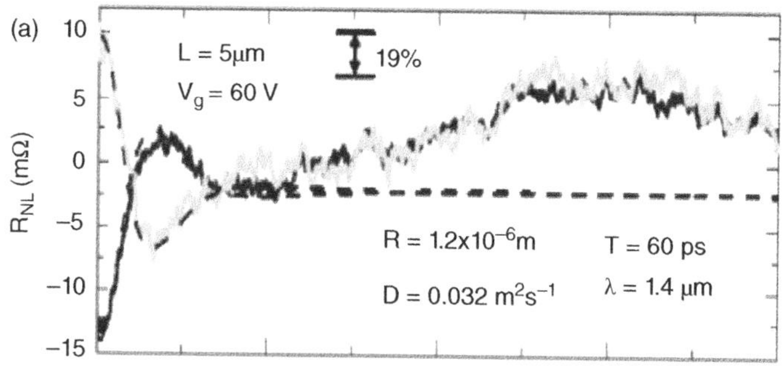

Figure 9.16 *An example of anisotropic spin relaxation in SLG*

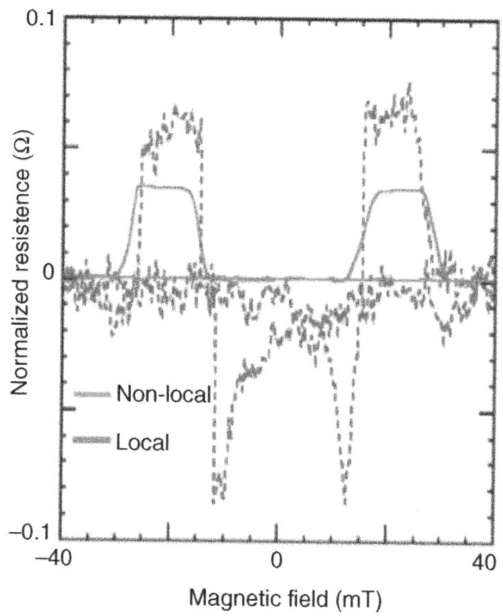

Figure 9.17 *Comparison of the spin signal intensities observed in MLG at RT in the local and the nonlocal geometries. The ratio of the intensities is almost equal to two*

attributable to the fact that the dominating spin relaxation mechanism is the Elliot–Yafet mechanism (Figure 9.16). Concerning the correspondence between spin signal intensities in the local and nonlocal schemes, which has been unclear in molecular spintronics as described above, it was clearly verified and steadfast basis of graphene spintronics was constructed in graphene spintronics (Figure 9.17) [53], and thus graphene is recognized as a model material for discussing spin transport in molecular materials. Spin relaxation in graphene is not a fully conclusive issue, since the Elliot–Yafet mechanism takes place in single-layer graphene but the D'yakonov–Perel mechanism governs the spin relaxation in double-layer graphene [66, 67]. Thus, further study on the origin of the different mechanisms in graphene is awaited.

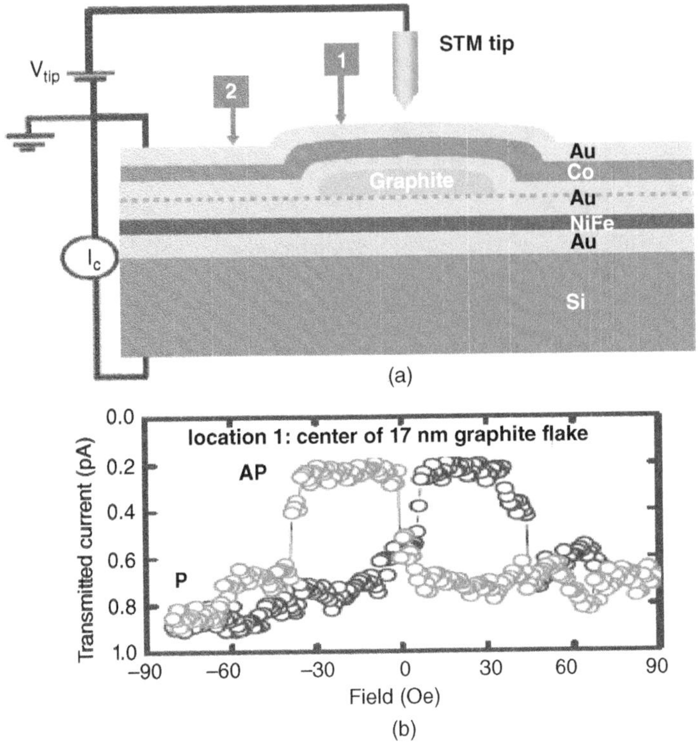

Figure 9.18 *(a) A schematic of the BEEM technique. (b) The observed local spin-valve signals at 150 K*

It is notable that a new approach for spin injection into graphitic materials was recently reported by Banerjee *et al.* [68], where perpendicular spin transport was realized by employing ballistic electron magnetic microscopy (Figure 9.18). As for a new approach, the dynamical spin pumping enables the generation of pure spin current and its transport is also quite noteworthy as mentioned in a previous section [30].

Contrary to studies on spintronics using graphene and carbon nanotubes, spintronics using fullerene has not been intensively studied until recently. The first report on spin-transport via C_{60} was implemented by fabricating a granular structure [69], in which Co particles (4–5 nm in diameter) were dispersed in various matrix species. Micklitz and co-worker observed MR up to 60 K, and the MR ratio was ca. 30% at 4 K. They concluded that the spin transport was governed by TMR. The interesting feature in their study was the TMR was observed in Co/CO or Co/Xe systems. However, it was not clarified that the MR was attributed to the magnetization of the Co and whether the MR was able to be observed up to higher temperature. Based on this motivation, Miwa *et al.* demonstrated the MR up to room temperature by using a C_{60}-Co nanocomposite system that has a similar structure with that of granular, and they also clarified that the observed MR was definitely due to the magnetization alignments of the Co nanoparticles of 2–3 nm in diameter [70]. The correspondence of the MR and the magnetization was clarified in the same sample, where the MR curve and the square of the normalized magnetization coordinated very well

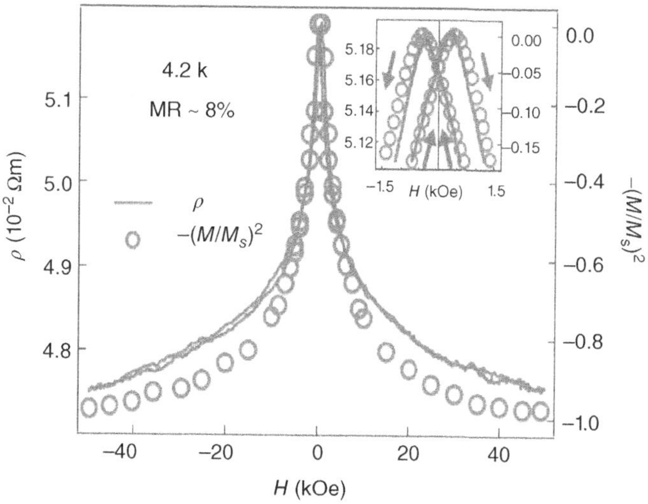

Figure 9.19 *The observed MR curve (the solid line) and the corresponding magnetization of the Co (open circles) in a* C_{60}*-Co nanocomposite spin device*

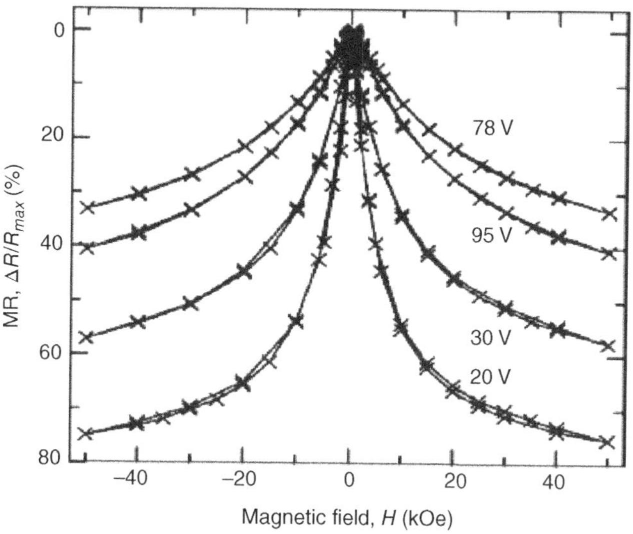

Figure 9.20 *An example of the large MR ratio in* C_{60}*-Co at 4.2 K*

(Figure 9.19). Sakai *et al.* also reported the similar effect by using the C_{60}-Co system but the observed MR ratio was extremely large (80% at 2 K in their definition, but ~300–500% in the conventional definition, see Figure 9.20) [71, 72]. They investigated the electronic structure of the C_{60}-Co covalent state by controlling the contents of the molecule and found that the spin polarization of the Co was strongly enhanced at the interface between C_{60} and Co by using X-ray magnetic circular dichroism [73]. It is noteworthy that recently Shiraishi and co-workers reported gigantic TMR ratio of 1.4 million % in the C_{60}-Co

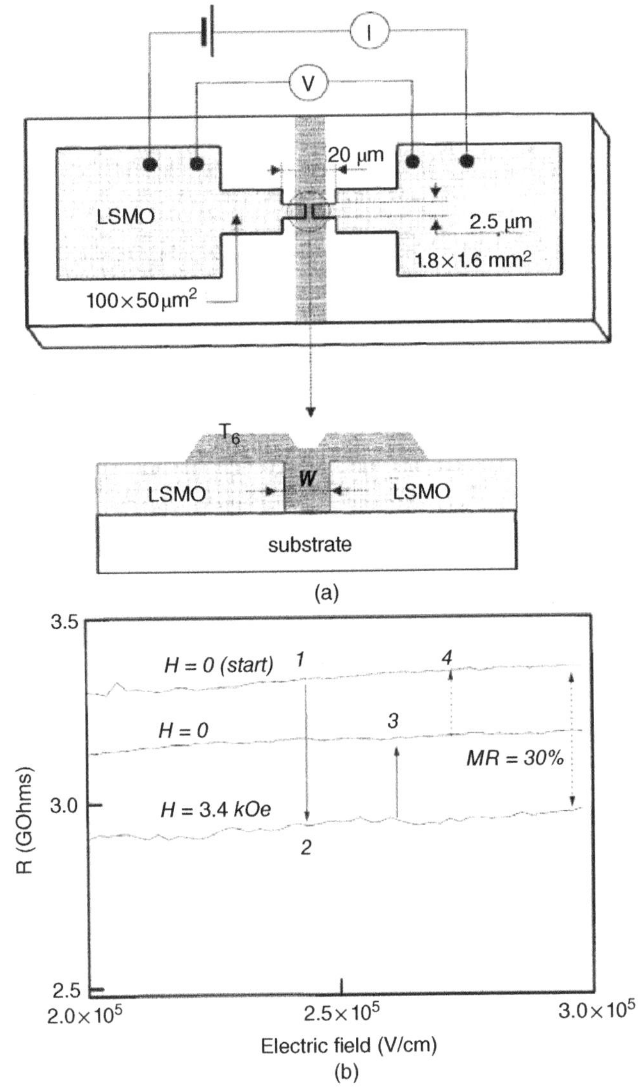

Figure 9.21 (a) Schematics of an organic spin valve using T6 and LSMO. (b) The observed resistance changes as a function of the external magnetic field

nanocomposite at 2 K [74]. This gigantic effect is ascribed to a novel magnetic switching effect due to a new magnetoelectric effect. It was not so much expected that a new physical feature would be found in spin-dependent tunneling transport via molecules at the first stage of such studies, but unexpected interesting issues have been observed, and nowadays, it is recognized that it is forth for implementing further investigation is needed. The other important findings in TMR via organic molecules will be described in the next subsection.

9.3.3 Molecular Spintronics using Organic Molecules

In this subsection, spin-dependent transport via organic molecules is introduced in detail. In 2002, Dediu and co-workers reported the change of the resistance of sexithienyl (T6)

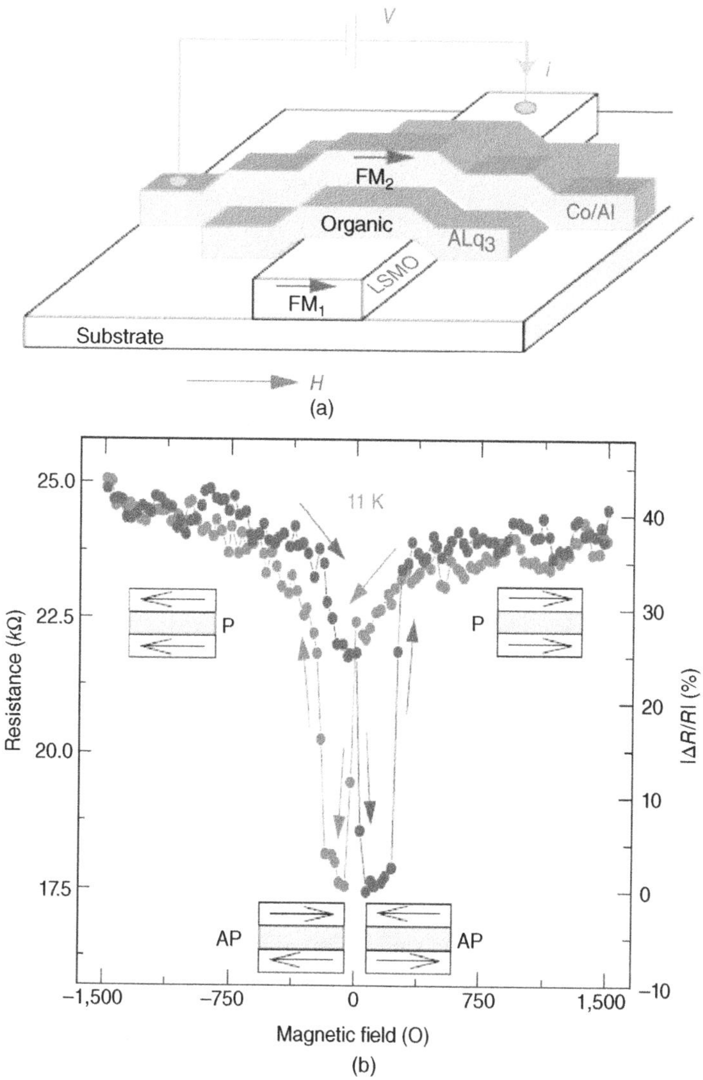

Figure 9.22 *(a) A schematic of a vertical spin valve using Alq3. (b) The observed MR effect in the organic spin valve at 11 K*

under an application of an external magnetic field by using LSMO electrodes (Figure 9.21) [75], which is recognized as the first report on an MR effect using an organic molecule. Following to this work, optical detection of spin coherent transport via molecularly bridged quantum dots was reported by Ouyang *et al.* in 2003 [76]. Because electrical detection of spin transport is comparably easier than optical detection in this early stage and Dediu used a lateral structure without the difference of the coercive force of the LSMO, Xiong *et al.* improved a device structure by introducing a sandwiched structure (Co/Alq3(~130 nm)/LSMO) and observed an MR effect with resistance hysteresis in 2004 (Figure 9.22) [77]. The MR effect was observed up to ~250 K which was close to the Curie temperature of LSMO. After these pioneering studies, a theoretical study in which large MR ratio of

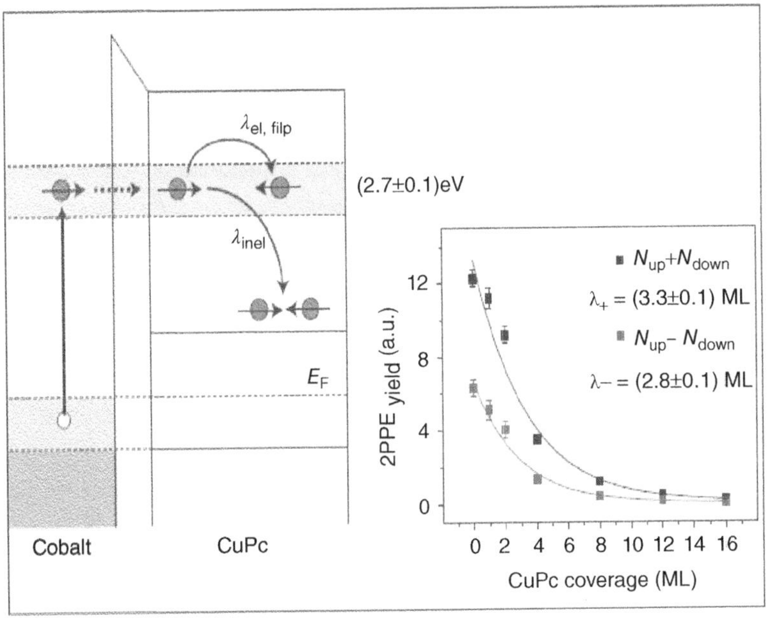

Figure 9.23 *A result on the 2PPE experiment in the purely spin injected case*

several hundreds percentage was predicted was carried out [78] and organic spintronics attracted much attention [79–81].

Contrary to nanocarbon spintronics, the nonlocal spin injection, generation of a pure spin current and the Hanle-type spin precession have not been achieved in organic spintronics. However, the unique investigations from the other stand-points were implemented [82, 83]. One is spin-resolved two-photon photoemission [82]. Cinchetti *et al.* fabricated a CuPc/Co heterojunction, and injected spin polarized photons from the Co into the lowest unoccupied level of the CuPc (the spin injection efficiency of 85%–90%). They reported the spin flip length of several tens nanometers in the CuPc at room temperature (Figure 9.23). Although this is not electrical detection of spin injection, it is valuable that this clarified the precise spin flip length in amorphous molecules. The other important study was carried out by using muon spin resonance [83], in which Drew and co-workers fabricated a NiFe/LiF/Alq3/TPD/FeCo vertical spin valve device. They observed MR in this device and simultaneously carried out depth resolved investigations for detecting spin polarization, which allows us to obtain a reliable proof of spin injection into organic molecules. According to their study, the spin diffusion length in the Alq3 was typically several tens nanometers at low temperature (Figure 9.24). These two studies pointed out that the spins which were injected both optically and electrically made relaxation less than 50 nm, and the results do not have accordance with previous studies using electrical techniques for spin injection and thicker (longer) spin transport channels. Here, it is noteworthy that spin transport even in a short organic spin channel using rubrene, Alq3 and CuPc (< 10 nm) is often tunneling spin transport (for example, see Figure 9.25) [84–88] and tunneling spin transport can be realized via ferromagnetic particles which were unintentionally formed in

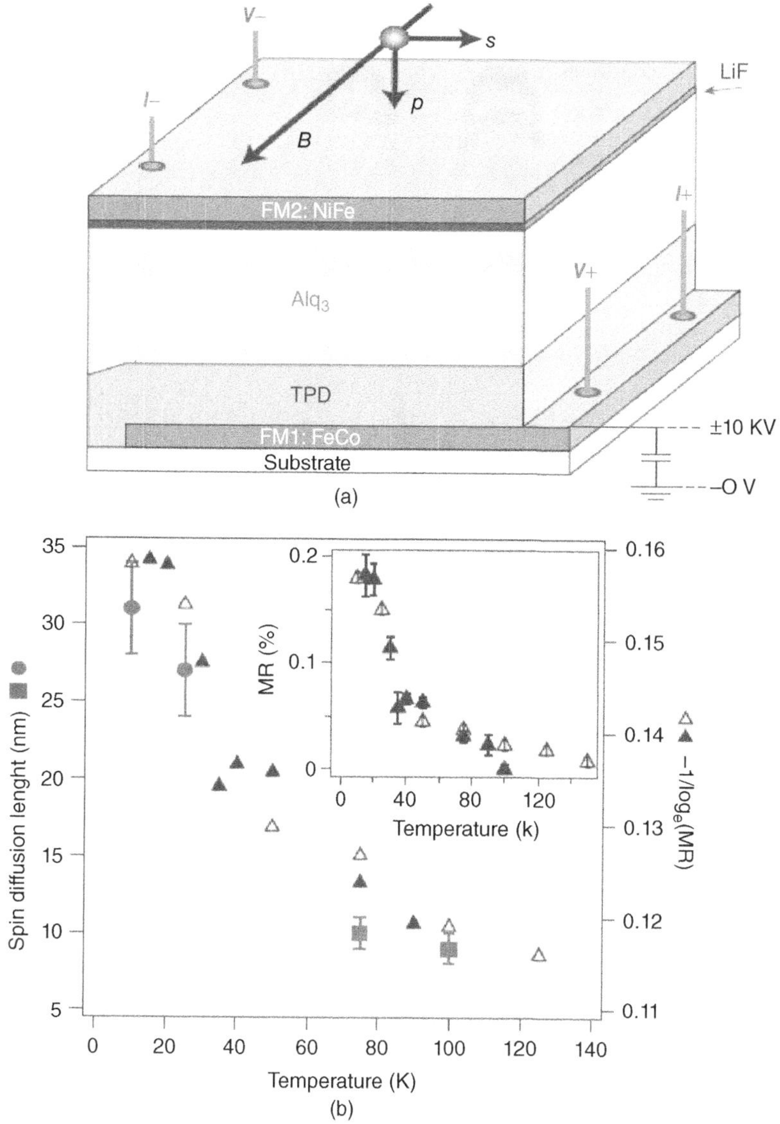

Figure 9.24 *(a) The spin valve structure. (b) The estimated spin diffusion length as a function of temperature*

organic spacers [89], and also that Jiang *et al.* reported the absence of the spin transport in the case of hole injection in Alq3 sandwiched by Co and LSMO [90]. On the other hand, Fahlman and Dediu carried out careful studies on an investigation of a band alignment of Co/Alq3/LSMO by introducing photoemission spectroscopy (Figure 9.26) [91–93] and clarified that spin transport in the lowest unoccupied molecular orbital can induce spin-dependent MR, because Alq3 is known as an appropriate electron transport material in a field of organic electroluminescence devices. Furthermore, they realized an optimum

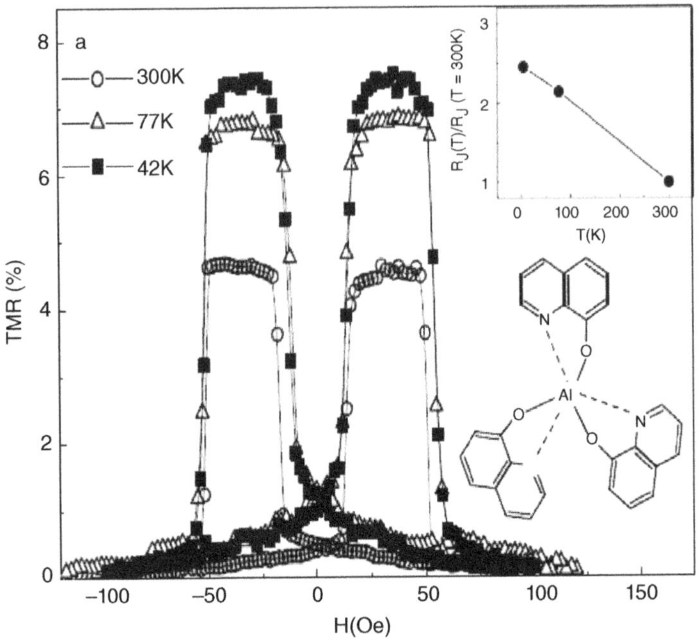

Figure 9.25 *An example of spin-dependent tunneling transport via molecules (here Alq3) in a Py/Alq3/Co spin valve*

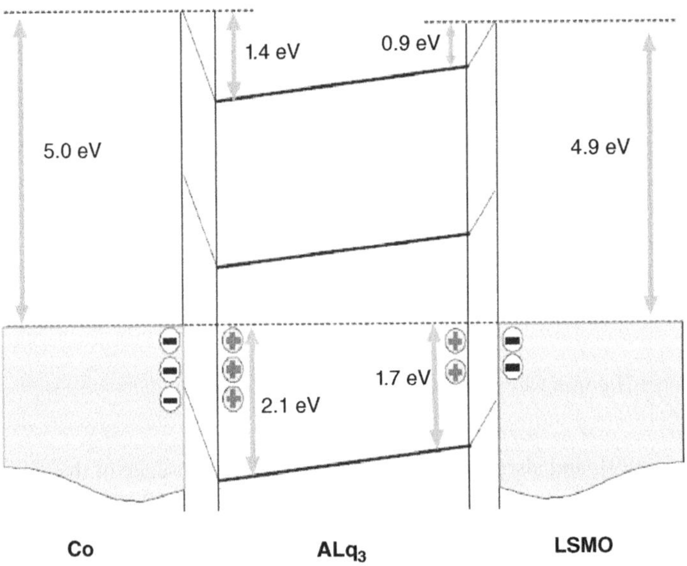

Figure 9.26 *A band diagram in a Co/Alq3/LSMO spin valve, which was obtained by using photoemission spectroscopy*

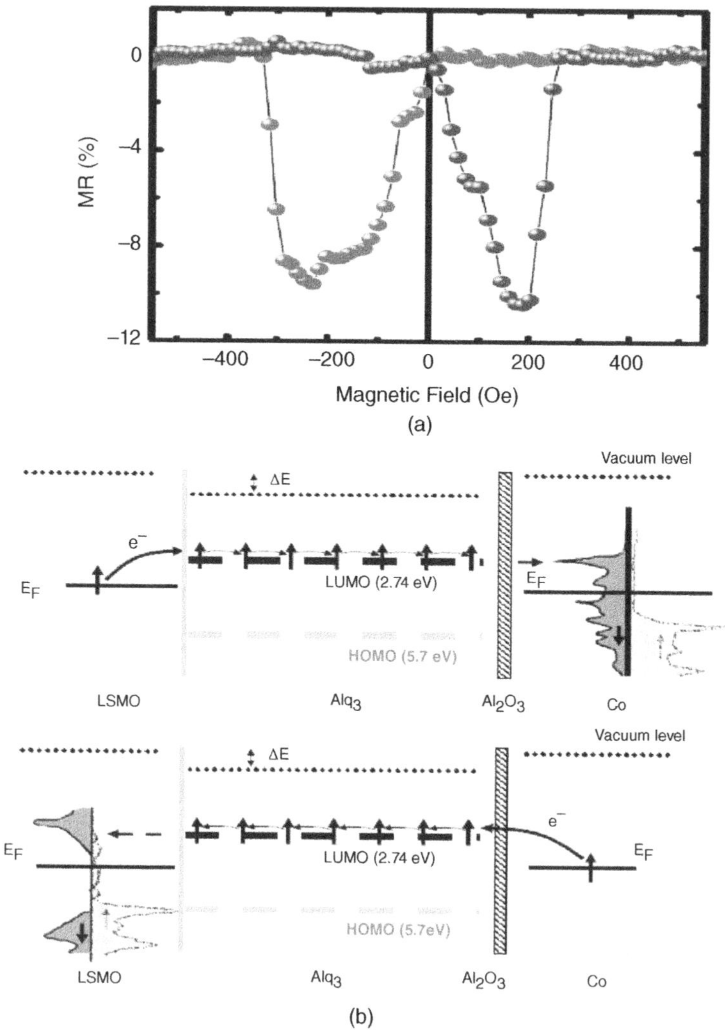

Figure 9.27 *(a) The observed MR effect in a Co/AlO/Alq3/LSMO spin valve at RT. (b) A model of spin transport in the Alq3. The injected spins transfer in the LUMO level of the Alq3 both in the case that the spins are injected from Co and from LSMO. This is thought to be a reason of the negative MR effect*

quality of an interface between Alq3 and Al_2O_3 which is a key issue to avoid the conductance mismatch for spin injection into organics and achieved spin transport via Alq3 at room temperature (Figure 9.27) [94]. Here, it should be emphasized that a hyperfine interaction plays an important role for spin relaxation in organics, and Bobbert theoretically clarified the role of the hyperfine interaction in Alq3 for the spin transport and the shape of the resistance hysteresis [95]. The author showed the good accordance of the theoretical fitting and the experimental data obtain by Dediu, which can be a proof of spin injection into Alq3 in their spin valves. Although negative MR was observed in their Alq3 spin valves, which does not seem usual based on a conventional spin transport model, they concluded in the article that the negative MR was ascribable to the combination of DOSs in Co and LSMO, namely, the DOS of the minority spins in the FMs was dominant in the spin transport (see Figure 9.27). It is noteworthy that a multifunctional spin device was fabricated by using this Alq3-based spin valve [96]. One remaining issue is how to rule out spin-polarized intergrain tunneling magnetoresistance appeared in LaSrMnO [97]. This effect easily induces an observation of magnetoresistance from a molecular spin valves, which is a spurious effect, i.e., an appearance of magnetoresistance without spin injection into molecules. Now, nobody has answered to this question, and a study is awaited.

As described above, it is not still conclusive whether GMR or TMR governs spin transport via organic molecules because there is no report on the Hanle-type spin precession. The missing of the Hanle-type spin precession in molecules is quite serious and its observation is strongly awaited in order to avoid imprecise estimation of spin coherence in molecules. In contrast to spin devices using nanocarbonaceous molecules, spin coherence in molecular spin devices has been merely estimated from resistance hysteresis, where strong offset signals were superimposed [98]. As pointed out in Ref. [98], this should be investigated and clarified as soon as possible for further progress in organic molecular spintronics.

It has been understood that TMR via organics includes interesting physical aspects. One example is an observation of large TMR ratio of ~300% at 2 K via Alq3 (Figure 9.28), where a vertical spin device was fabricated by a nanoindentation method [99]. Positive MR was found in their study, contrary to the previous work [94], and they clarified that the formation of a hybridized state between the Alq3 and the FM induced the positive MR. Currently, the MR ratio is much more enhanced when they utilize CuPc as a spacer [100]. In contrast, Sun *et al.* reported large negative MR of −300% at 10 K in Co/Alq3/LSMO vertical spin valves [101]. Concerning large MR ratio, Sakai *et al.* recently reported MR ratio of ~1,400,000% at 2 K by using a C_{60}-Co nanocomposite structure [74], and it was clarified that this large MR was ascribed to a novel magnetoelectric effect of the Co nanoparticles. The other interesting feature in TMR via molecules was reported by Hatanaka *et al.*, where Co nanoparticles were uniformly embedded in a rubrene matrix and higher order (at the maximum 5th-order) co-tunneling spin transport was realized [102]. Second order co-tunneling was already observed in insulating granular systems and theoretical explanation was clearly done [103], where MR ratio was enhanced due to Coulomb blockade and appearance of the co-tunneling effect. Hatanaka *et al.* observed the other type of the enhancement of the MR ratio inside of the Coulomb gap (see Figure 9.29), which was not able to be explained by the model by Takahashi and Maekawa [103]. They investigated I–V curves inside of the Coulomb gap and found that the power of the bias voltage was at maximum 9, which indicated that the 5th order co-tunneling occurred and this induced the further enhancement of the MR ratio up to 80%. It was also clarified by using a ^{57}Co spin echo method that the origin of such the large MR ratio observed in molecular nanocomposites was enhancement of spin polarization between rubrene and Co [104]. Whereas these studies were implemented by using the nanocomposite structure, in which ferromagnetic nanoparticles were dispersed, the similar structure was realized by using Fe_3O_4 nanoparticles. Wang *et al.* reported an observation of an MR effect in a combination of Fe_3O_4 and oleic acid molecules [105].

In the final part of this subsection, several interesting topics are introduced. The first topic is detection of circularly polarized light from an organic light emitting diode with FM electrodes [106]. The fusion of a spin degree of freedom and photon could give many

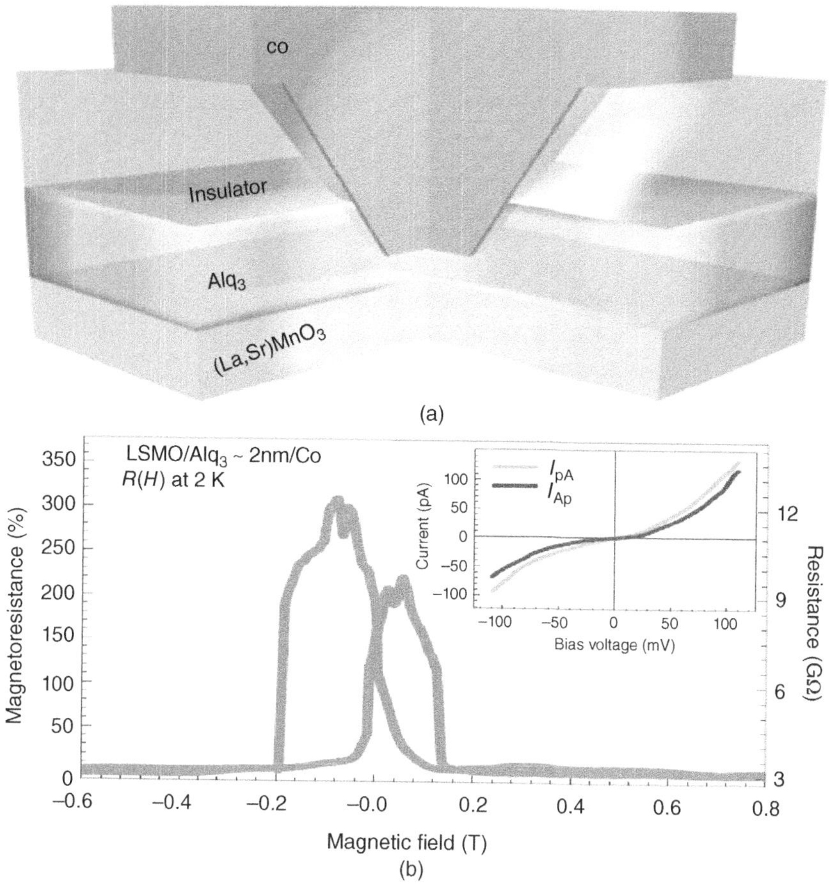

Figure 9.28 *(a) A schematic of a Alq3 spin valve fabricated by a nanoindentation method. (b) The observed MR effect at 2 K*

opportunities to fabricate novel spin devices using molecules. Recently, along this research direction, a group of Utah reported circularly polarized photoemission [107]. The second topic is utilization of molecular magnet, V(TCNE) for molecular spin devices with a short spin channel (~10 nm) [108]. The Curie temperature of V(TCNE) was reported to be ~400 K, which overcome the limitation of the Curie temperature of LSMO that is often used for molecular spin devices. In the conductance-mismatch-based point of view, FMs with low conductivity is a hopeful material for avoiding the conductance mismatch. Hence, the introduction of molecular magnets provides us a new point of view for the discussion of spin injection and transport in organics. But, the observed bias dependence was completely symmetric for the polarity change of the bias voltages, which is not consistent with a theory. Since the species of the electrodes are different, i.e., the conductivity of the electrodes are different, the bias dependence should be different as the polarity of the bias voltages. This is still an open question for the study. The third topic is an isotope effect [109]. Nguyen *et al.* fabricated vertical spin valves using a polymer material (DOO-PPV) with and without

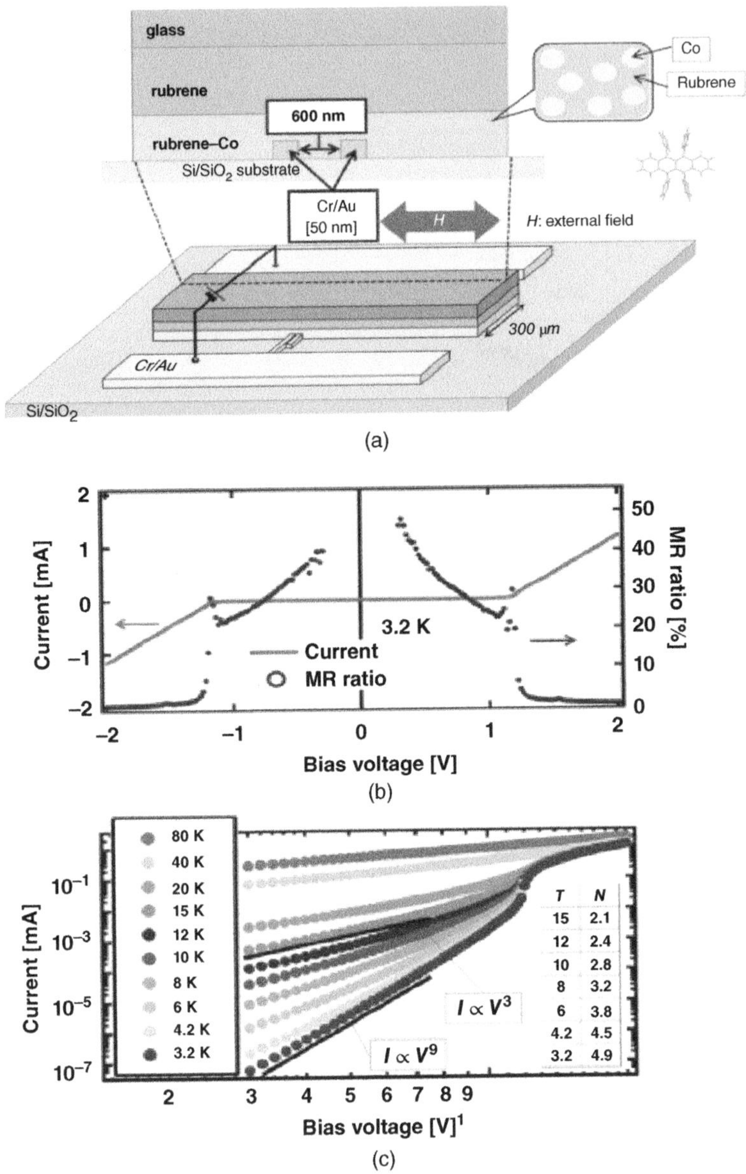

Figure 9.29 *(a) A device structure of a rubrene–cobalt nanocomposite spin device. (b) Correspondence between the I-V curve and the MR ratio at 3.2 K. (c) The I-V curves within the Coulomb gap. V2N-1 dependence was observed*

replacing hydrogen (H) to deuterium (D) in the polymer. The structure of the spin device was asymmetric, Co/DOO-PPV(25 nm)/LSMO. According to their experiments, MR ratio was increased up to ~40% at 10 K in the device with deuterium because the hyperfine interaction was thought to be suppressed, which is the isotope effect, and the bias voltage dependence of the MR ratio was symmetric.

9.4 Summary

The author introduced the current status of molecular spintronics. Spin injection, spin-dependent transport and spin-dependent binding effects, yielding magnetoresistance, were described in detail by introducing important achievements in the history of molecular spintronics. Several theories and important concepts, which are necessary to carry out studies of molecular spintronics, were also described. Whereas molecular spintronics includes many interesting physical issues, there are still some open questions and inclusive issues for concluding successful spin injection. Hence, further studies based on precise understandings of related physics is strongly awaited.

References

1. G. Binasch, P. Grünberg, F. Saurenbach and W. Zinn, *Phys. Rev. B* **39**, 4828 (1989).
2. M.N. Baibich, J.M. Broto, A. Fert, *et al., Phys. Rev. Lett.* **61**, 2472 (1989)
3. D.H. Mosca, F. Petroff, A. Fert, *et al., Appl. Phys. Lett.* **64**, 3500 (1994).
4. A. Berkowitz, J.R. Mitchell *et al., Phys. Rev. Lett.* **68**, 3745 (1992). J.Q. Xiao, J.S. Jiang and C.L. Chien, *Phys. Rev. Lett.* **68**, 3749 (1992).
5. T. Miyzaki and N. Tezuka, *J. Mag. Mag. Mater.* **139**, L231 (1995).
6. J.S. Moodera, L.R. Kinder, M. Wong and R. Meservey, *Phys. Rev. Lett.* **74**, 3273 (1995).
7. H. Fujimori, S. Mitani and S. Ohnuma, *Mater. Sci. Eng. B* **31**, 219 (1995); S. Mitani, S. Takahashi, K. Takanashi, K. Yakushiji, S. Maekawa and H. Fujimori, *Phys. Rev. Lett.* **81**, 2799 (1998).
8. P. Sheng, B. Abeles and Y. Arie, *Phys. Rev. Lett.* **31**, 44 (1973). J.S. Herman and B. Abeles, *Phys. Rev. Lett.* **37**, 1429 (1976).
9. S. Yuasa, T. Nagahama A. Fukushima, *et al., Nature Mater.* **3**, 868 (2004); S.S.P. Parkin, C. Kaiser, A. Panchula, P.M. Rice, B. Hughes, M. Samant and S-H. Yang, *Nature Mater.* **3**, 862 (2004).
10. S. Ikeda, J. Hayakawa, Y. Ashizawa, *et al., Appl. Phys. Lett.* **93**, 082508 (2008).
11. S. Datta and B. Das, *Appl. Phys. Lett.* **56**, 665 (1990).
12. For example, Y. Ohno, D.K. Young, B. Beschoten, *et al., Nature* **402**, 790 (1999), and I. Appelbaum, B. Huang and D.J. Monsma, *Nature* **447**, 295 (2007).
13. T. Matsuno, S. Sugahara and M. Tanaka, *Jpn. J. Appl. Phys.* **43**, 6032 (2003).
14. For detailed derivation, see *"Relativistic Quantum Mechanics"* by J.D. Bjorken and S.D. Drell (McGraw-Hill Inc., New York, 1964).
15. A. Fert and H. Jaffres, *Phys. Rev. B* **64**, 184420 (2001).
16. T. Kimura, Y. Haemle, and Y. Otani, *Phys. Rev. B* **71**, 014461, (2005).
17. T. Kimura, Y. Haemle, and Y. Otani, *Phys. Rev. Lett.* **96**, 03701, (2006).
18. T. Suzuki, T. Sasaki, T. Oikawa, *et al., Appl. Phys. Express* **4**, 023003, (2011).
19. T. Uemura, T. Akiho, M. Harada, *et al., Appl. Phys. Lett.* **99**, 082108, (2011).
20. M. Ohishi, M. Shiraishi, R. Nouchi, *et al., Jpn. J. Appl. Phys.* **46**, L605, (2007).
21. N. Tombros, C Jozsa, M. Popinciuc, *et al., Nature* **448**, 571, (2007).
22. N. Tombros, S.J. van der Molen and B.J. van Wees, *Phys. Rev. B* **73**, 233403 (2006).
23. S. Mizukami, Y. Ando, and T. Miyazaki, *Phys. Rev. B* **66**, 104413, (2002).

24. Y. Tserkovnyak, A. Brataas, and G. E. W. Bauer, *Phys. Rev. B* **66**, 224403, (2002).
25. E. Saitoh, M. Ueda, H. Miyajima, and G. Tatara, *Appl. Phys. Lett.* **88**, 182509, (2006).
26. K. Ando, S. Takahash, J. Ieda, *et al., Nature Mater.* **10**, 655, (2011).
27. K. Ando and E. Saitoh, *Nature Commun.* **3**, 629, (2012).
28. Y. Kajiwara, K. Harii, S. Takahashi, *et al., Nature* **464**, 262, (2010).
29. E. Shikoh, K. Ando, K. Kubo, *et al., Phys. Rev. Lett.* **110**, 127201, (2013).
30. Z. Tang, E. Shikoh, H. Ago, *et al., Phys. Rev. B* **87**, 140401(R) (2013).
31. A. Tsukahara, Y. Kitamura, Y. Ando, *et al.*, arXiv1301.3580.
32. A. K. Patra, S. Singh, B. Barin, *et al., Appl. Phys. Lett.* **101**, 162407 (2012).
33. For example, see J. Smit, *Physica* **17**, 612 (1995).
34. The local scheme is a method that we measure output voltages and inject an electric current in the same circuit. The detail will be shown in the section 3.2.
35. K. Tsukagoshi, B, Alphenaar and H. Ago, *Nature* **401**, 572 (1999).
36. A. Jensen, J.R. Hauptman, J. Nygard and P.E. Linderof, *Phys. Rev. B* **72**, 035419 (2005).
37. B. Zhao, I. Moench, H. Vinzelberg, *et al., Appl. Phys. Lett.* **80**, 3144 (2002).
38. R. Thamankar, S. Niyogi, B.Y. Yoo, *et al., Appl. Phys. Lett.* **89**, 033119 (2006).
39. S. Sahoo, T. Kontos, J. Furer, *et al., Nature Phys.* **1**, 99 (2005).
40. H.T. Man, I.J.W. Wever and A.F. Morpurgo, *Phys. Rev. B* **73**, 241401(R) (2006).
41. See for example M. Bockrath *et al., Science* **275**, 1922 (1997).
42. A. Javey, J. Guo, D.B. Farmer, *et al., Nano Lett.* **3**, 447 (2004).
43. S. Sahoo, T. Kontos, C. Schoenenberger and C. Suergers, *Appl. Phys. Lett.* **86**, 112109 (2005).
44. F.J. Jedema, H.B. Heersche, A.T. Filip, *et al., Nature* **416**, 713 (2002).
45. F.J. Jedema, M.S. Nijboer, A.T. Filip and B.J. van Wees, *Phys. Rev. B* **67**, 85319 (2003).
46. W. Hanle, *Z. Physik* **30**(1924) 93.
47. L.E. Hueso, A. Fert, N.D. Mathur *et al., Nature* **445**, 410 (2007).
48. K.S. Novoselov, A.K. Geim, S.V. Morozov, *et al., Nature* **438**, 197 (2005). Y. Zhang, Y-W. Tan, H. L. Stormer and P. Kim, *Nature* **438**, 201 (2005).
49. K. I. Bolotin, F. Ghaharil, M.D. Shulman, *et al., Nature* **462**, 196 (2009). X. Du. I. Skachko, F. Duerr, A. Luican and E.Y. Andrei, *Nature* **462**, 192 (2009).
50. H.B. Heersche, P. Jarillo-Herrero, J.B. Oostinga, *et al., Nature* **446**, 56 (2007).
51. K.I. Bolotin, K.J. Sikes, J. Hone, *et al., Phys. Rev. Lett.* **101**, 96802 (2008).
52. S. Cho, Y-F. Chen and M.S. Fuhrer, *Appl. Phys. Lett.* **91**, 123105 (2007).
53. M. Shiraishi, M. Ohishi, R. Nouchi, *et al., Adv. Func. Mater.* **19**, 3711 (2009).
54. T. Sasaki, T. Oikawa, T. Suzuki, *et al., IEEE Trans. Mag.* **46**, 1436 (2010).
55. N. Tombros, S. Tanabe, A. Veligura, *et al., Phys. Rev. Lett.* **101**, 046601 (2008).
56. C. Jozsa, M. Popinciuc, N. Tombros, *et al., Phys. Rev. Lett.* **100**, 236603 (2008).
57. C. Jozsa, M. Popinciuc, N. Tombros, *et al., Phys. Rev. B* **79**, 081402(R) (2009).
58. C. Jozsa, T. Massen, M. Popinciuc, *et al., Phys. Rev. B* **80**, 241403(R) (2009).
59. W. Han, W.H. Wang, K. Pi, *et al., Phys. Rev. Lett.* **102** 137205 (2009).
60. W. Han, K. Pi, K.M. McCreary, *et al., Phys. Rev. Lett.* **105**, 167202 (2010).
61. K. Pi, W. Han, K.M. McClreary, *et al., Phys. Rev. Lett.* **104**, 187201 (2010).
62. B. Dlubak, P. Seneor, A. Anane, *et al., Appl. Phys. Lett.* **97**, 092502 (2010).
63. H. Goto, A. Kanda, T. Sato, *et al., Appl. Phys. Lett.* **92**, 212110 (2008).

64. M. Popinciuc, C. Jozsa, P.J. Zomer, *et al.*, *Phys. Rev. B* **80**, 214427 (2009).
65. M. Shiraishi, Proc. SPIE Spintronics III, Vol. 7760, 77600H-1 (2010).
66. W. Han and R.K. Kawakami, *Phys. Rev. Lett.* **107**, 047207 (2011).
67. T.- Y. Yang, J. Balakrishnan, F. Volmer, *et al.*, *Phys. Rev. Lett.* **107**, 047206 (2011).
68. T. Banerjee, W.G. van der Wiel and R. Jansen, *Phys. Rev. B* **81**, 214409 (2010).
69. H. Zare-Kolsaraki and H. Micklitz, *Eur. Phys. J. B* **40**, 103 (2004).
70. S. Miwa, M. Shiraishi, M. Mizuguchi, *et al.*, *Jpn. J. Appl. Phys.* **45**, L717 (2006), and also S. Miwa, M. Shiraishi, M. Mizuguchi, T. Shinjo and Y. Suzuki, *Phys. Rev. B* **76**, 214414 (2007).
71. S. Sakai, K. Yakushiji, S. Mitani, *et al.*, *Appl. Phys. Lett.* **89**, 113118 (2006).
72. S. Sakai, I. Sugai, S. Mitani, *et al.*, *Appl. Phys. Lett.* **91**, 242104 (2007).
73. Y. Matsumoto and S. Sakai, *Chem. Phys. Lett.* **470**, 244 (2009).
74. Y. Sakai, E. Tamura, S. Toyokawa, *et al.*, *Adv. Func. Mater.* **22**, 3845 (2012).
75. V.A. Dediu, M. Murgia, F.C. Matacotta, *et al.*, *Solid State Commun.* **122**, 181 (2002).
76. M. Ouyang and D.D. Awschalom, *Science* **301**, 1074 (2003).
77. Z.H. Xiong, D. Wu, Z.V. Vardeny and J. Shi, *Nature* **427**, 821 (2004).
78. A.R. Rocha and S. Sanvito, *Nature Mater.* **4**, 335 (2005).
79. S. Majumdar, R. Laiho, P. Laukkanen, *et al.*, *Appl. Phys. Lett.* **89**, 122114 (2006).
80. S. Pramanik, C.-G. Stefanita, S. Patibandla, *et al.*, *Nature Nanotech.* **2**, 216 (2007).
81. T. Ikegami, I. Kawayama, M. Tonouchi, *et al.*, *Appl. Phys. Lett.* **92**, 153304 (2008).
82. M. Cinchetti and M. Aeschlimann, *Nature Mater.* **8**, 115 (2009).
83. A.J. Drew, W.P. Gillin, and J. Hoppler, *et al.*, *Nature Mater.* **8**, 109 (2009).
84. T.S. Santos, J.S. Lee, P. Migdal, *et al.*, *Phys. Rev. Lett.* **98**, 016601 (2007).
85. J.H. Shim, K.V. Raman, Y.J. Park, *et al.*, *Phys. Rev. Lett.* **100**, 226603 (2008).
86. H. Kusai, S. Miwa, M. Mizuguchi, *et al.*, *Chem. Phys. Lett.* **448**, 106 (2007).
87. S. Tanabe, S. Miwa, M. Mizuguchi, *et al.*, *Appl. Phys. Lett.* **91**, 063103 (2007).
88. Z. Tang, S. Tanabe, D. Hatanaka, *et al.*, *Jpn. J. Appl. Phys.* **49**, 033002 (2010).
89. H. Vinzelberg, J. Schumann, D. Elefant, *et al.*, *J. Appl. Phys.* **103**, 093720 (2008).
90. J.S. Jiang, J.E. Pearson and S.D. Bader, *Phys. Rev. B* **77**, 035303 (2008).
91. Y.Q. Zhan, I. Bergenti, L.E. Hueso, *et al.*, *Phys. Rev. B* **76**, 045406 (2007).
92. Y.Q. Zhan, M.P. de Jong, F.H. Li, *et al.*, *Phys. Rev. B* **78**, 045208 (2008).
93. F. Li, P. Graziosi, Q. Tang, *et al.*, *Phys Rev. B* **81**, 205415 (2010).
94. V.A. Dediu, L.E. Hueso, I. Bergenti, *et al.*, *Phys. Rev. B* **78**, 115203 (2009).
95. P.A. Bobbert, W. Wagemans, F.W.A. van Oost, *et al.*, *Phys. Rev. Lett.* **102**, 156604 (2009).
96. L. E. Hueso, I. Bergenti, A. Riminucci, *et al.*, *Adv. Mater.* **8**, 2639 (2007).
97. H.Y. Hwang, S.-W. Cheong, N.P. Ong and B. Batlogg, *Phys. Rev. Lett.* **77**, 2041 (1996). See also, X. Zhu, Z. Zhu, and C. Chen, *et al.*, "Hysteresis and vertical anisotropy of magnetoresistance in La0.67A0.33MnOz (A=Ca, Sr) polycrystalline films deposited on amorphous quartz substrates", appeared in Ceramics International, where a nice resistance hysteresis from simple LSMO film.
98. C. Boehme and J.M. Lupton, *Nature Nanotech.* **8**, 612 (2013).
99. C. Barraud, P. Seneor, R. Mattana, *et al.*, *Nature Phys.* **6**, 615 (2010).
100. C. Barruad *et al.*, The 3rd Topical Meeting on Spins in Organic Semiconductor (Amsterdam, 2010), PB-01.
101. D. Sun, L. Yin, C. Sun, *et al.*, *Phys. Rev. Lett.* **104**, 236602 (2010).

102. D. Hatanaka, S. Tanabe, H. Kusai, *et al.*, *Phys. Rev. B* **79**, 235402 (2009).
103. S. Takahashi and S. Maekawa, *Phys. Rev. Lett.* **80**, 1758 (1998)
104. M. Shiraishi, H. Kusai, R. Nouchi, *et al.*, *Appl. Phys. Lett.* **93**, 053103 (2008).
105. S. Wang, F.J. Yue, D. Wu, *et al.*, *Appl. Phys. Lett.* **94**, 012507 (2009).
106. E. Shikoh, A. Fujiwara, Y. Ando and T. Miyazaki, *Jpn. J. Appl. Phys.* **45**, 6897 (2006).
107. T.D. Nguyen, E. Ehrenfreund, and Z. V. Varderny, *Science* **337**, 204 (2012).
108. J-W. Yoo and A.J. Epstein, *Nature Mater.* **9**, 638 (2010).
109. Tho D. Nguyen and Z.V. Vardeny, *Nature Mater.* **609**, 345 (2010).

Appendix

As shown in Equation (9.11), the Dirac equation with an electromagnetic field can be written as

$$i\frac{\partial}{\partial t}\psi' = \left[e^{iS}\left(H - i\frac{\partial}{\partial t}\right)e^{-iS}\right]\psi' = H'\psi'.$$

Here, the unitary transformation can be written as

$$\exp(+iS)H\exp(-iS) = H + i[S,H] + \frac{i^2}{2!}[S,[S,\dot{S}]] + \ \ldots, \tag{A.1}$$

where the Baker–Hausdorff formula is used. Hence

$$\begin{aligned} H' &= H + i[S,H] - \frac{1}{2}[S,[S,H]] - \frac{i}{6}[S,[S,[S,H]]] \\ &- \frac{1}{24}[S,[S,[S,[S,\beta m]]]] - \dot{S} - \frac{i}{2}[S,\dot{S}] + \frac{1}{6}[S,[S,\dot{S}]]. \end{aligned} \tag{A.2}$$

The above Hamiltonian is assumed to correspond to

$$H' = \beta m + \varepsilon + \theta + i[S,\beta]m,$$

and the odd terms in H' should be eliminated. For this purpose, we set S as

$$S = -\frac{i\beta\theta}{2m}$$

Then, we carry out the following calculation, $[S, H]$, $[S, [S, H]]$, $[S, [S, [S, H]]]$ and so on. Because β and θ are anticommutable and $\beta^2 = \mathbf{1}$, then

$$\begin{aligned} Si[S,H] &= i\left[-\frac{i\beta\theta}{2m}, \beta m + \theta + \varepsilon\right] \\ &= -\theta + \frac{\beta}{m}\theta^2 + \frac{\beta}{2m}[\theta,\varepsilon], \\ \frac{i^2}{2!}[S,[S,H]] &= \frac{i^2}{2}\left[S, i\theta - i\frac{\beta}{m}\theta^2 - i\frac{\beta}{2m}[\theta,\varepsilon]\right] \\ &= -\frac{\beta}{2m}\theta^2 - \frac{\beta}{2m^2}\theta^3 - \frac{1}{8m^2}[\theta,[\theta,\varepsilon]], \end{aligned}$$

$$\frac{i^3}{3!}[S,[S,[S,H]]] = \frac{1}{6m^2}\theta^3 - \frac{1}{6m^3}\beta\theta^4,$$
$$\frac{i^4}{4!}[S,[S,[S,[S,H]]]] = \frac{\beta\theta^4}{24m^3},$$
$$-\dot{S} = \frac{i\beta\dot{\theta}}{2m},$$
$$-\frac{i}{2}[S,\dot{S}] = -\frac{i}{8m}[\theta,\dot{\theta}].$$

Hence,

$$H' = \beta\left(m + \frac{\theta^2}{2m} - \frac{\theta^4}{8m^3}\right) + \varepsilon - \frac{1}{8m^2}[\theta,[\theta,\varepsilon]] - \frac{i}{8m}[\theta,\dot{\theta}]$$
$$+ \frac{\beta}{2m}[\theta,\varepsilon] - \frac{\theta^3}{3m^2} + \frac{i\beta\dot{\theta}}{2m} = m + \varepsilon' + \theta',$$

where it is noteworthy that the odd term is included only in the term of $O(1/m)$. In order to rewrite the above Hamiltonian as

$$H'' = \beta m + \varepsilon' + \theta'', \quad \left(\theta'' \sim O\left(\frac{1}{m^2}\right)\right)$$

the unitary transformation is once again implemented by setting S' as

$$S' = \frac{-i\beta}{2m}\theta' = \frac{-i\beta}{2m}\left(\frac{\beta}{2m}[\theta,\varepsilon] - \frac{\theta^3}{3m^2} + \frac{i\beta\dot{\theta}}{2m}\right).$$

Then the Hamiltonian can be rewritten as

$$H'' = \beta m + \varepsilon - \frac{\beta}{2m}[\theta',\varepsilon'] - \frac{i\beta\theta'}{2m}$$
$$\equiv \beta m + \varepsilon' + \theta''.$$

Furthermore, we set S'' as

$$S'' = \frac{-i\beta}{2m}\theta'',$$

and thus,

$$H'' = \beta\left(m + \frac{\theta^2}{2m} - \frac{\theta^4}{8m^3}\right) + \varepsilon - \frac{1}{8m^2}[\theta,[\theta,\varepsilon]] - \frac{i}{8m}[\theta,\dot{\theta}],$$

is the transformed Hamiltonian, where the odd term becomes small enough ($\sim (1/m^2)$). The third and the fourth terms in Equation (A.8) can be written as

$$-\frac{1}{8m^2}[\theta,[\theta,\varepsilon] + i\dot{\theta}],$$

and we carry out the calculation of the above term explicitly as follows:

$$\frac{\theta^2}{2m} = \frac{\{\alpha \cdot (p - eA)\}^2}{2m} = \frac{1}{2m}\{(p - eA)^2 + i\alpha \cdot (p - eA) \times (p - eA)\}$$
$$= \frac{1}{2m}\{(p - eA)^2 - e(\sigma \cdot B)\},$$
$$\frac{1}{8m^2}([\theta, \varepsilon] + i\dot{\theta}) = \frac{1}{8m^2}([\alpha \cdot (p - eA), e\phi] + i\dot{\theta})$$
$$= \frac{1}{8m^2}(-ie\alpha\nabla\phi + ie\nabla\dot{A})$$
$$= \frac{ie}{8m^2}\alpha \cdot E$$
$$\left[\theta, \frac{i}{8m^2}\alpha \cdot E\right] = \frac{ie}{8m^2}[\alpha \cdot p, \alpha \cdot E]$$
$$= \frac{ie}{8m^2}\sum_{i,j}\alpha_i\alpha_j\left(-i\frac{\partial E^j}{\partial x^i}\right) + \frac{e}{4m^2}\sigma(E \times p)$$
$$= \left(\frac{e}{8m^2}\sigma \cdot rot\,(E) + \frac{ie}{8m^2}div(E)\right) + \frac{e}{2m^2}\sigma(E \times p).$$

In the above calculation, remember that the time derivative of momentum in the Schrödinger representation is equal to zero. From the above calculations, we can obtain Equation (9.12) as

$$H' = \beta\left\{m + \frac{(p - eA)^2}{2m} - \frac{p^4}{8m^3}\right\} + e\phi - e\frac{1}{2m}\beta(\sigma \cdot B)$$
$$- \frac{ie}{8m^2}\sigma \cdot \text{rot}(E) - \frac{e}{4m^2}\sigma \cdot (E \times p) - \frac{e}{8m^2}\,\text{div}(E).$$

10

Silicon Spintronics for Next-Generation Devices

Kohei Hamaya
Department of Systems Innovation, Graduate School of Engineering Science
Osaka University, Japan

10.1 Recent Progress in Silicon Spintronics

Spin-based electronics (spintronics) is expected to markedly improve the device performance because of its nonvolatility, reconstructibility, low power consumption, and so forth [1]. In particular, combining the spintronics with silicon (Si)-based semiconductor industry will become important to overcome the ultimate scaling limits of the shrinking of silicon (Si)-based conventional complementary metal-oxide-semiconductor (CMOS) transistors [2]. Fortunately, Si has been predicted to be a semiconductor with enhanced spin lifetime and spin transport length due to its low spin–orbit scattering and lattice inversion symmetry [1, 2].

In 2006, Appelbaum *et al.* demonstrated spin transport across more than micrometer-order Si channels in spin-dependent ballistic hot-electron transport devices [3]. Since then, they demonstrated lots of very nice works on spin transport in undoped [4–7] and doped [8, 9] Si channels. Recently, the useful data extracted were discussed theoretically in detail, and then the phonon-induced spin relaxation in conduction band in Si was proposed [10, 11]. Unfortunately, the observed spin transport using the ballistic hot-electron transport devices was limited at low temperatures.

In 2009, even at room temperature, electrical detections of spin accumulation with the three-terminal technique for heavily doped Si channels with NiFe/AlO_x contacts were

Spintronics for Next Generation Innovative Devices, First Edition. Edited by Katsuaki Sato and Eiji Saitoh.
© 2015 John Wiley & Sons, Ltd. Published 2015 by John Wiley & Sons, Ltd.

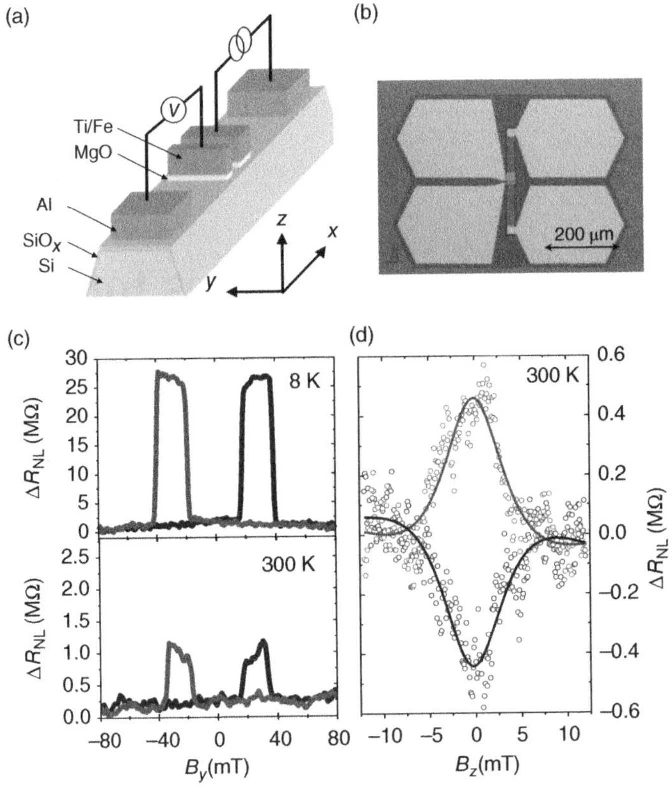

Figure 10.1 (Color online) (a) Schematic and (b) optical micrograph of the four-terminal nonlocal spin-valve device. (c) Nonlocal spin signals at 8 and 300 K in an Fe/MgO/Si device. (d) Hanle-type spin dephasing detected by nonlocal geometry at 300 K in the Fe/MgO/Si device

reported by Jansen *et al.* [12]. Since their reports were very impressive for the room-temperature applications of Si spintronics, lots of researchers used the three-terminal technique in order to create the spin accumulation in similar Si-based device structures [12–16]. At the same time, Sasaki *et al.* demonstrated electrical detection of pure spin current through the degenerated Si channels using a four-terminal nonlocal spin-valve technique in lateral devices [17–20]. And then, they reported room-temperature spin transport in the degenerated Si channels in 2011 [17]. Schematic and an optical micrograph image of their device structures are shown in Figure 10.1(a) and (b), respectively [17]. The device structure consisted of a heavily doped Si channel ($n \sim 10^{20}$ cm^{-3}) and Fe/MgO tunnel contacts [17–20]. Figure 10.1(c) shows nonlocal spin-valve signals measured at 8 and 300 K [17]. By sweeping B_y, the relative magnetization orientation of two wire-shaped Fe electrodes was controlled, leading to the abrupt changes in the nonlocal magnetoresistance. They also observed Hanle-type spin precession at 300 K, as shown in Figure 10.1(d), by applying perpendicular magnetic field (B_z) [17]. By other groups, thermal [21] and dynamical [22] creations of spin accumulation in Si were recently developed.

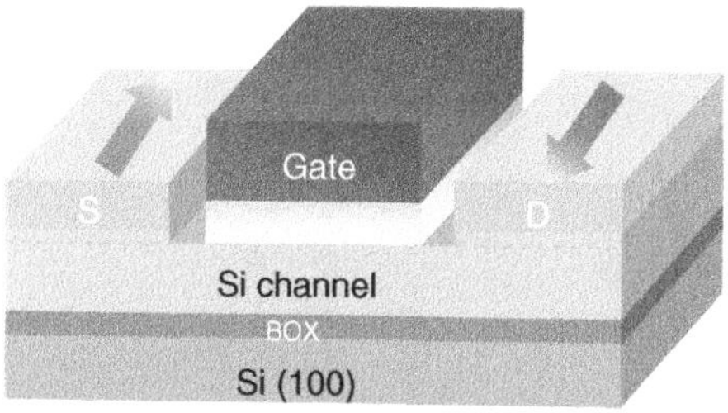

Figure 10.2 *(Color online) Schematic diagram of spin MOSFET with metallic Schottky-tunnel contacts*

Unfortunately, there are still many issues for developing next-generation devices compatible with Si, in particular, spin metal-oxide-semiconductor field-effect transistors (MOSFETs) suggested [23, 24]. Figure 10.2 shows a schematic diagram of a spin-MOSFET with metallic source and drain (S/D) contacts. For scalable spin MOSFETs with ultralow power consumption, reducing the parasitic resistance between S/D contacts will be strongly required. However, S/D contacts with insulating tunnel barriers have so far been utilized for electrical spin injection techniques for Si devices [3, 12–20, 25] because the tunnel barriers provide the solution of the impedance mismatch [26–28] and of the silicidation reaction between ferromagnetic metal (FM) and Si. From now on, one needs to simultaneously realize highly efficient spin injection and low-resistance S/D contacts in Si-based devices. Thus, the spin injection and detection without insulating tunnel barriers at S/D structures should be explored.

Until now, the author and co-workers have individually explored two important technologies, high-quality epitaxial growth of FM alloys including half-metallic Heusler alloys on Si [29–33] and spin injection into nondegenerated Si channel across the Schottky-tunnel barrier [34–40]. If the above two technologies are simultaneously realized in a single device, a way for highly efficient spin injection and detection with low parasitic resistance will be able to be opened in Si-based devices in the future. Recently, the author and co-workers developed electrical detection of spin-accumulation signals in Si with a MOSFET structure with a high-quality CoFe/n^+-Si contact [38, 40]. Interestingly, the spin signals can be modulated by applying the gate voltage even at room temperature [38, 40].

Here this text presents recent research of the author's group for silicon spintronics for next-generation devices. First, high-quality epitaxial growth of spin-injector materials on Si is shown. Next, electrical spin injection and detection across the high-quality Schottky-tunnel barriers are established in Si-based devices. By combining these technologies with a MOSFET structure, room-temperature creation and detection of spin accumulation in the nondegenerated Si channels can be demonstrated. These results are expected to build the basis of a key technology toward developing next-generation Si spintronic devices.

10.2 High-Quality Schottky Tunnel Contact

As a spin injector and detector, bcc-type FMs are special candidates for Si spintronics. Since there is only less than 5% lattice mismatch between some of the bcc-type FMs (0.564–0.570 nm) and Si (0.543 nm), it will be possible to realize epitaxial growth of

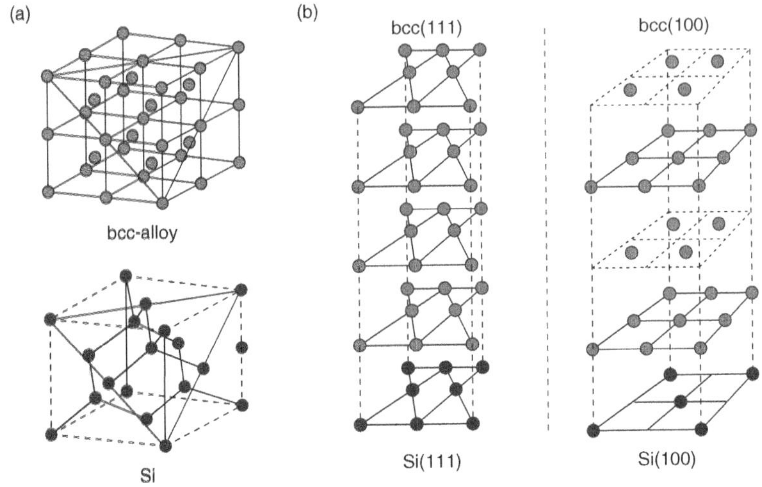

Figure 10.3 (Color online) (a) Crystal structures of a bcc-type FM alloy and Si. (b) The atomic arrangements of the bcc-type FM alloy on Si(100) (left) and on Si(111) (right)

high-quality FM films on Si. Figure 10.3(a) illustrates the crystal structures for a bcc-type FM alloy and Si. Looking at the (111) plane, denoted by triangles (⟨111⟩ planes), there is very good atomic arrangement matching between FM and Si in the left of Figure 10.3(b). In contrast, at the (100) plane, there are atomic mismatches at the interface between Fe_3Si and Ge, as depicted in the right of Figure 10.3(b). In the following, the fabrication procedures of the atomically controlled FM/Si(111) heterojunctions are explained. Basically, the bcc-type ferromagnetic films were grown by low-temperature molecular beam epitaxy (LT-MBE) at 60–200°C [29–33]. An undoped Si wafer with (111) orientation was used as the substrate. After cleaning the substrates with an aqueous HF solution ($HF : H_2O = 1 : 40$), a heat treatment was conducted at 450°C for 20 min in an MBE chamber with a base pressure of $\sim 10^{-7}$ Torr. Prior to the growth of FM films, the substrate temperature was reduced to 60–200°C. Reflection high-energy electron diffraction (RHEED) patterns of Si surface show atomically smooth as shown in the bottom picture of Figure 10.4(a). Using Knudsen cells, the elements of FM alloys were coevaporated, leading to precisely tuning the chemical composition of these elements. For example, we co-evaporated Co and Fe for CoFe alloys using Knudsen cells, leading to precisely tuning the chemical composition [31]. During the growth, two-dimensional epitaxial growth was confirmed by the observation of RHEED patterns, such as the top picture of Figure 10.4(a). By the way, two-dimensional epitaxial growth could not be achieved for the substrates with smooth (100) surface, as shown in the pictures of Figure 10.4(b). Thus, it is very important for the epitaxial growth of bcc-type FM alloys on Si to utilize the atomic arrangement matching at the (111) plane between FM and Si.

Figure 10.5(a) shows a cross-sectional transmission electron microscopy (TEM) image of $Co_{45}Fe_{55}$ on Si(111). There is almost no roughness at the CoFe/Si interface and no interfacial reaction layer at the entire region observed. A high-resolution TEM image near the interface is shown in Figure 10.5(b). Almost atomically smooth interface can be seen and single-crystalline CoFe can be achieved. The influence of the atomic composition on

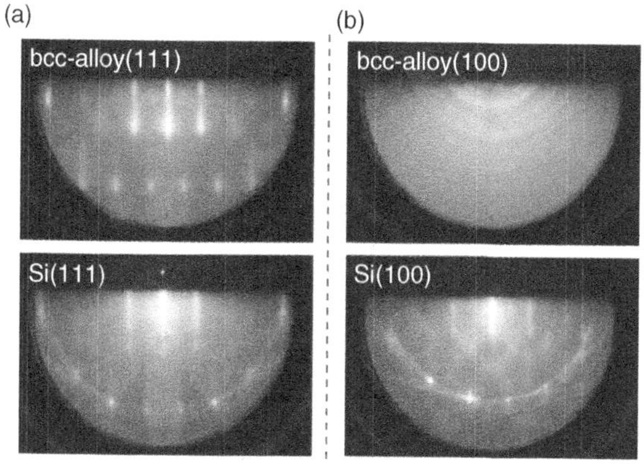

Figure 10.4 *(Color online) RHEED patterns during the growth of the FM layers (top figures) on Si(100) (left) and Si(111) (right). The bottom figures are the RHEED patterns for the Si(100) and Si(111) surfaces just before the growth of FM layers*

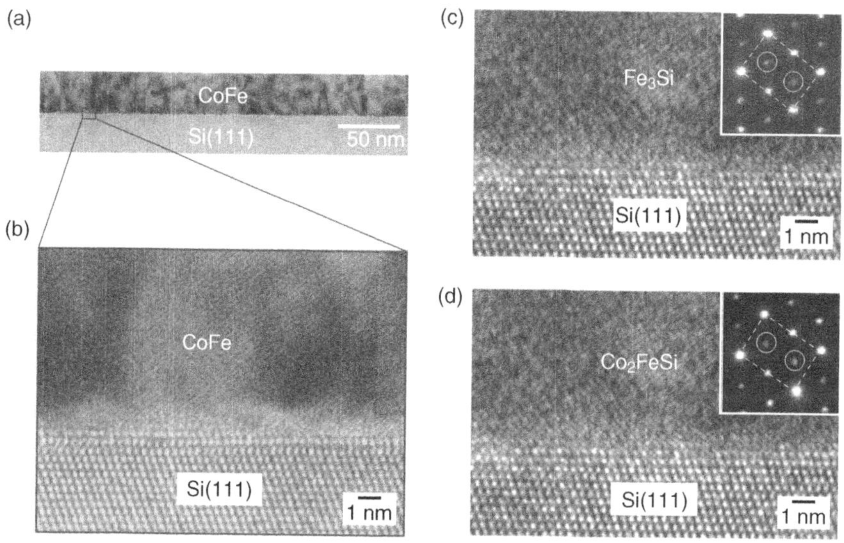

Figure 10.5 *(Color online) Cross-sectional (a) low-magnification and (b) high-resolution TEM images of CoFe/Si(111) interface. (c) and (d) show high-resolution TEM images of* Fe_3Si*/Si(111) and* Co_2FeSi*/Si(111) interfaces, respectively, together with the nanobeam ED patterns in the FM layers. The axis of the incident electron beam is parallel to the* $[1\bar{1}0]$ *direction*

the interfacial reaction for $Co_{100-x}Fe_x$ alloys was also examined. As a result, high-quality films on Si(111) from $x = 25$ to 55 were formed [31]. When Fe or Co single-element films were deposited on Si(111), such high-quality films could not be obtained. It can be speculated that this special correlation between bcc-alloys(111) and Si(111) can realize lowering crystallization energy. Actually, many Heusler alloys which have the same bcc-type crystal

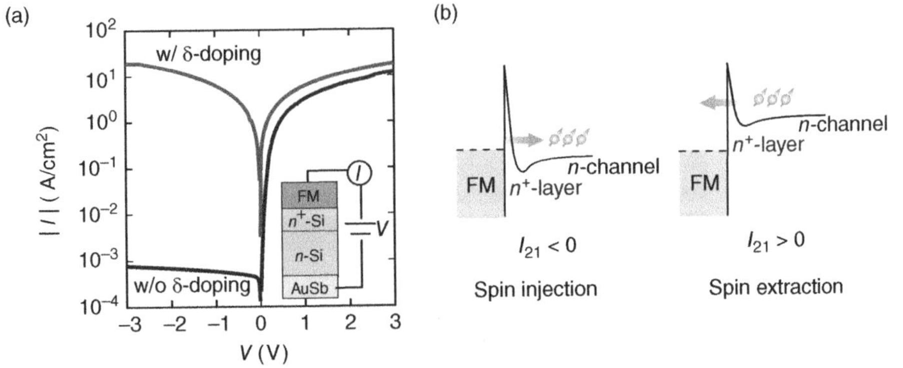

Figure 10.6 (Color online) (a) $|I| - V$ characteristics of the fabricated Schottky diodes for Si at 300 K. The insets show schematic illustrations of the FM/n^+-Si/n-Si structure. (b) Schematic illustration of spin accumulation in spin-injection and spin-extraction conditions

structure can be achieved on Si(111) [29, 30, 32]. Figure 10.5(c) and (d) shows representative high-quality epitaxial films such as Fe_3Si and Co_2FeSi, respectively, grown at ~100°C. Even for extremely low-temperature growth, highly-ordered Fe_3Si and Co_2FeSi with $D0_3$ and $L2_1$, respectively, were obtained on Si(111) [29, 30]. These techniques can be extend to the Ge-based technologies [41–48].

In order to achieve tunneling conduction across the FM/Si Schottky barriers, a heavily doped n^+-Si layer should be inserted between FM and Si. The inserted n^+-Si layer was formed by a combination of Si epitaxy using an MBE process with an Sb δ-doping technique [49, 50]. To evaluate the electrical properties of the FM/Si contact, two different vertical devices (~1 mm in diameter) with and without the n^+-Si layer (~10^{20} cm^{-3}) between FM and n-Si (~10^{15} cm^{-3}) were fabricated. For both devices, the absolute value of the room-temperature current density, $|I|$, as a function of bias voltage (V) is shown in Figure 10.6(a). These characteristics were reproduced for ten devices. Typical rectifying behavior of conventional Schottky diodes can be seen for devices without n^+- layers, while almost symmetric behavior with respect to V polarity is found for devices with the n^+-layer. Note that the reverse bias $|I|$ is extremely enhanced by inserting the n^+-layer. These results indicate that the transport mechanism was changed from thermionic emission to tunneling. As a result, the realistic conduction band should have an energy profile with a Schottky-tunnel barrier, as schematically illustrated in Figure 10.6(b). Since the doped Sb atoms can easily diffuse toward the surface because of the segregation during the growth, the carrier density near the interface can become ~10^{18} cm^{-3} with some distributions.

10.3 Si-MOSFET Structure for Detecting Spin Accumulation

A (111)-oriented silicon on insulator (SOI), where thicknesses of the SOI and buried oxide (BOX) layers were about ~75 and 200 nm, respectively, was used as a channel for MOSFET structures. The carrier density of the SOI layer was ~4.5×10^{15} cm^{-3} (1–5 Ω cm) at room temperature. To obtain tunneling conduction of electrons for spin

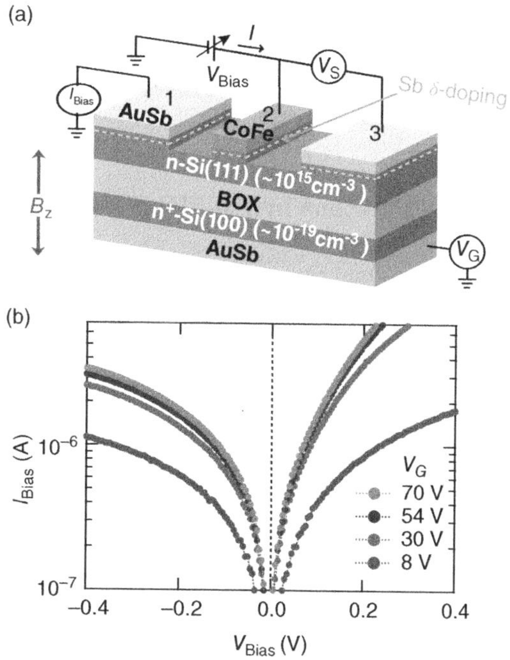

Figure 10.7 *(Color online) (a) Schematic diagram of a Si-MOSFET structure with a CoFe/n*$^+$*-Si Schottky-tunnel contact fabricated. (b)* I_{Bias}–V_{Bias} *characteristics for various* V_G *at room temperature*

injection and detection, n$^+$-Si layer was inserted between CoFe and SOI by a combination of the Si epitaxy using an MBE process with an Sb δ-doping technique [49, 50], where the doping density of Sb atoms is $\sim 1 \times 10^{19}$ cm^{-3}. The Sb δ-doped n$^+$-Si layer on the channel region was removed by the Ar$^+$ ion milling. An ohmic contact (AuSb) for backside heavily doped Si was formed at less than 300°C. Conventional processes with electron-beam lithography, Ar$^+$ ion milling, and reactive ion etching were used to fabricate three-terminal lateral devices with a backside gate electrode, illustrated in Figure 10.7(a). The CoFe/n$^+$-Si contact (contact 2) and AuSb ohmic contacts (contact 1 and 3) have lateral dimensions of 1×100 μm^2 and 100×100 μm^2, respectively. The distance between the contacts 2 and 1 or 3 is ~30 μm.

With increasing gate voltage (V_G), the bias current (I_{Bias}) value with respect to the bias voltage (V_{Bias}) gradually increases (Figure 10.7(b)). This means that the conduction channel is formed from the vicinity of the interface between SOI and BOX by the V_G applications, clearly indicating that this device can operate as a MOSFET. By evaluating detailed electrical properties at the interface under the application of V_G, we found that there are almost no changes in the properties of interface resistance (R_{Int}) values (105–120 kΩ), being independent on V_G. The resistance area product of the CoFe/n$^+$-Si contact is estimated to be $\sim 10^7$ Ω μm^2, which is sufficient large value for the spin injection and detection in nondegenerated Si channels with a carrier density of $\sim 10^{15}$ cm^{-3} [2, 27]. Also, electrical properties of the Si channel region were evaluated under the application of V_G. Contrary to the interface properties, the significant modulation of the channel resistance values was observed by applying V_G. For this device, electrical properties were largely dominated by the change in the characteristics of the Si channel. By applying V_G, sufficiently large current flows of several μA can be demonstrated for spin injection from the CoFe contact to the Si channel.

10.4 Spin Injection and Detection in a Si-MOSFET Structure

The three-terminal Hanle measurements were performed by a dc method with the current–voltage configuration shown in Figure 10.7(a) at room temperature, where a small magnetic field perpendicular to the plane, B_Z, was applied after the magnetic moment of the contact 2 aligned parallel to the plane along the long axis of the contact. If the spin accumulation in the Si channel is obtained, the spin splitting ($\Delta\mu$) of the electrochemical potential in Si can be generated, leading to the change in the resistance of the contact. When the spins are manipulated by inducing spin precession with a perpendicular magnetic field to the direction of injected spins, a voltage change (ΔV) with a characteristic Lorentzian line-shape can be seen [2].

Figure 10.8(a) and (b) shows ΔV_S–B_Z curves for $V_G = 8.0$ V at $I_{Bias} = -2.0$ and -0.1 μA, respectively, at room temperature, where a quadratic background voltage depending on B_Z is subtracted from the raw data. Here in this condition ($I_{Bias} < 0$) the electrons are injected from the spin-polarized states of CoFe into the conduction band of Si, as shown in Figure 10.6(b). When B_Z increases from zero to ±20 mT, clear ΔV_S changes are observed for both I_{Bias} even at room temperature. The presence of the changes in ΔV_S is caused by the depolarization of the accumulated spins [2]. It should be noted that the magnitude of ΔV_S reaches ~33 μV at $I_{Bias} = -2.0$ μA, showing the resistance change in more than 15 Ω. With decreasing I_{Bias} down to 0.1 μA, the magnitude of ΔV_S is markedly reduced. This feature is consistent with the decrease in the injection of spin-polarized electrons.

Also, ΔV_S–B_Z curves were examined for various V_G at a constant I_{Bias} of -2.0 μA. Because of the large channel resistance and the presence of the large electrical noise,

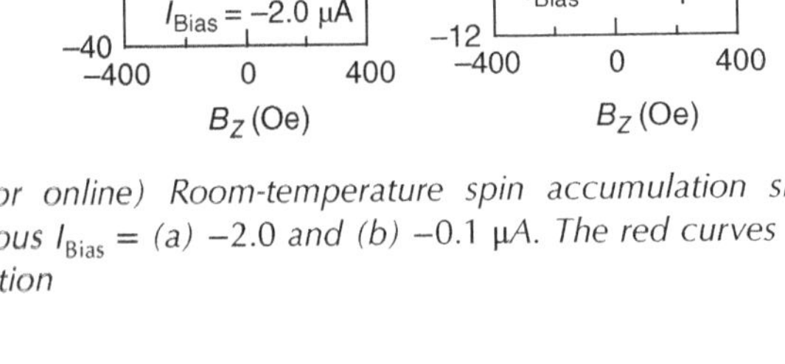

Figure 10.8 (Color online) Room-temperature spin accumulation signals measured at $V_G = 8.0$ V for various I_{Bias} = (a) −2.0 and (b) −0.1 μA. The red curves are fitting results by the Lorentzian function

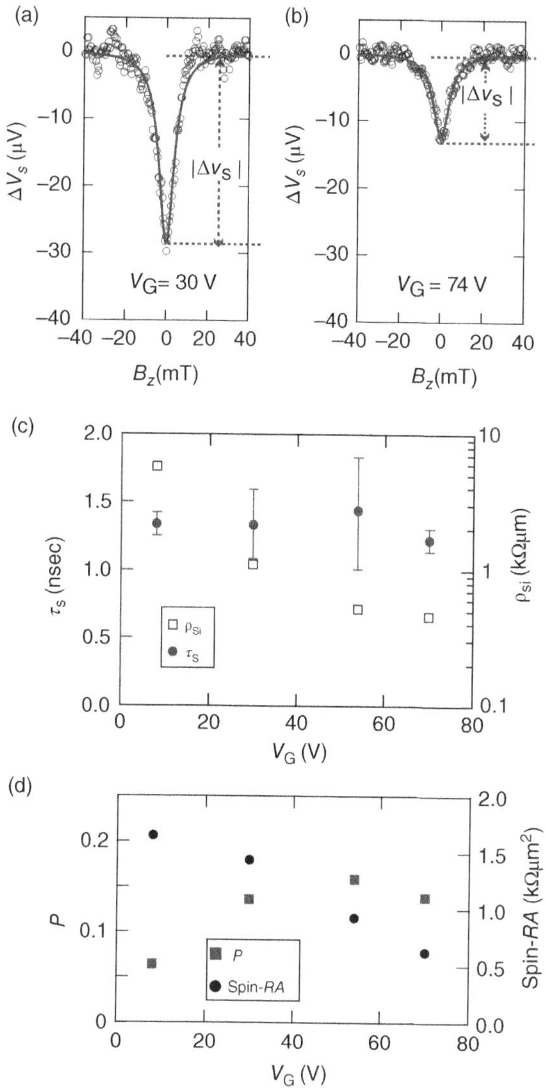

Figure 10.9 *(Color online) Room-temperature spin accumulation signals measured at (a) $V_G = 30$ V and (b) = 74 V at $I_{Bias} = -2.0$ μA. The red curves are fitting results by the Lorentzian function. (c) Spin lifetime (τ_S) and channel resistivity (ρ_{Si}) as a function of V_G at room temperature. (d) The estimated tunnel spin polarization (P) and the spin RA as a function of V_G at room temperature*

Hanle-like shapes could not be seen in $V_G < 8.0$ V. At $V_G = 30$ V, a relatively large change in ΔV_S of ~29 μV is also obtained in Figure 10.9(a), but it is found that the magnitude of ΔV_S, $|\Delta V_S|$, is slightly reduced from ~33 to ~29 μV by applying V_G from 8.0 to 30 V. Interestingly, the magnitude of ΔV_S is further decreased to ~12 μV when the gate voltage is further applied up to $V_G = 70$ V in Figure 10.9(b).

Hereafter the mechanism of the reduction in ΔV_S by applying V_G should be discussed. First, the spin relaxation induced by the application of V_G should be considered. The lower limit of spin lifetime (τ_S) can be extracted from the obtained Hanle-effect curves [2]. By using the Lorentzian function, $\Delta V_S(B_Z) = \Delta V_S(0)/[1 + (\omega_L \tau_S)^2]$ [2, 12], the width of the Hanle-effect curve can be regarded as the value of τ_S, where $\omega_L = g\mu_B B_Z/\hbar$ is the Lamor frequency, g is the electron g-factor ($g = 2$), μ_B is the Bohr magneton. The fitting results were denoted by the red solid curves in Figures 10.8(a), (b) and 10.9(a), (b). Figure 10.9(c) shows the extracted τ_S versus V_G at room temperature, together with the change in ρ_{Si} with V_G. Note that the τ_S values are ranging from 1.0 to 2.0 ns, being nearly constant. In general, the τ_S values are affected by the doping elements and its doping density [2]. For these devices, the spin accumulation is created in the P-doped channel beneath the contact with Sb δ-doped interface. In Figure 10.9(c), τ_S is almost constant whereas ρ_{Si} largely changes more than one ordered of magnitude. This means that there is almost no correlation between τ_S and the change of the carrier density induced by V_G in the Si channel, largely different from the dependence of the impurity doping density on the τ_S value presented in previous works [2, 12, 15]. Therefore, at room temperature, the spin diffusion length (λ_{Si}) can be regarded as almost constant value for our devices. Assuming $D \sim 40$ cm^2 s^{-1} ($n \sim 10^{15}$ cm^{-3}) [51], $\lambda_{Si} \sim 2.3$ μm at room temperature can be extracted from the relationship of $\lambda_N = \sqrt{D\tau_S}$ ($\tau_S \sim 1.3$ ns).

Second, the components of the magnitude of spin signals should be considered in detail. When the current density (J) is assumed to be $|I_{Bias}|/A$, where A is contact area (100 μm^2) of this device, the experimental spin resistance area-product (spin-RA) can roughly be estimated to be $\frac{|\Delta V_{23}|}{J}$ (Ωμm^2). For example, the spin-RA at $V_G = 8.0$ V and $I_{Bias} = -2.0$ μA is roughly ~1.65 kΩμm^2. On the basis of the simple spin diffusion model [27, 28], the theoretical spin-RA can be expressed as follows [2].

$$\text{Spin-}RA = P^2 \times \lambda_{Si} \times \rho_{Si} \times \frac{\lambda_{Si}}{d}, \tag{10.1}$$

where P is the tunnel spin polarization, λ_{Si} and ρ_{Si} are the spin diffusion length and resistivity of the Si channel used, respectively. $\frac{\lambda_{Si}}{d}$ is the geometrical factor of our devices [2] (d is the channel thickness ~0.075 μm). By comparing Equation (10.1) with experimental spin-RA values, experimental P values can roughly be extracted. Here $\lambda_{Si} = 2.3$ μm is assumed and ρ_{Si} values at various V_G are experimentally obtained in previous Figure 10.9(c). In Figure 10.9(d), the extracted P as a function of V_G is summarized, together with the experimental spin-RA at $I_{Bias} = -2.0$ μA. The P values are less than 0.2 for all V_G. Recently, the room-temperature spin polarization of our epitaxial CoFe films on Si(111) was estimated to be less than 0.25 by nonlocal spin-signal measurements using metallic lateral spin valves [52]. The experimental P extracted here are smaller than 0.25. From these considerations, at least, these experimental results can be understood within the framework of the theoretical spin diffusion model [27, 28]. It should be noted that P ~0.065 at $V_G = 8.0$ V is relatively small compared with P ~0.15 at other V_G.

As shown in Figure 10.9(c), ρ_{Si} at $V_G = 8.0$ V is one order of magnitude larger than those at other V_G. Since the resistance area product for our device is almost constant (~10^7 Ωμm^2) despite changing V_G, it can be understood that the spin injection efficiency for $V_G = 8.0$ V is relatively low compared with that for other V_G due to influence of the impedance mismatch problem [2, 27, 28]. Note that spin-RA is linearly reduced in

Figure 10.9(d) despite the logarithmic decrease in ρ_{Si} in Figure 10.9(c). Although it can tentatively be speculated that the change in spin-*RA* with V_G originates from the change in ρ_{Si} from Equation (10.1), the correlation between P and ρ_{Si} with changing V_G should be considered. From $V_G = 8$ to 30 V, P is enhanced from 0.065 to 0.14, leading to the enhancement in spin-*RA*, whereas ρ_{Si} is logarithmically reduced, giving rise to the decrease in spin-*RA*. In this case, since the change in ρ_{Si} is much larger that in P from $V_G = 8$ to 74 V, the change in spin-*RA* is dominantly affected by the decrease in ρ_{Si}. Phenomenologically, the density of state in Si at the Fermi level should be varied by the application of V_G, resulting in the reduction in spin accumulation ($\Delta\mu$), even if the same I_{Bias} is used for spin injection into the Si channel [38]. Since the data shown in Figures 10.9 can be understood within the framework of the theoretical model [27, 28], the reduction in $|\Delta V_S|$ with V_G can qualitatively be explained by the change in ρ_{Si}. In order to precisely discuss the magnitude of spin signals, the correlation between P and ρ_{Si} with changing V_G should be considered.

10.5 Summary

In this text, the author has presented recent progresses of silicon spintronics, particularly compatible with spin MOSFET. Using LTMBE techniques, the authors have developed high-quality epitaxial growth of bcc-type FM alloys on Si(111). Even for Heusler alloys, atomically smooth heterointerfaces and highly ordered structures can be achieved. In order to obtain tunneling conduction of electrons through such high-quality FM/Si interfaces, we have also developed an Sb δ-doping technique with a Si epitaxy. In a MOSFET structure with a CoFe/n$^+$-Si Schottky tunnel contact, room-temperature spin-accumulation signals were detected in a nondegenerated Si channel with a carrier density of $\sim 10^{15}$ cm^{-3}. The observed spin accumulation signals were modulated by the application of the gate voltage. These features can be explained within a framework of the theoretical spin diffusion model. The change in the spin-accumulation signals was attributed dominantly to the change in the resistivity of the Si channel, indicating reliable evidence for the spin injection into the nondegenerated Si channel at room temperature. Now, a room-temperature spin injection technique for the nondegenerated Si channel without insulating tunnel barriers can be established, indicating a technological progress for Si-based spintronic applications with gate electrodes.

References

1. Žutić I, Fabian J, and Das Sarma S 2004 *Rev. Mod. Phys.* **76**, 323.
2. Jansen R 2012 *Nature Mater.* **11**, 400; Jansen R, Dash SP, Sharma S and Min BC 2012 *Semicond. Sci. Technol.* **27**, 083001.
3. Appelbaum I, Huang B and Monsma D J 2007 *Nature* **447**, 295.
4. Huang B, Monsma DJ and Appelbaum I 2007 *Phys. Rev. Lett.* **99**, 177209.
5. Huang B, Zhao L, Monsma DJ and Appelbaum I 2007 *Appl. Phys. Lett.* **91**, 052501.
6. Huang B, Jang H-J and Appelbaum I 2008 *Appl. Phys. Lett.* **93**, 162508.
7. Li J and Appelbaum I 2012 *Appl. Phys. Lett.* **100**, 162408.

8. Jang HJ, Xu J, Li J, *et al.* 2008 *Phys. Rev. B* **78**, 165329.
9. Lu Y, Li J and Appelbaum I 2011 *Phys. Rev. Lett.* **106**, 217202.
10. Cheng JL, Wu MW and Fabian J 2010 *Phys. Rev. Lett.* **104**, 016601.
11. Song Y and Dery H 2012 *Phys. Rev. B* **86**, 085201.
12. Dash SP, Sharma S, Patel RS, *et al.* 2009 *Nature (London)* **462**, 491.
13. Jansen R, Min BC, Dash SP, *et al.* 2010 *Phys. Rev. B* **82**, 241305(R).
14. Jeon KR, Min BC, Shin IJ, *et al.* 2011 *Appl. Phys. Lett.* **98**, 262102.
15. Li CH, van't Erve OMJ and Jonker BT 2011 *Nat. Commun.* **2**, 245.
16. Ishikawa M, Sugiyama H, Inokuchi T, *et al.* 2012 *Appl. Phys. Lett.* **100**, 252404.
17. Suzuki T, Sasaki T, Oikawa T, *et al.* 2011 *Appl. Phys. Express* **4**, 023003.
18. Sasaki T, Oikawa T, Suzuki T, *et al.* 2009 *Appl. Phys. Express* **2**, 053003.
19. Sasaki T, Oikawa T, Suzuki *et al.* 2010 *IEEE Trans. Mag.* **46**, 1436.
20. Sasaki T, Oikawa T, Shiraishi M, *et al.* 2011 *Appl. Phys. Lett.* **98**, 012508.
21. Le Breton JC, Sharma S, Saito H, *et al.* 2011 *Nature* **475**, 82.
22. Ando K and Saitoh E 2012 *Nature Commun.* **3**, 629.
23. Sugahara S and Tanaka M 2004 *Appl. Phys. Lett.* **84**, 2307.
24. Dery H, Dalal P, Cywiński Ł and Sham LJ 2007 *Nature (London)* **447**, 573.
25. van't Erve OMJ, Hanbicki AT, Holub M, *et al.* 2007 *Appl. Phys. Lett.* **91**, 212109.
26. Rashba EI 2000 *Phys. Rev. B* **62**, R16267.
27. Fert A and Jaffrès H 2001 *Phys. Rev. B* **64**, 184420.
28. Takahashi S and Maekawa S 2003 *Phys. Rev. B* **67**, 052406.
29. Hamaya K, Ueda K, Kishi Y, *et al.* 2008 *Appl. Phys. Lett.* **93** 132117.
30. Yamada S, Hamaya K, Yamamoto K, *et al.* 2010 *Appl. Phys. Lett.* **96**, 082511.
31. Maeda Y, Hamaya K, Yamada S, *et al.* 2010 *Appl. Phys. Lett.* **97**, 192501.
32. Oki S, Yamada S, Murakami T, *et al.* 2012 *Thin Solid Films* **520**, 3419.
33. Tanikawa K, Oki S, Yamada S, *et al.* 2013 *Phys. Rev. B* **88**, 014402.
34. Ando Y, Hamaya K, Kasahara K, *et al.* 2009 *Appl. Phys. Lett.* **94**, 182105.
35. Ando Y, Kasahara K, Yamane K, *et al.* 2010 *Appl. Phys. Express* **3**, 093001.
36. Hamaya K, Ando Y, Sadoh T and Miyao M 2011 *Jpn. J. Appl. Phys.* **50**, 010101.
37. Ando Y, Kasahara K, Yamane K, *et al.* 2011 *Appl. Phys. Lett.* **99**, 012113.
38. Ando Y, Maeda Y, Kasahara K, *et al.* 2011 *Appl. Phys. Lett.* **99**, 132511.
39. Ando Y, Kasahara K, Yamada S, *et al.* 2012 *Phys. Rev. B* **85**, 035320.
40. Hamaya K, Ando Y, Masaki K, *et al.* 2013 *J. Appl. Phys.* **113**, 17C501.
41. Ueda K, Hamaya K, Yamamoto K, *et al.* 2008 *Appl. Phys. Lett.* **93**, 112108.
42. Ando Y, Hamaya K, Kasahara K, *et al.* 2009 *J. Appl. Phys.* **105**, 07B102.
43. Hamaya K, Itoh H, Nakatsuka O, *et al.* 2009 *Phys. Rev. Lett.* **102**, 137204.
44. Kasahara K, Yamamoto K, Yamada S, *et al.* 2010 *J. Appl. Phys.* **107**, 09B105.
45. Hamaya K, Murakami T, Yamada S, *et al.* 2011 *Phys. Rev. B* **83**, 144411.
46. Yamada S, Hamaya K, Murakami T, *et al.* 2011 *J. Appl. Phys.* **109**, 07B113.
47. Yamada S, Sagar J, Honda S, *et al.* 2012 *Phys. Rev. B* **86**, 174406.
48. Yamada S, Tanikawa K, Miyao M and Hamaya K 2012 *Cryst. Growth Des.* **12**, 4703.
49. Nakagawa K, Miyao M and Shiraki Y 1989 *Thin Solid Films* **183**, 315.
50. Miyao M and Nakagawa K, 1994 *Jpn. J. Appl. Phys.* **33**, 3791.
51. Sze SM (ed.) 1981 *Physics of Semiconductor Devices*, 2nd edn., pp. 27–30. Wiley.
52. Oki S, Yamada S, Hashimoto N, *et al.* 2012 *Appl. Phys. Express* **5**, 063004.

11

Electric-Field Control of Magnetism in Ferromagnetic Semiconductors

Tomoteru Fukumura
Department of Chemistry, The University of Tokyo, Japan

Electric field effect enables us to control charge carrier accumulation/depletion and is an indispensable technology in semiconductor electronics. By applying the electric field effect to ferromagnetic semiconductor, not only the charge quantity but also the magnetism can be controlled. For a decade, the electric field effect of ferromagnetic semiconductors has been extensively studied. Quite recently, the electric field induced ferromagnetism at room temperature in a ferromagnetic oxide semiconductor was reported. In this chapter, current status about the electric field effect of ferromagnetic semiconductors is described.

11.1 Introduction

11.1.1 Ferromagnetic Semiconductor

Recently, semiconductor spintronics has been extensively studied, and one of the promising semiconductor spintronic materials is the ferromagnetic semiconductor, a compound which possesses properties of semiconductor and ferromagnet [1, 2]. In ferromagnetic semiconductor electron (or hole) carriers possess information of both charge and spin, in which the latter is nonvolatile and various interplay between charge and spin is expected. Properties of the semiconductor such as electronic states have to be maintained sufficiently in ferromagnetic semiconductor even if possessing the ferromagnetic character. The diluted magnetic semiconductor (DMS) is a semiconductor doped with small amount of magnetic element. The magnetic atoms (cations) substitute partially for nonmagnetic atoms in the semiconductor, and provide the localized spins (Figure 11.1). The electronic states are approximately the same as that of the host semiconductor, hence ferromagnetic DMS is

Spintronics for Next Generation Innovative Devices, First Edition. Edited by Katsuaki Sato and Eiji Saitoh.
© 2015 John Wiley & Sons, Ltd. Published 2015 by John Wiley & Sons, Ltd.

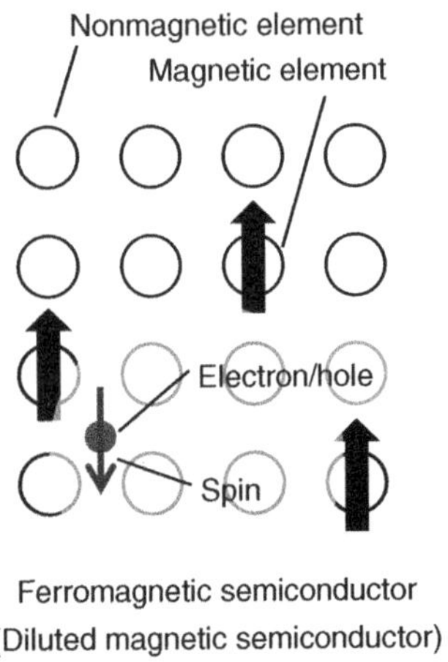

Figure 11.1 *Schematic structure of ferromagnetic semiconductor*

suitable for semiconductor spintronics. Although Mn-doped II–VI group DMSs are usually paramagnetic (or spin glass) [3], Mn-doped III–V group DMSs are ferromagnetic because of the ferromagnetic exchange interaction between spatially separated localized spins intermediated by itinerant hole carriers (Figure 11.1) [4, 5]. Thus, Mn-doped III–V group DMS has been extensively studied in the field of semiconductor spintronics.

In general, the magnetization in ferromagnet is functions of temperature and magnetic field. The magnetization is monotonically decreasing function of the temperature, and is monotonically increasing function of the magnetic field until the saturation. In case of the ferromagnetic semiconductor, the magnetization also depends on the carrier density because the itinerant carriers intermediate the ferromagnetic exchange coupling. By changing the carrier density via, e.g., chemical doping, the ferromagnetism (i.e. the magnetization amplitude and the Curie temperature) can be controlled.

11.1.2 Electric Field Effect of Ferromagnetic Semiconductor

In ordinary semiconductors, the carrier density can be controlled not only by chemical doping but also by different means: electric field effect. By applying gate voltage to the semiconducting channel through gate insulator, the carriers are accumulated or depleted depending on the polarity of carriers and electric field. This principle can be applied to ferromagnetic semiconductor. By applying the electric field, accumulated or depleted carriers enhance or suppress the ferromagnetic exchange coupling, respectively, resulting in the variation of the magnetism such as the magnetization and the Curie temperature. This variation is usually difficult to be achieved in ferromagnetic metals, because the carrier density is not significantly changed by any means because of the much higher carrier density in the metal than that in ferromagnetic semiconductor.

In fields of spintronics, the generation, the transfer, the control, and the detection of spin are fundamental technology. In case of ferromagnetic semiconductor, the generation of spin can be performed by using electric field effect without spin injection from ferromagnetic electrode. Also, the magnetization reversal by current injection is expected to be performed with much smaller current density when the magnetization amplitude is reduced by electric

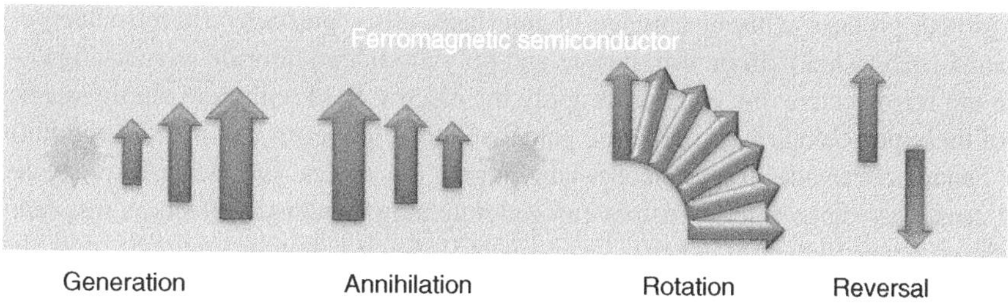

Figure 11.2 Possible electric field effects on spins in ferromagnetic semiconductor

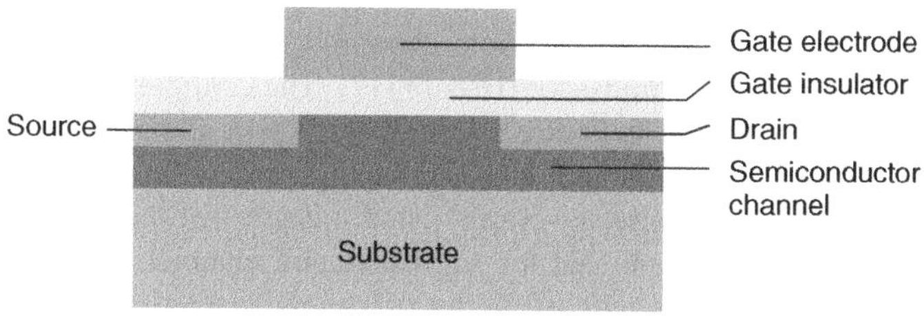

Figure 11.3 Metal-insulator (oxide)-semiconductor field effect transistor

field effect. In addition, the magnetic anisotropy can be modulated by using electric field effect. Accordingly, the electric field effect is one of the important techniques to manipulate spins in ferromagnetic semiconductor (Figure 11.2), and is useful to advance available technology in metal spintronics.

11.2 Experimental Techniques of Electric Field Effect

11.2.1 Field Effect Transistor

For electric field experiments, the standard and established device structure is field effect transistor [6]. The structure is basically trilayer composed of semiconductor channel, gate insulator, and gate electrode (Figure 11.3). In case of electric field effect of ferromagnetic semiconductor, high carrier accumulation is needed to control the ferromagnetism. To this end, the gate insulator has to be suited for the application of high electric field without dielectric breakdown. Thus, highly insulating and chemically stable materials are usually used as the gate insulator, such as SiO_2, Al_2O_3, and HfO_2, among which HfO_2 is advantageous due to its high dielectric constant. In addition to the gate insulator material itself, the deposition process is important to form high quality gate insulator and its interface with semiconductor channel. Spin coating method is technically easy without using any expensive equipment, for SiO_2 and insulating polymers such as polyimide [7, 8]. Physical vapor deposition methods such as sputtering method and pulsed laser deposition method are versatile for the deposition of various insulators [9, 10]. Recently developed atomic layer deposition method is capable of atomic layer-by-layer deposition of the gate insulator materials, which is one of the promising methods for the high electric field experiments, because of the high coverage of homogeneously grown film and the mild growth process without damage of interface. Also, piezoelectric/ferroelectric gate insulators such as lead zirconium titanate and polyvinylidene fluoride were used [11–14], where the ferroelectric one enables to apply the electric field with nonvolatile manner. In case of these piezoelectric/ferroelectric gate insulators, however, it is noted that additional effect can appear in addition to electrostatic doping of carriers. For example, piezoelectric effect caused by these gate insulators can generate magnetoelastic effect, as was reported in lead zirconate titanate/(Ga,Mn)As system [15]. Such behavior could be classified as a multiferroic effect [16] rather than electrostatic charge accumulation/depletion. In this chapter, the electrostatic gating is mainly described.

11.2.2 Electric Double Layer Transistor

Recently, another technique for electric field experiment has been developed, which is called the electric double layer transistor (EDLT) [17]. The configuration of EDLT is significantly different from that of field effect transistor, because the gate insulator is liquid electrolyte. One of the EDLT structure is shown in Figure 11.4(a). The gate electrode is located adjacent to the semiconductor channel, and a droplet of the liquid electrolyte is in contact with the gate electrode and the semiconductor channel. Figure 11.4(b) illustrates the operation principle in case of the ferromagnetic semiconductor channel. With the application of gate voltage, the cations and anions move according to the electric field. With the application of the positive gate voltage, the cations are distributed above the channel forming an electric double layer with $\sim$1 nm thick. Several voltages along the ultrathin electric double layer result in the high electric field with an order of $\geq 10^1$ MV/cm, which is approximately equivalent to the electron carrier accumulation of $\sim 10^{14}$ cm^{-2}, resulting in the electrically induced ferromagnetism. EDLT is very useful to apply high electric field by using the simple configuration without preparation of high quality interface in trilayer structure like field effect transistor. Accordingly, high electric field effect for various forms of samples such as single crystal, layered crystal, and epitaxial thin film has been studied, yielding in, e.g., the electrically induced superconductivity and the insulator to metal transition, as a result of high carrier accumulation [18–20]. In case of some compounds, electrochemical reaction hinders the application of high gate voltage, thus chemical stability of samples is an important factor for EDLT.

11.2.3 Probe of Ferromagnetism

In order to investigate the electric field effect of ferromagnetism, probing methods of ferromagnetism as well as the application of high electric field are important. The most fundamental quantity is the magnetization, which is measured with standard superconducting quantum interference device (SQUID) magnetometer under the application of gate voltage (Figure 11.5(a)). The sensitivity of SQUID magnetometer is so superior that any change in the magnetization can be measured quantitatively. However, its magnetization signal comes not only from the sample itself but also from any external sources such as electrodes and gate insulator, where the separation of those components is not easy. Hence, some alternative methods are sometimes needed for the electric field experiments.

Ferromagnetic semiconductor has an energy gap in contrast with ferromagnetic metal. In general, magnetic circular dichroism (MCD) of ferromagnetic semiconductor is proportional to the energy derivative of the absorption spectrum [23], so that MCD signal is

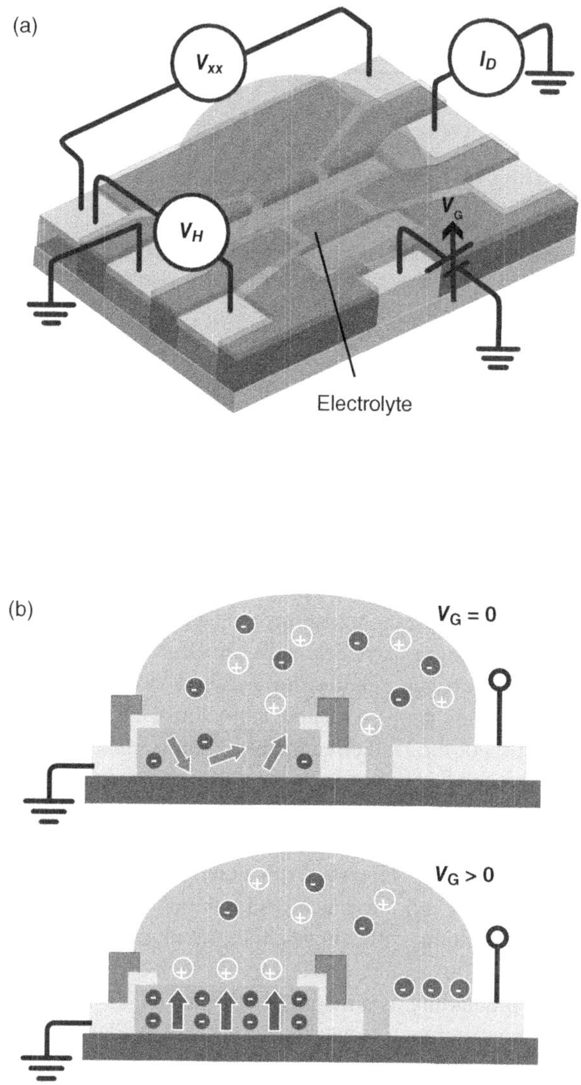

Figure 11.4 (a) Schematic diagram of an electric double layer transistor (EDLT). The liquid electrolyte contacts with the semiconductor channel and the planar gate electrode [21]. (b) Operation principle of electric double layer transistor with the channel of ferromagnetic semiconductor. The formation of electric double layer with the application of positive gate voltage accumulates high density electron carriers, resulting in an electrically induced ferromagnetism owing to the carrier-mediated ferromagnetism [21].

sufficiently large around at the absorption edge, and its magnetic field dependence reflects that of the magnetization (Figure 11.5(b)). But it is noted that the sample has to be optically accessible under electric field effect experiment, thus semitransparent gate electrode is often used.

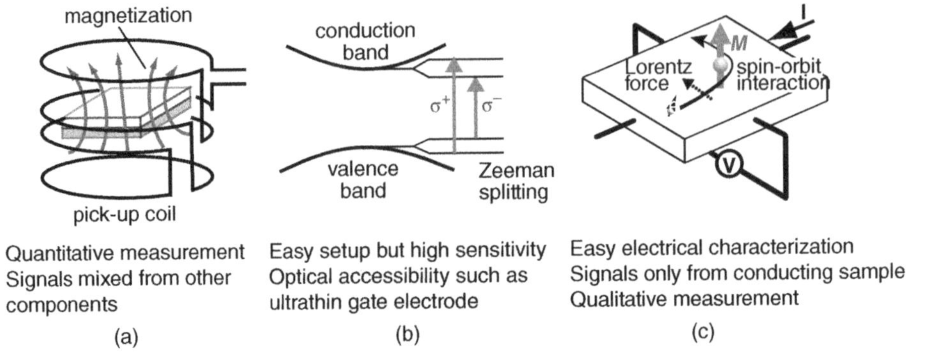

Figure 11.5 *Schematic experimental setups for (a) magnetization, (b) magneto-optical effect, and (c) anomalous Hall effect measurements modified from [22]. Merits and demerits of each measurement are listed.*

Another fundamental property is anomalous Hall effect, which is observed generally in ferromagnetic semiconductors and metals (Figure 11.5(c)). The Hall effect is the sum of ordinary and anomalous Hall effects. The former is proportional to the magnetic field caused by the Lorentz force, and the latter is proportional to the magnetization caused by the spin–orbit interaction. Thus, the magnetization can be evaluated from the anomalous Hall effect measurement, whereas the carrier density can be evaluated from the ordinary Hall effect measurement. Therefore, the relationship between the magnetization and the carrier density can be obtained when the ordinary and anomalous Hall effects are separated for the Hall effect measurement. For the Hall effect measurements, the electric current does probe the magnetization in the conductor but does not probe the magnetization in the insulator, because the current flows only in the conductive region. Hence, the magnetization except the sample, e.g., the gate insulator and substrate, is not detected in contrast with the standard magnetization measurement. However, it is noted that the Hall effect measurement is not always versatile. Laterally inhomogeneous sample may show a Hall voltage due to the meandering current path without reflecting neither the magnetization nor the carrier density. Vertically inhomogeneous nonmagnetic sample such as bilayer may show nonlinear anomalous Hall effect pretending as a magnetic sample [24, 25].

In addition to these measurements, other techniques are available such as direct observation of magnetic domain by magneto-optical means [14]. Also, simpler characterization of the ferromagnetism was carried out by temperature dependence of resistivity [26], because ferromagnetic semiconductors often show a maximum of resistivity around at the Curie temperature reflecting an insulator to metal transition accompanied with a paramagnetic to ferromagnetic transition.

11.3 Electric Field Control of Ferromagnetism in Ferromagnetic Semiconductors

11.3.1 (III,Mn)V Ferromagnetic Semiconductors

One of the important properties of ferromagnetic semiconductor is the capability of electrical control of the ferromagnetism, as a result of the carrier-mediated mechanism. Ohno *et al.* have demonstrated for the first time the electrical control of the ferromagnetism

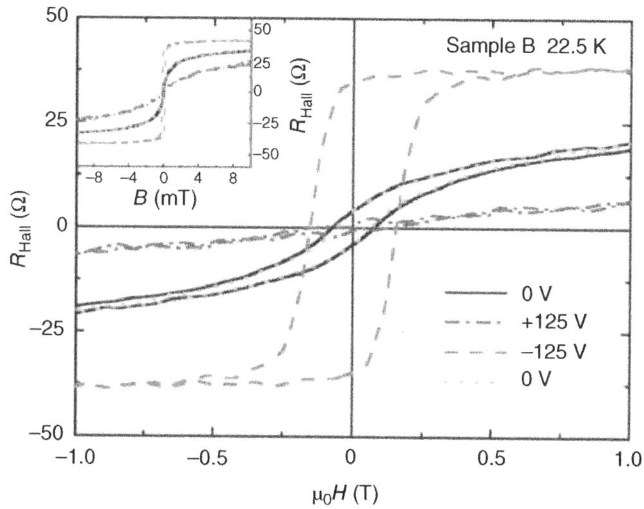

Figure 11.6 Hall resistance (R_{Hall}) vs. magnetic field curves for (In,Mn)As at different gate voltages at 22.5 K. With the positive gate voltage, a paramagnetic behavior is observed due to depleted holes. With the negative gate voltage, clear ferromagnetic hysteresis is observed due to accumulated holes. Inset shows the data for wider range of magnetic fields [7]. (Reprinted by permission from MacMillan Publishers Ltd. Copyright (2000))

in (In,Mn)As by means of a MISFET structure with polyimide gate insulator [7]. The accumulated hole carriers with the application of negative gate voltage enhances the magnetization and vice versa observed by the anomalous Hall effect (Figure 11.6). Subsequently, the magnetic field assisted magnetization reversal was demonstrated in (In,Mn)As MOSFET with spun-on SiO_2 gate insulator by using the result that the coercive force is monotonically decreasing function of the electric field (i.e., the gate voltage) (Figure 11.7(a)) [8]. When the coercive force is decreased by the electric field below an external magnetic field, the magnetization can be reversed even setting back the electric field (Figure 11.7(b)). In addition, the magnetization rotation was demonstrated in (Ga,Mn)As MOSFET with atomic layer deposited ZrO_2 gate insulator as a result of the carrier density dependence of magnetic anisotropy (Figure 11.8) [27]. All these results are technologically important.

However, the electric field effect experiment is very useful for the investigation of fundamentals of the ferromagnetism. One of the advantages to the electric field effect experiments is that the carrier density can be varied continuously while unchanging the other sample parameters such as the amount of chemical dopants, in contrast with chemical doping experiment. The hole carrier density dependence of the Curie temperature in (Ga,Mn)As, which is fundamental quantity in relation with the microscopic mechanism of the ferromagnetism, was studied by systematically varied hole carrier density from both anomalous Hall effect [28] and magnetization measurements [29]. In addition, a peculiar sign reversal of anomalous Hall coefficient was observed from the electric field effect of (Ga,Mn)As ultrathin films [30]. The EDLT for (Ga,Mn)As was also developed indicating that lower gate voltage than that of MOSFET is sufficient for wider tuning of carrier density, although the electrochemical reaction hampers the application of high gate voltage [31].

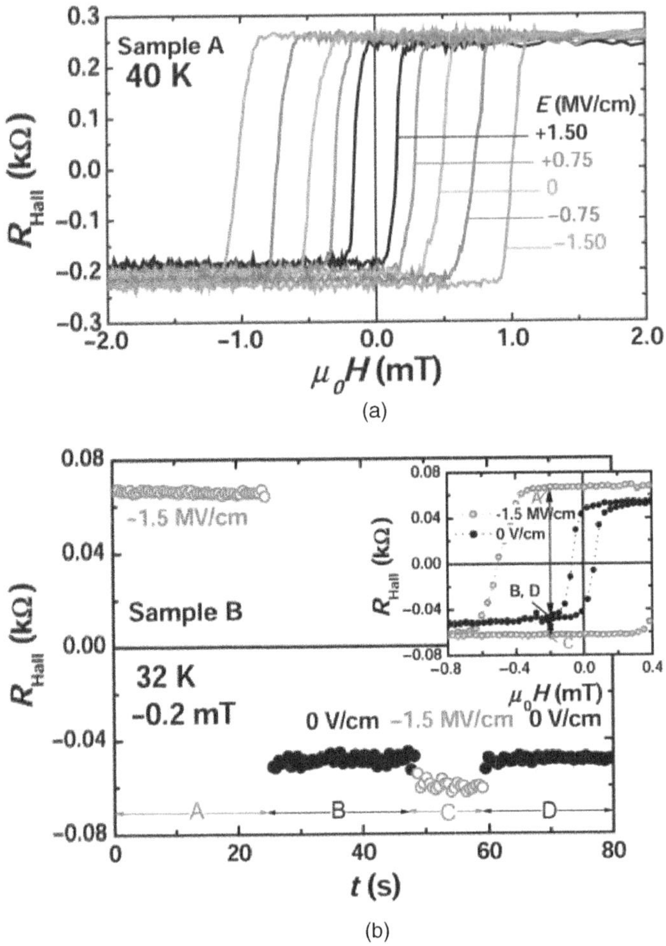

Figure 11.7 (a) Hall resistance (R_{Hall}) vs. magnetic field curves at 40 K at different electric fields for (In,Mn)As [8]. (b) Time evolution of R_{Hall} under different electric fields at 32 K in −0.2 mT, representing magnetic field-assisted magnetization reversal by electric field effect. Inset shows corresponding states A–D in magnetization curves [8]. (Reprinted by permission from American Association for the Advancement of Science. Copyright (2003))

11.3.2 High-T Ferromagnetic Oxide Semiconductor (Ti,Co)O_2

As described in previous section, (Ga,Mn)As is a prototypical ferromagnetic semiconductor so that various spintronic devices were successfully demonstrated. However, the Curie temperature, ~200 K at present, hampers the device operation at room temperature. Accordingly, room temperature ferromagnetic semiconductors have been extensively searched so far. Various semiconductors have been investigated as a host compound of ferromagnetic semiconductors. One of the promising candidates was wide gap oxide semiconductors such as ZnO and TiO_2. Various transition metal doped oxide semiconductors, which is called magnetic oxide semiconductors, have been reported to be ferromagnetic

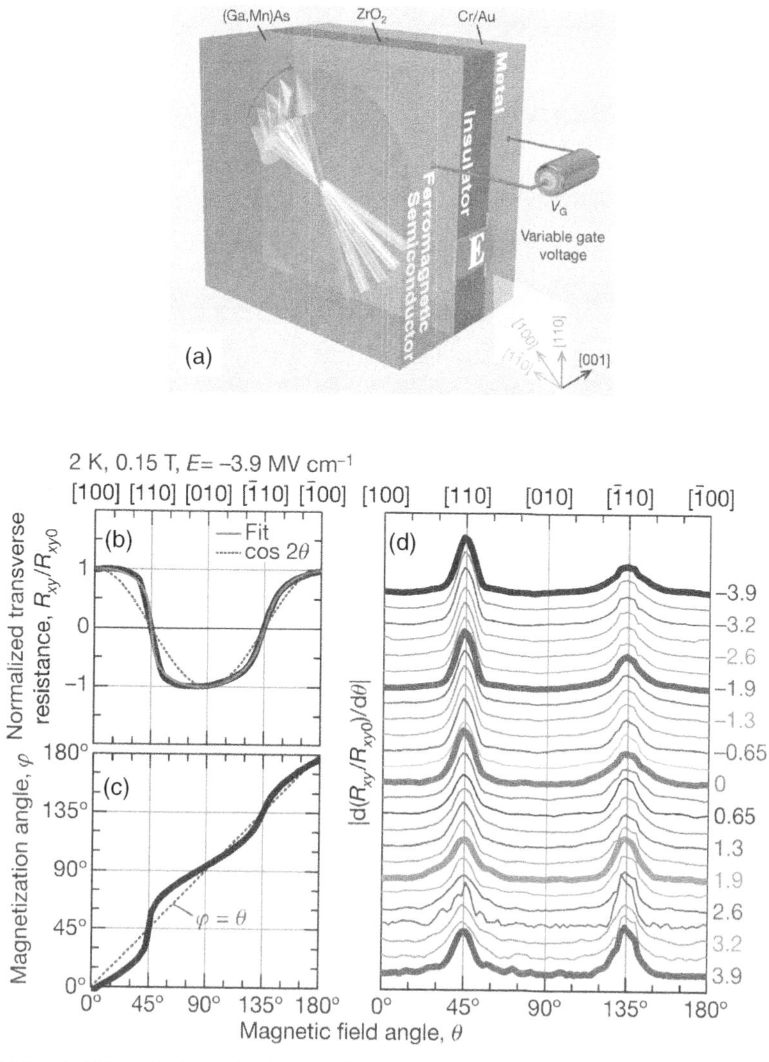

Figure 11.8 *(a) Electrical manipulation of magnetization vector in (Ga,Mn)As MOSFET. The magnetization variation is caused by the change in magnetic anisotropy with the hole density [27]. (b) Magnetic field angle (θ) dependence of normalized planar Hall resistance R_{xy}/R_{xy0} [27]. (c) Magnetization angle (ϕ) as a function of θ. (d) $|R_{xy}/R_{xy0}|$ vs. θ at different electric fields, in which the higher peak denotes the harder magnetization axis [27]. (Reprinted by permission from MacMillan Publishers Ltd. Copyright (2008))*

at room temperature [32–37]. The strong concern is why their Curie temperatures are so high, typically 500–700 K, since standard carrier-mediated interaction predicts the Curie temperature to be ~350 K [38]. Thus, several groups claimed the defect-mediated ferromagnetism due to the presence of oxygen vacancy, while the other groups claimed the carrier-mediated ferromagnetism [39, 40]. Only for the latter mechanism, the high temperature ferromagnetism could be utilized for semiconductor spintronics. In case of magnetic

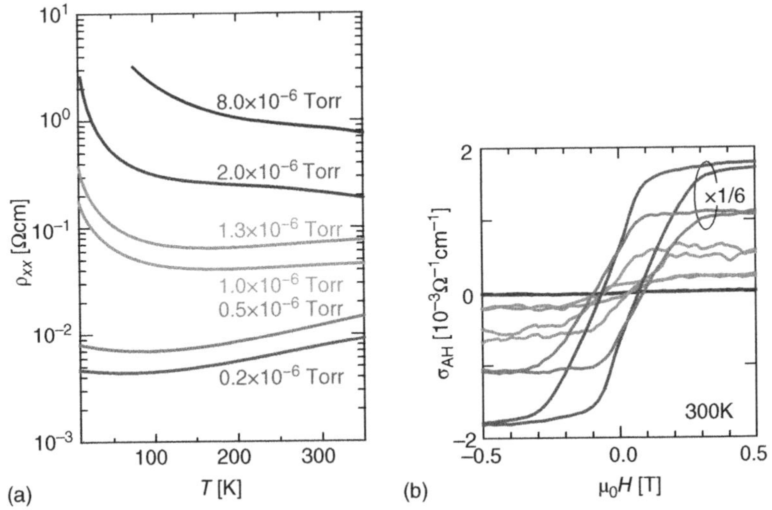

Figure 11.9 *(a) Resistivity (ρ_{xx}) vs. temperature curves and (b) anomalous Hall conductivity σ_{AH} vs. magnetic field curves at 300 K for $Ti_{0.90}Co_{0.10}O_{2-\delta}$ films grown in various oxygen pressures. Colors denote each sample [41]. (Reprinted by permission from American Institute of Physics. Copyright (2011))*

oxide semiconductors, the oxygen vacancy served as the electron dopant, hence these two mechanisms are difficult to be distinguished by using the oxygen vacancy as electron dopant. By using electric field effect, on the other hand, the carrier density can be varied without changing the amount of oxygen vacancy. Thus, the electric field effect is expected to settle the debate about these mechanisms. In addition, the electric field control of the ferromagnetism at room temperature is an important milestone for semiconductor spintronics.

$(Ti,Co)O_2$ is ferromagnetic at room temperature both for anatase and rutile structures [42, 43]. Among these, anatase $(Ti,Co)O_2$ shows clearer paramagnetic to ferromagnetic and insulator to metal transitions by varying the lower range of carrier density (together with the amount of oxygen vacancy) via chemical doping method (Figure 11.9) [41]. The optimal Curie temperature is as high as ~600 K [44]. Hence, anatase $(Ti,Co)O_2$ is a suitable system to investigate the electric field effect of the ferromagnetism at room temperature.

EDLT with the channel of anatase $(Ti,Co)O_2$ was developed by Yamada *et al.* [21]. With the application of positive gate voltage, anatase $(Ti,Co)O_2$ showed an insulator to metal transition (Figure 11.10(a)). The change in the resistivity was mainly caused by the increased electron carrier density evaluated from the ordinary Hall term (Figure 11.10(b)). From the anomalous Hall term simultaneously measured, the hysteresis curve of the anomalous Hall conductivity was enlarged with increasing the carrier density (Figure 11.11(a)). Since the anomalous Hall conductivity is monotonically increasing functions of both the magnetization and the carrier density on the verge of the paramagnetic to ferromagnetic transition, this result represents that the electrostatically accumulated electron carriers induced the ferromagnetic magnetization, i.e. electrically

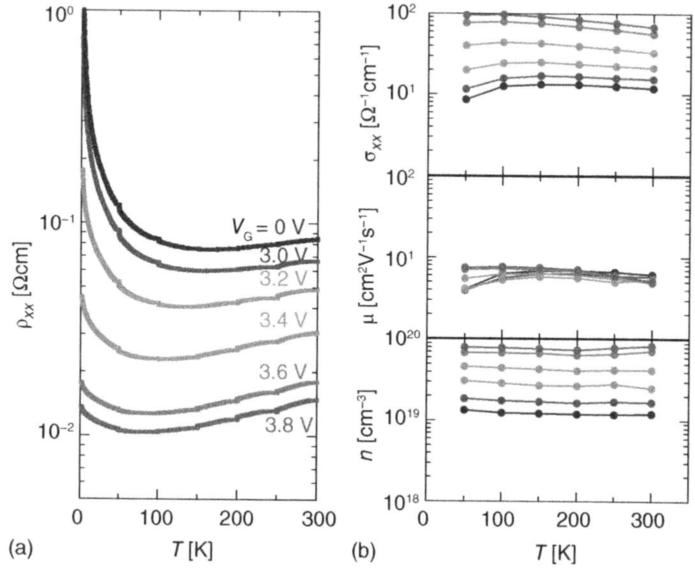

Figure 11.10 *(a) Resistivity (ρ_{xx}) vs. temperature curves in $Ti_{0.90}Co_{0.10}O_2$ with the application of gate voltages (V_G) [21]. (b) Conductivity (ρ_{xx}; upper panel), mobility (μ; center panel), and electron density (n; bottom panel) vs. temperature curves at each V_G. Colors denote each sample [21].*

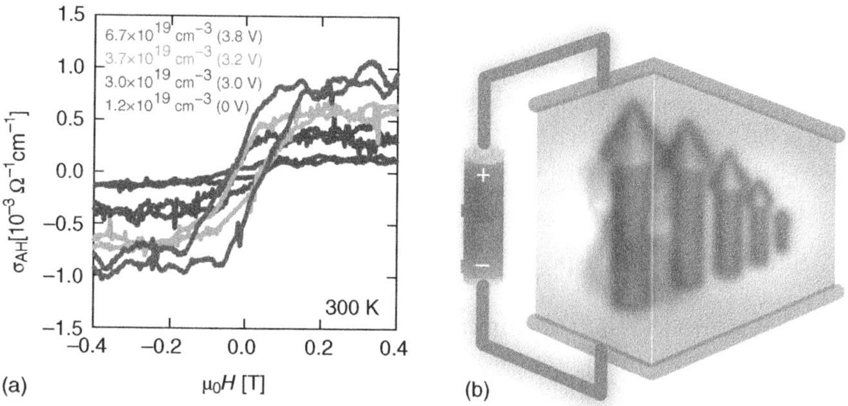

Figure 11.11 *(a) Anomalous Hall conductivity (σ_{AH}) vs. magnetic field curves at 300 K at each V_G. The values of electron density (n) at each V_G are shown [21]. (b) A schematic of electrically induced ferromagnetism. The accumulated electron carriers (green circle) by electric field effect induce ferromagnetism. Red arrow and blue trajectory denote the magnetization and the flow of electron carriers, respectively [21].*

induced ferromagnetism at room temperature (Figure 11.11(b)). A possibility of magnetic anisotropy change was ruled out from magnetization measurements, in which the magnetic anisotropy was not significantly changed among the samples with different carrier density. This electrically induced ferromagnetism in (Ti,Co)O_2 represents that the carrier-mediated mechanism governs its high temperature ferromagnetism and that ferromagnetic semiconductor spintronic devices can be operated at room temperature.

11.3.3 Other Ferromagnetic Semiconductors and Related Compounds

Electric field effect of various ferromagnetic semiconductors and oxides has been extensively studied (Table 11.1), although some of them could be related not to the electrostatic carrier doping but to the multiferroic effects or the extrinsic effects. For compound semiconductors, Boukari *et al.* studied photoluminescence of modulation-doped p-type $Cd_{0.96}Mn_{0.04}Te$ quantum well [45]. Park *et al.* studied the anomalous Hall effect of MnGe [46]. Nepal *et al.* studied the magnetization and the Hall effect of GaMnN/*p*-GaN/n-GaN multilayer structures at 300 K [47]. Xiu *et al.* studied the magnetization of MnGe quantum dots at up to 400 K [48]. For oxides, various compounds have been studied. Mathews *et al.* studied ferroelectric field effect of $La_{0.7}Ca_{0.3}MnO_3$ with $PbZr_{0.2}Ti_{0.8}O_3$ gate insulator [49]. Zhao *et al.* studied the magnetization of anatase (Ti,Co)O_2 with $PbZr_{0.2}Ti_{0.8}O_3$ gate insulator [50]. Kanki *et al.* studied MCD of $La_{0.85}Ba_{0.15}MnO_3$ with $Pb(Zr_{0.2}Ti_{0.8})O_3$ gate insulator [51]. Thiele *et al.* studied the magnetization and magnetoelectric effect of $La_{0.7}A_{0.3}MnO_3$ (A = Ca, Sr) with $Pb(Mg_{1/3}Nb_{2/3})_{0.72}Ti_{0.28}O_3$ gate insulator [52]. Chou *et al.* studied the magnetization of $[Co/ZnO]_{25}\delta$-doped multilayers at room temperature [53]. Dhoot *et al.* studied the resistance of $La_{0.8}Ca_{0.2}MnO_3$ with EDLT [26]. Ichimura *et al.* studied the resistance of $(Fe,Zn)_3O_4$ with parylene gate insulator [54]. Metal–insulator transitions of $NdNiO_3$ and $(Sm,Ca)MnO_3$ were studied with EDLT by several groups [55, 56].

Recently, the electric field effect of even the ferromagnetic metals and alloys become possible and is extensively studied. In the form of ultrathin film, the electric field effect resulted in the control of magnetic anisotropy [57–63], the ferromagnetic to paramagnetic transition [64], and the control of magnetic domain wall velocity [65].

11.4 Summary and Prospect

Through the electric field effect experiments of ferromagnetic semiconductors, particularly (In,Mn)As and (Ga,Mn)As, various semiconductor spintronic devices have been demonstrated. Recent progress on the techniques of electric field application and magnetic probing enables to perform sophisticated experiments with the application of very high electric field, leading to the implementation of highly functional spintronic devices and the discovery of new phase of matter that was inaccessible ever. At present, EDLT attracts much attention as a powerful tool for high electric field effect on various functionalities. However, the development of all-solid-state field effect transistor for high electric field effect is desired for the device integration and the controllability at low temperature. In addition, recent development of room temperature ferromagnetic semiconductors such as (Ti,Co)O_2 enables to conduct such device operation at room temperature. Both the development of

Table 11.1 *Electric field effect studies on ferromagnetic semiconductors and oxides*

Channel compound	Sample type	Gate insulator	Method for gate insulator	T_C (K)	$\|V_G\|$ (V)	Carrier type	Probe of M	Ref.
(In,Mn)As	Thin film	Polyimide	Spin coating	≤27.5	≤125	p	AHE	[7]
(In,Mn)As	Thin film	SiO_2	Spin coating	≤53.5	≤135	p	AHE	[8]
(Ga,Mn)As	Thin film	Al_2O_3	ALD	≤70	≤25	p	AHE	[10]
(Ga,Mn)As	Thin film	ZrO_2	ALD	46	≤12	p	AHE	[27]
(Ga,Mn)As	Thin film	HfO_2	ALD	≤41	≤15	p	*M*	[29]
(Ga,Mn)As	Thin film	$KClO_4$ + PEO	dropping	≤109	≤3	p	AHE	[31]
(Ga,Mn)As	Thin film	Al_2O_3, HfO_2, ZrO_2	ALD	≤165	≤2.5	p	AHE	[30]
(Ga,Mn)As	Thin film	Al_2O_3, HfO_2	ALD	≤165	≤3.2	p	AHE	[28]
(Ga,Mn)As	Thin film	P(VDF-TrFE)	Spin coating	≤84	≤30	p	AHE	[11]
(Ga,Mn)As	Thin film	Piezoelectric actuator	Contact	≤80	≤200	p	*R*	[12]
(Ga,Mn)As	Thin film	PZT	Contact	80	≤200	p	*R*	[13]
(Ga,Mn)As	Thin film	P(VDF-TrFE)	Spin coating	≤60	≤10	p	*R*	[66]
(Ga,Mn)As	Thin film	Piezoelectric actuator	Contact	≤85	≤200	p	*R*	[15]
Mn δ-doped GaAs	Thin film	SiO_2	Spin coating	≤117	≤18	p	AHE	[67]
(Ga,Mn)(As,P)	Thin film	P(VDF-TrFE)	Spin coating	≤44	≤40	p	AHE	[14]
(Cd,Mn)Te	Thin film	(Cd,Mg,Zn)Te	MBE	≤2.5	≤1	p	PL	[45]
MnGe	Thin film	SiN	Sputtering	>50	≤10	p	AHE	[46]
MnGe	Quantum dots	Al_2O_3	ALD	>100	≤40	p	*M*	[48]
GaMnN	Thin film	GaMnN	MOCVD	>300	≤6	p	*M*, AHE	[47]
$(Ti,Co)O_2$	Thin film	PZT	PLD	>300	≤30	n	*M*	[50]
$(Ti,Co)O_2$	Thin film	DEME-TFSI $CsClO_4$ + PEO	Dropping	>300	≤3.8	n	AHE	[21]
(Zn,Co)O	Thin film	Al_2O_3	Sputtering	>300	≤10	n	*M*	[53]
$(Fe,Zn)_3O_4$	Thin film	parylene	Pyrolyzing	>300	≤150	n	AHE	[54]
$(La,Ca)MnO_3$	Thin film	EMIM-TFSI $KClO_4$ + PEO	Dropping	≤190	≤3	p	*R*	[26]
$(La,Ba)MnO_3$	Thin film	PZT	PLD	>300	~4	p	*R*, MCD	[51]
$(La,Ca)MnO_3$	Thin film	PZT	PLD		≤15	p	*R*	[49]
$NdNiO_3$	Thin film	EMI-TFSI	Dropping		≤4	p	*R*	[55]
$(La,Sr)MnO_3$	Thin film	PMN-PT	Substrate	≤346	≤500	p	*M*	[52]
$SmCoO_3$	Thin film	$DEME\text{-}BF_4$	Dropping		≤2	p	*R*	[56]

T_C: Curie temperature under electric field; V_G: gate voltage; *M*: magnetization; PEO: polyethylene-oxide; P(VDF-TrFE): polyvinylidene fluoride-trifluoroethylene; PZT: $PbZr_{0.2}Ti_{0.8}O_3$; DEME-TFSI: *N*,*N*-diethyl-*N*-(2-methoxyethyl)-*N*-methylammonium-bis(trifluoromethylsulfonyl)imide; EMIM-TFSI: 1-ethyl-3-methylimidazolium bis(trifluoromethylsulfonyl)imide; EMI-TFSI: 1-ethyl-3-methylimidazolium-bis(trifl uoromethylsulfonyl)imide; PMN-PT: $Pb(Mn_{1/3}Nb_{2/3})_{0.72}Ti_{0.28}O_3$; $DEME\text{-}BF_4$: *N*, *N*-diethyl-*N*-(2-methoxyethyl)-Nmethylammonium-tetrafluoroborate; ALD: atomic layer deposition; MBE: molecular beam epitaxy; MOCVD: metal-organic chemical-vapor deposition; PLD: pulsed laser deposition; p: hole carrier; n: electron carrier; *R*: resistance; AHE: anomalous Hall effect; PL: photoluminescence; MCD: magnetic circular dichroism.

novel techniques of electric field effect and the discovery/optimization of new ferromagnetic semiconductors are indispensable to advance the science and applications in the electric field effect of ferromagnetic semiconductors.

Acknowledgments

The author acknowledges T. S. Krasienapibal and J. Wei for technical assistance. This work was based in part on the collaboration with Y. Yamada, K. Ueno, M. Kawasaki, H. T. Yuan, H. Shimotani, Y. Iwasa, L. Gu, S. Tsukimoto, and Y. Ikuhara. This work was supported in part by JST-PRESTO and JSPS through NEXT Program initiated by CSTP.

References

1. I. Žutić, J. Fabian, and S. Das Sarma. Spintronics: Fundamentals and applications. *Rev. Mod. Phys.*, **76**:323–410, 2004.
2. J. Fabian, A. Matos-Abiaguea, C. Ertlera, P. Stanoa, and I. Žutić. Semiconductor spintronics. *Acta Phys. Slov.*, **57**:565–907, 2007.
3. J. K. Furdyna. Diluted magnetic semiconductors. *J. Appl. Phys.*, **64(4)**:R29–R64, 1988.
4. H. Ohno. Making nonmagnetic semiconductors ferromagnetic. *Science*, **281(5379)**: 951–956, 1998.
5. T. Jungwirth, J. Sinova, J. Mašek, J. Kučera, and A. H. MacDonald. Theory of ferromagnetic (III,Mn)V semiconductors. *Rev. Mod. Phys.*, **78**:809–864, 2006.
6. S. M. Sze. *Physics of Semiconductor Devices*. Wiley, New York, 1987.
7. H. Ohno, D. Chiba, F. Matsukura, T. Omiya, E. Abe, T. Dietl, Y. Ohno, and K. Ohtani. Electric-field control of ferromagnetism. *Nature*, **408(6815)**:944–946, 2000.
8. D. Chiba, M. Yamanouchi, F. Matsukura, and H. Ohno. Electrical manipulation of magnetization reversal in a ferromagnetic semiconductor. *Science*, **301(5635)**: 943–945, 2003.
9. M. J. Biercuk, D. J. Monsma, C. M. Marcus, J. S. Becker, and R. G. Gordon. Low-temperature atomic-layer-deposition lift-off method for microelectronic and nanoelectronic applications. *Appl. Phys. Lett.*, **83(12)**:2405–2407, 2003.
10. D. Chiba, F. Matsukura, and H. Ohno. Electric-field control of ferromagnetism in (Ga,Mn)As. *Appl. Phys. Lett.*, **89(16)**:162505, 2006.
11. I. Stolichnov, S. W. E. Riester, H. J. Trodahl, N. Setter, A. W. Rushforth, K. W. Edmonds, R. P. Campion, C. T. Foxon, B. L. Gallagher, and T. Jungwirth. Non-volatile ferroelectric control of ferromagnetism in (Ga,Mn)As. *Nat. Mater.*, **7(6)**: 464–467, 2008.
12. S. T. B. Goennenwein, M. Althammer, C. Bihler, A. Brandlmaier, S. Geprägs, M. Opel, W. Schoch, W. Limmer, R. Gross, and M. S. Brandt. Piezo-voltage control of magnetization orientation in a ferromagnetic semiconductor. *Phys. Stat. Sol.*, **2(3)**:96–98, 2008.
13. M. Overby, A. Chernyshov, L. P. Rokhinson, X. Liu, and J. K. Furdyna. GaMnAs-based hybrid multiferroic memory device. *Appl. Phys. Lett.*, **92(19)**:192501, 2008.

14. E. Mikheev, I. Stolichnov, E. De Ranieri, J. Wunderlich, H. J. Trodahl, A. W. Rushforth, S. W. E. Riester, R. P. Campion, K. W. Edmonds, B. L. Gallagher, and N. Setter. Magnetic domain wall propagation under ferroelectric control. *Phys. Rev. B*, **86**:235130, 2012.
15. C. Bihler, M. Althammer, A. Brandlmaier, S. Geprägs, M. Weiler, M. Opel, W. Schoch, W. Limmer, R. Gross, M. S. Brandt, and S. T. B. Goennenwein. $Ga_{1-x}Mn_xAs$/piezoelectric actuator hybrids: A model system for magnetoelastic magnetization manipulation. *Phys. Rev. B*, **78**:045203, 2008.
16. C. A. F. Vaz. Electric field control of magnetism in multiferroic heterostructures. *J. Phys. Condens. Matter*, **24(33)**:333201, 2012.
17. M. Kawasaki and Y. Iwasa. Electronics: 'Cut and stick' ion gels. *Nature*, **489(7417)**: 510–511, 2012.
18. K. Ueno, S. Nakamura, H. Shimotani, A. Ohtomo, N. Kimura, T. Nojima, H. Aoki, Y. Iwasa, and M. Kawasaki. Electric-field-induced superconductivity in an insulator. *Nat. Mater.*, **7(11)**:855–858, 2008.
19. J. T. Ye, S. Inoue, K. Kobayashi, Y. Kasahara, H. T. Yuan, H. Shimotani, and Y. Iwasa. Liquid-gated interface superconductivity on an atomically flat film. *Nat. Mater.*, **9(2)**:125–128, 2010.
20. H. Shimotani, H. Asanuma, A. Tsukazaki, A. Ohtomo, M. Kawasaki, and Y. Iwasa. Insulator-to-metal transition in ZnO by electric double layer gating. *Appl. Phys. Lett.*, **91(8)**:082106, 2007.
21. Y. Yamada, K. Ueno, T. Fukumura, H. T. Yuan, H. Shimotani, Y. Iwasa, L. Gu, S. Tsukimoto, Y. Ikuhara, and M. Kawasaki. Electrically induced ferromagnetism at room temperature in cobalt-doped titanium dioxide. *Science*, **332(6033)**:1065–1067, 2011.
22. T. Fukumura, H. Toyosaki, and Y. Yamada. Magnetic oxide semiconductors. *Semicond. Sci. Technol.*, **20(4)**:S103–S111, 2005.
23. S. Sugano and N. Kojima. *Magneto-Optics*. Springer-Verlag, New York, 2000.
24. H. H. Wieder. Transport coefficients of InAs epilayers. *Appl. Phys. Lett.* **25(4)**:206–208, 1974.
25. W. A. Beck and J. R. Anderson. Determination of electrical transport properties using a novel magnetic fielddependent Hall technique. *J. Appl. Phys.*, **62(2)**:541–554, 1987.
26. A. S. Dhoot, C. Israel, X. Moya, N. D. Mathur, and R. H. Friend. Large electric field effect in electrolyte-gated manganites. *Phys. Rev. Lett.*, **102**:136402, 2009.
27. D. Chiba, M. Sawicki, Y. Nishitani, Y. Nakatani, F. Matsukura, and H. Ohno. Magnetization vector manipulation by electric fields. *Nature*, **455(7212)**:515–518, 2008.
28. Y. Nishitani, D. Chiba, M. Endo, M. Sawicki, F. Matsukura, T. Dietl, and H. Ohno. Curie temperature versus hole concentration in field-effect structures of $Ga_{1-x}Mn_xAs$. *Phys. Rev. B*, **81**:045208, 2010.
29. M. Sawicki, D. Chiba, A. Korbecka, Y. Nishitani, J. A. Majewski, F. Matsukura, T. Dietl, and H. Ohno. Experimental probing of the interplay between ferromagnetism and localization in (Ga,Mn)As. *Nat. Phys.*, **6(1)**:22–25, 2010.
30. D. Chiba, A. Werpachowska, M. Endo, Y. Nishitani, F. Matsukura, T. Dietl, and H. Ohno. Anomalous Hall effect in field-effect structures of (Ga,Mn)As. *Phys. Rev. Lett.*, **104**:106601, 2010.

31. M. Endo, D. Chiba, H. Shimotani, F. Matsukura, Y. Iwasa, and H. Ohno. Electric double layer transistor with a (Ga,Mn)As channel. *Appl. Phys. Lett.*, **96(2)**: 2010.
32. S. J. Pearton, C. R. Abernathy, M. E. Overberg, G. T. Thaler, D. P. Norton, N. Theodoropoulou, A. F. Hebard, Y. D. Park, F. Ren, J. Kim, and L. A. Boatner. Wide band gap ferromagnetic semiconductors and oxides. *J. Appl. Phys.*, **93(1)**:1–13, 2003.
33. W. Prellier, A. Fouchet, and B. Mercey. Oxide-diluted magnetic semiconductors: a review of the experimental status. *J. Phys. Condens. Matter*, **15**:1583, 2003.
34. R. Janisch, P. Gopal, and N. A. Spaldin. Transition metal-doped TiO_2 and ZnO–present status of the field. *J. Phys. Condens. Matter*, **17(27)**:R657, 2005.
35. C. Liu, F. Yun, and H. Morkoc. Ferromagnetism of ZnO and GaN: A review. *J. Mater. Sci.: Mater. Electron.*, **16(9)**:555–597, 2005.
36. F. Pan, C. Song, X. J. Liu, Y. C. Yang, and F. Zeng. Ferromagnetism and possible application in spintronics of transition-metal-doped ZnO films. *Mater. Sci. Eng. R*, **62(1)**:1–35, 2008.
37. S. B. Ogale. Dilute doping, defects, and ferromagnetism in metal oxide systems. *Adv. Mater.*, **22(29)**:3125–3155, 2010.
38. J. M. D. Coey, M. Venkatesan, and C. B. Fitzgerald. Donor impurity band exchange in dilute ferromagnetic oxides. *Nat. Mater.*, **4(2)**:173–179, 2005.
39. T. Fukumura, Y. Yamada, K. Ueno, H. T. Yuan, H. Shimotani, Y. Iwasa, L. Gu, S. Tsukimoto, Y. Ikuhara, and M. Kawasaki. Electron carrier-mediated room temperature ferromagnetism in anatase $(Ti, Co)O_2$. *Spin*, **02(04)**:1230005, 2012.
40. T. Fukumura and M. Kawasaki. Magnetic oxide semiconductors: On the high-temperature ferromagnetism in TiO_2-and ZnO-based compounds. in *Functional Metal Oxides*, pages S. B. Ogale, T. V. Venkatesan, and M. Blamire (Eds.), Wiley–VCH, Weinheim, 2013.
41. Y. Yamada, T. Fukumura, K. Ueno, and M. Kawasaki. Control of ferromagnetism at room temperature in $(Ti,Co)O_{2-\delta}$ via chemical doping of electron carriers. *Appl. Phys. Lett.*, **99(24)**:242502, 2011.
42. Y. Matsumoto, M. Murakami, T. Shono, T. Hasegawa, T. Fukumura, M. Kawasaki, P. Ahmet, T. Chikyow, S. Koshihara, and H. Koinuma. Room-temperature ferromagnetism in transparent transition metal-doped titanium dioxide. *Science*, **291(5505)**:854–856, 2001.
43. Y. Matsumoto, R. Takahashi, M. Murakami, T. Koida, X.-Juan Fan, T. Hasegawa, T. Fukumura, M. Kawasaki, S. Koshihara, and H. Koinuma. Ferromagnetism in Co-doped TiO_2 rutile thin films grown by laser molecular beam epitaxy. *Jpn. J. Appl. Phys.*, **40**:L1204–L1206, 2001.
44. K. Ueno, T. Fukumura, H. Toyosaki, M. Nakano, T. Yamasaki, Y. Yamada, and M. Kawasaki. Anomalous Hall effect in anatase $Ti_{1-x}Co_xO_{2-\delta}$ above room temperature. *J. Appl. Phys.*, **103(7)**:07D114, 2008.
45. H. Boukari, P. Kossacki, M. Bertolini, D. Ferrand, J. Cibert, S. Tatarenko, A. Wasiela, J. A. Gaj, and T. Dietl. Light and electric field control of ferromagnetism in magnetic quantum structures. *Phys. Rev. Lett.*, **88(20)**:207204, 2002.
46. Y. D. Park, A. T. Hanbicki, S. C. Erwin, C. S. Hellberg, J. M. Sullivan, J. E. Mattson, T. F. Ambrose, A. Wilson, G. Spanos, and B. T. Jonker. A group-IV ferromagnetic semiconductor: Mn_xGe_{1-x}. *Science*, **295(5555)**:651–654, 2002.

47. N. Nepal, M. Oliver Luen, J. M. Zavada, S. M. Bedair, P. Frajtag, and N. A. El-Masry. Electric field control of room temperature ferromagnetism in III-N dilute magnetic semiconductor films. *Appl. Phys. Lett.*, **94**(**13**):132505, 2009.
48. F. Xiu, Y. Wang, J. Kim, A. Hong, J. Tang, A. P. Jacob, J. Zou, and K. L. Wang. Electric-field-controlled ferromagnetism in high-curie-temperature $Mn_{0.05}Ge_{0.95}$ quantum dots. *Nat. Mater.*, **9**(**4**):337–344, 2010.
49. S. Mathews, R. Ramesh, T. Venkatesan, and J. Benedetto. Ferroelectric field effect transistor based on epitaxial perovskite heterostructures. *Science*, **276**(**5310**):238–240, 1997.
50. T. Zhao, S. R. Shinde, S. B. Ogale, H. Zheng, T. Venkatesan, R. Ramesh, and S. Das Sarma. Electric field effect in diluted magnetic insulator anatase $Co{:}TiO_2$. *Phys. Rev. Lett.*, **94**(**12**):126601, 2005.
51. T. Kanki, H. Tanaka, and T. Kawai. Electric control of room temperature ferromagnetism in a $Pb(Zr_{0.2}Ti_{0.8})O_3/La_{0.85}Ba_{0.15}MnO_3$ field-effect transistor. *Appl. Phys. Lett.*, **89**(**24**):242506, 2006.
52. C. Thiele, K. Dörr, O. Bilani, J. Rödel, and L. Schultz. Influence of strain on the magnetization and magnetoelectric effect in $La_{0.7}A_{0.3}MnO_3$/PMN-Pt(001)(A= Sr, Ca). *Phys. Rev. B*, **75**(**5**):054408, 2007.
53. H. Chou, C. P. Lin, J.-C. A. Huang, and H. S. Hsu. Magnetic coupling and electric conduction in oxide diluted magnetic semiconductors. *Phys. Rev. B*, **77**(**24**):245210, 2008.
54. T. Ichimura, K. Fujiwara, T. Kushizaki, T. Kanki, and H. Tanaka. Unstrained epitaxial zn-substituted Fe_3O_4 films for ferromagnetic field-effect transistors. *Jpn. J. Appl. Phys.*, **52**(**6R**):068002, 2013.
55. R. Scherwitzl, P. Zubko, I. G. Lezama, S. Ono, A. F. Morpurgo, G. Catalan, and J.-M. Triscone. Electric-field control of the metal-insulator transition in ultrathin $NdNiO_3$ films. *Adv. Mater.*, **22**(**48**):5517–5520, 2010.
56. P.-H. Xiang, S. Asanuma, H. Yamada, H. Sato, I. H. Inoue, H. Akoh, A. Sawa, M. Kawasaki, and Y. Iwasa. Electrolyte-gated $SmCoO_3$ thin-film transistors exhibiting thickness-dependent large switching ratio at room temperature. *Adv. Mater.*, **25**(**15**):2158–2161, 2013.
57. M. Weisheit, S. Fähler, A. Marty, Y. Souche, C. Poinsignon, and D. Givord. Electric field-induced modification of magnetism in thin-film ferromagnets. *Science*, **315**(**5810**):349–351, 2007.
58. T. Maruyama, Y. Shiota, T. Nozaki, K. Ohta, N. Toda, M. Mizuguchi, A. A. Tulapurkar, T. Shinjo, M. Shiraishi, S. Mizukami, et al. Large voltage-induced magnetic anisotropy change in a few atomic layers of iron. *Nat. Nanotech.*, **4**(**3**):158–161, 2009.
59. M. Endo, S. Kanai, S. Ikeda, F. Matsukura, and H. Ohno. Electric-field effects on thickness dependent magnetic anisotropy of sputtered $MgO/Co_{40}Fe_{40}B_{20}/Ta$ structures. *Appl. Phys. Lett.*, **96**(**21**):212503, 2010.
60. X. He, Y. Wang, N. Wu, A. N. Caruso, E. Vescovo, K. D. Belashchenko, P. A. Dowben, and C. Binek. Robust isothermal electric control of exchange bias at room temperature. *Nat. Mater.*, **9**(**7**):579–585, 2010.
61. T. Seki, M. Kohda, J. Nitta, and K. Takanashi. Coercivity change in an FePt thin layer in a Hall device by voltage application. *Appl. Phys. Lett.*, **98**(**21**):212505, 2011.
62. W.-G. Wang, M. Li, S. Hageman, and C. L. Chien. Electric-field-assisted switching in magnetic tunnel junctions. *Nat. Mater.*, **11**(**1**):64–68, 2011.

63. K. Kita, D. W. Abraham, M. J. Gajek, and D. C. Worledge. Electric-field-control of magnetic anisotropy of $Co_{0.6}Fe_{0.2}B_{0.2}$/oxide stacks using reduced voltage. *J. Appl. Phys.*, **112(3)**:033919, 2012.
64. D. Chiba, S. Fukami, K. Shimamura, N. Ishiwata, K. Kobayashi, and T. Ono. Electrical control of the ferromagnetic phase transition in cobalt at room temperature. *Nat. Mater.*, **10(11)**:853–856, 2011.
65. D. Chiba, M. Kawaguchi, S. Fukami, N. Ishiwata, K. Shimamura, K. Kobayashi, and T. Ono. Electric-field control of magnetic domain-wall velocity in ultrathin cobalt with perpendicular magnetization. *Nat. Commun.*, **3**:888, 2012.
66. S. W. E. Riester, I. Stolichnov, H. J. Trodahl, N. Setter, A. W. Rushforth, K. W. Edmonds, R. P. Campion, C. T. Foxon, B. L. Gallagher, and T. Jungwirth. Toward a low-voltage multiferroic transistor: Magnetic (Ga,Mn)As under ferroelectric control. *Appl. Phys. Lett.*, **94(6)**:063504, 2009.
67. A. M. Nazmul, S. Kobayashi, S. Sugahara, and M. Tanaka. Electrical and optical control of ferromagnetism in III-V semiconductor heterostructures at high temperature (100 K). *Jpn. J. Appl. Phys.*, **43**:233, 2004.

12

Quantum Information Processing Using Nitrogen-Vacancy Centres in Diamond

Norikazu Mizuochi [1,2]
[1] *Graduate School of Engineering Science, Osaka University, Toyonaka-city, Osaka, Japan*
[2] *CREST, Japan Science Technology, Kawaguchi, Saitama, Japan*

12.1 Introduction

The smallest size of features in state-of-the-art technological devices is a few tens of nanometres. These devices are electrically controlled, which is superior to optical control where diffraction limits the smallest size of features to be an order of magnitude larger. Nowadays, even electrical control of single atoms is possible; for example, for single P dopants in silicon (Si:P) [1, 2] and for quantum dots [3, 4]. The quantum nature of these systems opens up novel fields such as quantum electronics, spintronics and metrology. A severe limitation of these systems, however, is the requirement of cryogenic conditions for operation.

Recently, nitrogen-vacancy (NV) defect centres in diamond (Figure 12.1) have attracted significant attention. A NV centre is a joint defect in the carbon lattice of diamond and consists of a substitutional nitrogen atom and an adjacent vacancy. Its spin triplet (spin 1) ground state can be polarized and read out optically, so that electron-spin resonance (ESR) experiments can be performed on a single spin at ambient conditions, such as room temperature.

In 1997, a single NV centre was optically observed at room temperature for the first time [5]. A single spin state was also observed by optically detected magnetic

Spintronics for Next Generation Innovative Devices, First Edition. Edited by Katsuaki Sato and Eiji Saitoh.
© 2015 John Wiley & Sons, Ltd. Published 2015 by John Wiley & Sons, Ltd.

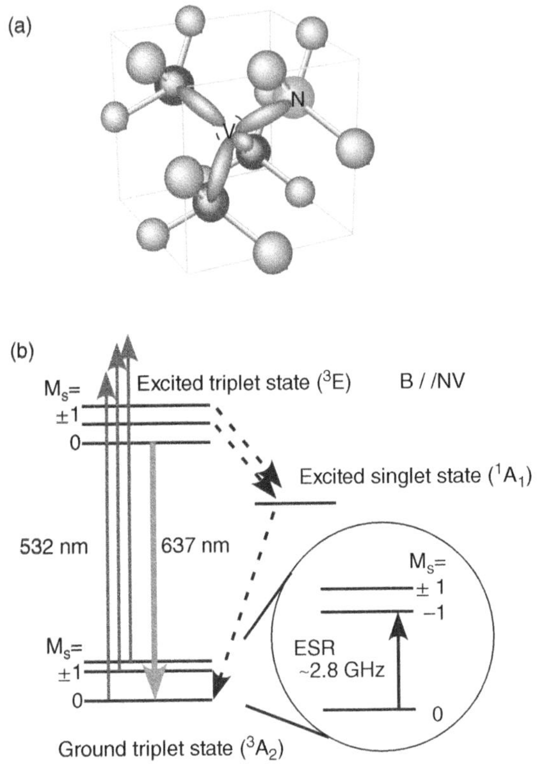

Figure 12.1 (a) Atomic structure of NV center. (b) Energy levels of NV center

resonance (ODMR). In 2004, Rabi oscillation of a single electron spin and nuclear spin was observed [6, 7]. After these pioneering studies, a lot of important research was done on the manipulation of single electron and nuclear spin. NV centres are thus promising candidates as solid-state room-temperature quantum bits and as quantum sensors with various applications in quantum information [8–12], magnetometry [13–18] and biosensing [19–21]. Recently, purely electrical operations yielded the first room-temperature single-photon source [10, 22]. In particular, the important topics are the isotopical engineering of diamond, with its extended coherence time T_2, the decoherence mechanisms and associated applications for magnetometry. Significant research has shown the excellent spin characteristics of diamond compared with other solid-state systems. From this viewpoint, we summarize our previous research and review on these topics.

12.2 Longitudinal Electron-Spin Relaxation (T_1) of NV Centre

To begin, we introduce the longitudinal electron-spin relaxation because, in principle, its time constant T_1 must be longer than the coherence time T_2. In longitudinal ESR, the magnetic quantum number M_S changes. For example, for a NV centre, T_1 is the relaxation time for $M_S = \pm 1$ to change to $M_S = 0$. In the solid state, longitudinal ESR is usually caused by absorption or stimulated emission of phonons, so the longitudinal ESR is usually called a spin lattice relaxation. Therefore, T_1 strongly depends on temperature.

As far as we know, the longest T_1 measured for a single NV centre at room temperature is 6 ms [12]. For low temperature, T_1 measured for an ensemble of NV centres is reported to be 199 s at 10 K [23] and 265 s at 1.9 K [24].

Measurements of the temperature dependence of T_1 revealed that T_1 is predominantly determined by a Raman process at room temperature [23]. Below room temperature, it is reported that Orbach-type processes dominate. At yet lower temperatures, sample-dependent cross-relaxation processes dominate, resulting in a temperature-independent T_1 which can range from milliseconds to minutes [23].

12.3 Coherence Time (T_2) of NV Centre in Diamond with Natural Abundance of ^{13}C

In the last decade, the coherence time T_2 of NV centres was significantly increased. In 1989, the longest T_2 of an ensemble of NV centres was 80 μs at 1.4 K. In 2003, T_2 of an ensemble NV centres was reported to be 58 μs at room temperature [25]. In 2006, T_2 for a single NV centre was reported to be 242 μs [26] and 350 μs [27].

In N-rich type-Ib diamond, it was shown in 2008 that T_2 of an ensemble of NV centres was dominated by the dipolar interaction among electron spins of nitrogen (P1 centres) and NV centres [28]. The diamond studied was made by a high-temperature high-pressure technique, where the concentration of high-density nitrogen was $10^{19}-10^{20}$ cm^{-3} in the type-Ib diamond. It was demonstrated that the decoherence of NV-centre spin by the electron-spin bath can be eliminated by polarizing the electron-spin bath, which in turn is done by lowering the temperature to 1.3 K and by using a high-frequency magnetic field (240 GHz and 8 T) [28]. It was shown that below 2 K, T_2 saturates at ~250 μs, where the polarization of the electron-spin bath exceeds 99%. This result shows that T_2 is greatly increased when the spin–spin interaction with the electron-spin bath was eliminated. Therefore, T_2 can be much longer if the paramagnetic impurities and defects are removed.

For high-quality diamond, where the nitrogen concentration is less than 10^{12} to 10^{14} cm^{-3}, it was revealed that T_2 is longer than 250 μs, even at room temperature. Furthermore, in 2009, T_2 for a single NV centre was reported to be 0.65 ms which, to the best of our knowledge, is the longest room-temperature T_2 for diamond with a natural abundance of ^{13}C (1.1%) [29]. The longer T_2 is attributed to the removal of nitrogen, other impurities and defects by the synthesis technique of chemical vapour deposition (CVD) [30].

Given that $T_2 = 0.65$ ms is the longest coherence time available for high-quality diamond, where impurities and defects are removed, we now discuss the dominant mechanism which determines T_2. In the next section, we see how the ^{13}C concentration depends on the spin-relaxation times [29, 31].

12.4 T_2^* Free-Induction Decay Time

In addition to the effect of the nuclear-spin bath on individual ^{13}C spins, the static interaction between single NV electron spins and the environment for different ^{13}C concentrations

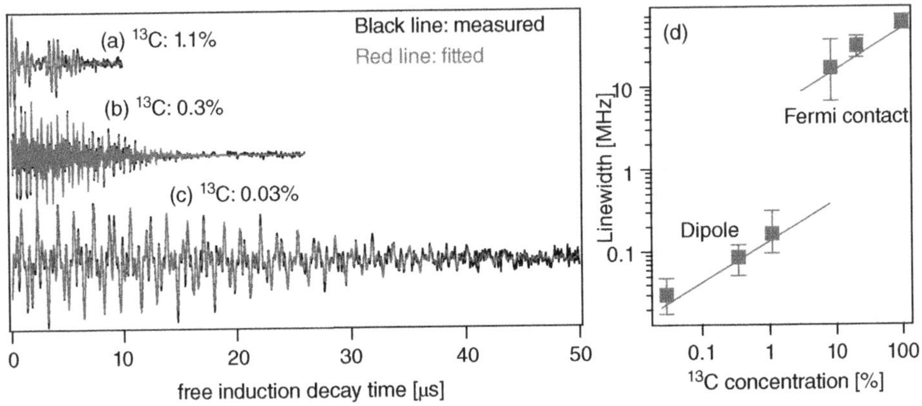

Figure 12.2 *Free induction decays of single NV in (a) 1.1%, (b) 0.3%, (c) 0.03% ^{13}C diamond. (d) Inhomogeneous linewidth as a function of ^{13}C concentration. Error bars indicate the distributions measured. The details of the simulated lines are shown in the text*

was investigated by measuring T_2^* (i.e., the inhomogeneous ESR linewidth). It was observed that with decreasing ^{13}C concentration, T_2^* increases and the linewidth narrows, as shown in Figure 12.2. In diamond with 0.03% ^{13}C, an extremely long T_2^* of 30 μs was found. In the region of low ^{13}C concentration (≤1.1%), the linewidth W (full width at half maximum) is derived from T_2^* by

$$W = \frac{2\sqrt{ln\ 2}}{\pi T_2^*}. \tag{12.1}$$

In the region of high ^{13}C concentration (>1.1%), the linewidth is obtained by fitting the ESR line of a single NV centre with a Gaussian lineshape. Average linewidths are plotted in Figure 12.2.

A likely cause for the inhomogeneous ESR linewidth is the hyperfine coupling constant (HFC) to ^{13}C nuclear spins. For example, in Si, the dependence of the inhomogeneous linewidth of P donors in ^{29}Si is well fit by an isotropic HFC (a_l) because of the Fermi-contact interaction with ^{29}Si nuclear spins with a concentration f [32, 33]:

$$W = 2\sqrt{2\ ln\ 2}\left[f\sum_l\left(\frac{a_l}{2}\right)^2\right]^{1/2}. \tag{12.2}$$

The sum runs over all nuclear-spin sites l. In Figure 12.2, the solid line for high ^{13}C concentrations is calculated by summing over all nine sites in the third shell with $a_l = 14$ MHz (see above for assignment of sites and HFC constants). Note that contributions from ^{13}C in the first shell were not considered in the linewidth calculations because they contribute to an observable splitting but not to the linewidth. As seen from Figure 12.2, it fits the experimental results well for high ^{13}C concentration.

For lower ^{13}C concentration, experimental data deviate from this behaviour owing to the probability that any ^{13}C located close to a NV centre decreases upon reducing the ^{13}C concentration. Furthermore, the density of unpaired electron spins rapidly decreases with distance from the three dangling bonds around the vacancy. This is known from the HFC parameters [34, 35], indicating that almost 100% spin density is localized on the C sites in the first and third shell. Therefore, in such a situation, the most prominent contribution to the inhomogeneous linewidth is the weaker dipole–dipole interaction between electron spin and ^{13}C nuclear spin at distant sites. The lower line in Figure 12.2 is the linewidth

$$W = \sqrt{\left(\frac{\mu_0 \mu_e \mu_n g_e g_n}{4\pi h}\right)^2 (3.195 \times 10^{46}) n}, \tag{12.3}$$

calculated from the second moment [36] with more than 3000 lattice sites for each ^{13}C concentration (n). Contributions from ^{13}C in the first and second shell are not considered. As seen from Figure 12.2, W fits the experimental results in the low ^{13}C concentration (<1%) quite well. At low ^{13}C concentration, the linewidth is dominated by dipole–dipole interactions.

Recently, the longest T_2^* reported was more than 30 and 470 μs in diamond with a natural abundance of ^{13}C [37] and in ^{12}C-enriched diamond [9], respectively, under μ-metal shielded environments. The ^{12}C concentration of the latter sample is reported to be 99.99% [9]. In Ref. [37], the dependence of T_2^* of NV centres on the axial component of the magnetic field was reported. The coherence time T_2^* clearly attained a maximum when the applied axial magnetic field was reduced to zero. The results for T_2^* deviate from that predicted by Equation (12.3) mainly owing to the μ-metal shielded environment.

12.5 Coherence Time T_2 of Electron and Nuclear Spin in ^{12}C-Enriched Diamond

The dephasing time T_2 of the electron spin is measured by two-pulse Hahn echo-decay curves (see Figure 12.3). We analysed T_2 for diamond made by CVD and excluded diamond with 0.03% ^{13}C concentration. In 1.1% diamond, we found $T_2 = 0.65$ ms, which is the longest T_2 measured so far for diamond with a natural abundance of ^{13}C. For the lower ^{13}C concentration of 0.3%, an even longer T_2 of 1.8 ms was reported [31]. This is the longest T_2 at room temperature. T_2 is found to be inversely proportional to the ^{13}C concentration, as shown in Figure 12.3(d). This benchmark may allow the observation of coherent coupling between spins separated by a few tens of nanometres. Single electron spins in the same isotopically engineered CVD diamond have been used to detect external magnetic fields with a sensitivity reaching 4 nT Hz$^{-1/2}$ and with sub-nanometre spatial resolution.

A theoretical analysis of T_2 by the disjoint-cluster approach [38] proposes the relationship $T_2 \sim (\bar{C}A_c)^{-1/2}$, where $\bar{C}$ is the averaged nuclear–nuclear dipole interaction in the bath and Ac is a characteristic value for the electron–nuclear dipole interaction. Since both interactions scale linearly with ^{13}C concentration n, T_2 decreases approximately as $1/n$ in this model. The fit to the data shown in Figure 12.3 supports this inverse proportionality and fits our data. Note that our data are also fit by the values calculated in Ref. [38] within 30% deviation.

The above results indicate that even if the ^{13}C concentration increases to about 10%, T_2 should be about 120 μs, which thus indicates that the threshold (~10^4 operation) for quantum-error-correction schemes [39] can be exceeded even in ^{13}C-enriched diamond at room temperature with typical single-qubit flips of several ns. As for the gate-operation

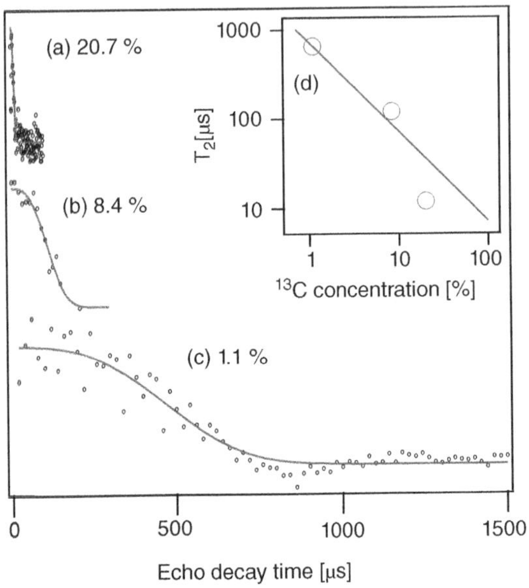

Figure 12.3 *Echo decays of electron spin in (a) 20.7%, (b) 8.4%, and (c) 1.1%* ^{13}C *diamond. The MW pulse sequence is* $\pi/2 - \tau - \pi - \tau - \pi/2$ *where* τ *is interpulse delay. Red lines are fits to* $\exp[-(t/T_2)^3]$. *(d)* T_2 *over* ^{13}C *concentration n. The solid line is fitted with a* $1/n$ *dependence*

speed for electron spin, a π pulse is realized with a time of less than 1 ns in the 'strong-driving' regime [40] and of about 3.5 ns by using strip-line wires [41].

The coherence time T_2 for nuclear spin is potentially much longer than that for electron spin. However, T_2 for nuclear spin states is measured to be around ~6 ms because it is limited by T1 for electron spin (~6 ms) [12]. Recently, T_2 for nuclear spin was significantly extended to over 1 s by decoupling the single nuclear spin from its local environment [9]. In experiments, the decoupling was done by irradiating with a high-power laser (532 nm, > 10 mW). The report shows that the decoupling is caused by rapid ionization and deionization, the rate of which is proportional to the laser intensity. When these transition rates exceed the hyperfine coupling strength, the interaction between the nuclear and electronic spins is strongly suppressed, owing to a phenomenon analogous to motional averaging. In the experiment, the use of even higher laser intensities is limited by heating of the diamond sample, which causes drifts in the ESR transition. In the analysis, further improvement can be achieved by decreasing the hyperfine and nuclear-nuclear interaction strengths by reducing the ^{13}C concentration, potentially resulting in hour-long storage times.

12.6 Spin and Optical Properties of NV Centres Close to Surface

Engineering surface-proximate NV centres with long spin coherence times is of significant interest because a long T_2 enables the detection of weak coupling between single NV centres and other quantum degrees of freedom, which would be useful for a wide variety of applications such as nanoscale probes, including magnetometry, magnetic imaging and electric-field sensing. However, the resulting implanted-spin-coherence times are generally short and widely dispersed, even in isotopically purified substrates, presumably because of crystal damage intrinsic to the impact of accelerated ions. Therefore, engineering surface-proximate NV centres with long spin-coherence times remains a challenge.

Nanometre-precision depth control of NV-centre creation near the surface of synthetic diamond was demonstrated by using an *in-situ* nitrogen delta-doping technique during plasma-enhanced CVD. Despite their proximity to the surface, doped NV centres with depths d ranging from 5 to 100 nm display long spin-coherence times ($T_2 > 100\,\mu s$ at $d = 5$ nm and $T_2 > 600\,\mu s$ at $d = 5$ nm) [42]. NV centres shallow implanted by epitaxial overgrowth result in an increase of the coherence times by up to an order of magnitude ($T_2 = 250\,\mu s$), and ms decoherence times were realized by using dynamical decoupling [43]. For isotopically purified 12C-single-crystal diamond layers, where NV centres were positioned within 100 nm of the surface, enhanced spin dephasing times are reported ($T_2^* \approx 90\,\mu s, T_2 \approx 1.7$ ms).

The spin and optical properties of NV centres within 1 to 10 nm were systematically investigated by Ofori-Okai *et al.* [44]. They measured the photoluminescence intensity both as a function of energy and ion fluence and determined the relative concentrations of NV^- and NV^0. They observed a monotonic decrease in NV^- in a very shallow region and deduced the presence of a depth threshold below which the negative-charge state becomes unstable. Furthermore, they found a small but systematic ODMR spectral broadening for defects shallower than about 2 nm. This broadening is consistent with the presence of a surface paramagnetic impurity layer [45] largely decoupled by motional averaging.

12.7 Magnetometry

For applications in magnetic sensing, NV centres are of significant interest because of their ultralong coherence time. Nanoscale imaging magnetometry with diamond spins was demonstrated by two groups [46, 47]. They detected 3 nT ac magnetic fields at kHz frequencies after 100 s of averaging and a sensitivity of $0.5\,\mu T\,Hz^{-1/2}$ for a diamond nanocrystal with a diameter of 30 nm [46]. It was shown that single electron spins in isotopically engineered CVD diamond can be used to detect external magnetic fields with a sensitivity reaching $4\,nT\,Hz^{-1/2}$ and sub-nanometre spatial resolution [31].

For magnetometry in the high-density regime, high-density n improves the sensitivity to fields because the projection noise per unit volume decreases as $1/\sqrt{n}$ [48, 49]. The performance of a magnetometer based on electronic spins is, in principle, limited by the spin-projection noise. The minimum-detectable magnetic field δB is

$$\delta B \cong \frac{1}{g\mu}\frac{1}{R\sqrt{\eta}}\frac{1}{\sqrt{NtT_2^*}}, \tag{12.4}$$

where R is the measurement contrast, η is the detection efficiency, N is the number of spin centres and t is the integration time. From this equation, a shot-noise-limited sensitivity of less than $1\,pT/Hz^{-1/2}$ is estimated theoretically under ideal conditions.

One of the promising applications of magnetic-field sensing by NV centres is nuclear magnetic resonance (NMR) spectroscopy of nanoscale samples. Much effort has been directed to this goal and NMR detection of a (four nanometre)^3 voxel of protons has been achieved with magnetic resonance force microscopy – a challenging experimental technique that requires ultralow temperatures in vacuum. By using NV centres, NMR signals from a (five nanometre)^3 voxel of various fluid and solid organic samples were detected under ambient conditions. NV centres were embedded approximately 7 nm under the surface of a bulk diamond to record NMR spectra of various samples placed on the diamond surface. Its detection volume consisted of only 104 nuclear spins [50].

12.8 Summary

We reviewed recent research into the dependence on ^{13}C concentration of the coherence times T_2^* and T_2 of the electron spin of NV centres in diamond. The results point toward ^{13}C nuclei as being the main cause for dephasing in otherwise clean diamond. The fit with the calculated T_2 is very important for elucidating the dephasing mechanism and for increasing T_2 for quantum information devices and ultraprecise magnetometry. In future, we consider that electrical control of qubits, such as carried by spin and photon [50–53], and role of the quantum interface between qubits [54–57] will become important. These topics will be reviewed elsewhere.

References

1. J. J. Pla, K. Y. Tan, J. P. Dehollain, *et al.*, *Nature* **489**, 541 (2012).
2. A. Morello, J. J. Pla, F. A. Zwanenburg, *et al.*, Single-shot readout of an electron spin in silicon, *Nature* **467**, 687 (2010).
3. T. Hayashi, T. Fujisawa, H. D. Cheong, *et al.*, Coherent manipulation of electronic states in a double quantum dot, *Phys. Rev. Lett.* **91**, 226804 (2003).
4. T. Fujisawa, T. Hayashi, R. Tomita, and Y. Hirayama, Bidirectional counting of single electrons, *Science* **312**, 1634 (2006).
5. A. Gruber, A. Dräbenstedt, C. Tietz, *et al.*, *Science,* **276**, 2012 (1997).
6. F. Jelezko, T. Gaebel, I. Popa, *et al.*, *Phys. Rev. Lett.*, **92**, 076401 (2004).
7. F. Jelezko, T. Gaebel, I. Popa, *et al.*, *Phys. Rev. Lett.*, **93**, 130501 (2006).
8. H. Bernien, B. Hensen, W. Pfaff, *et al.*, *Nature* **497**, 86 (2013).
9. P. C. Maurer, G. Kucsko, C. Latta, *et al.*, *Science* **336**, 1283 (2012).
10. N. Mizuochi, T. Makino, H. Kato, *et al.*, *Nat. Photon.* **6**, 299 (2012).
11. T. M. Babinec, B. J. M. Hausmann, M. Khan, *et al.*, *Nat. Nanotechnol.* **5**, 195 (2010).
12. P. Neumann, N. Mizuochi, F. Rempp, *et al.*, *Science* **320**, 1326 (2008).
13. J. R. Maze, P. L. Stanwix, J. S. Hodges, *et al.*, *Nature* **455**, 644 (2008).
14. G. Balasubramanian, I. Y. Chan, R. Kolesov, *et al.*, *Nature* **455**, 648 (2008).
15. P. Maletinsky, S. Hong, M. S. Grinolds, *et al.*, *Nat. Nanotechnol.* **7**, 320 (2012).
16. V. M. Acosta, E. Bauch, M. Ledbetter, *et al.*, *Phys. Rev. B* **80**, 115202 (2009).
17. H. J. Mamin, M. Kim, M. H. Sherwood, *et al.*, *Science,* **339**, 557 (2013).
18. T. Staudacher, F. Shi, S. Pezzagna, *et al.*, *Science,* **339**, 561 (2013).

19. L. T. Hall, C. D. Hill, J. H. Cole, *et al.*, *PNAS* **107**, 18777 (2010).
20. L. P. McGuinness, Y. Yan, A. Stacey, *et al.*, *Nat. Nanotechnol.* **6**, 358 (2011).
21. D. Le Sage, K. Arai, D. R. Glenn, *et al.*, *Nature* **496**, 486 (2013).
22. A. Lohrmann, S. Pezzagna, I. Dobrinets, *et al.*, *Appl. Phys. Lett.* **99**, 251106 (2011).
23. A. Jarmola, V. M. Acosta, K. Jensen, *et al.*, *Phys. Rev. Lett*, **108**, 197601 (2012).
24. J. Harrison, M. J. Sellars, and N. B. Manson, *Diam. Relat. Mater.* **15**, 586 (2006).
25. T. A. Kennedy, J. S. Colton, J. E. Butler, *et al.*, *Appl. Phys. Lett.* **83**, 4190 (2003).
26. L. Childress, M.V.G. Dutt, J. M. Taylor, *et al.*, *Science,* **314**, 281 (2006).
27. T. Gaebel, M. Domhan, I. Popa, *et al.*, *Nat. Phys.*, **2**, 408 (2006).
28. S. Takahashi, R. Hanson, J. van Tol, *et al.*, *Phys. Rev. Lett*, **101**, 047601 (2008).
29. N. Mizuochi, P. Neumann, F. Rempp, *et al.*, *Phys. Rev. B* **80**, 041201(R) (2009).
30. J. Isberg, J. Hammersberg, E. Johansson, *et al.*, *Science,* **297**, 1670 (2002).
31. G. Balasubramanian, P. Neumann, D. Twitchen, *et al.*, *Nat. Mater.*, **8**, 383 (2009).
32. W. Kohn, in *Solid State Physics* (Academic Press, New York, 1957), vol. **5**, p. 257.
33. E. Abe, A. Fujimoto, J. Isoya, *et al.*, *Phys. Rev. B* **82**, 121201 (2010).
34. J. H. N. Loubser and J. A. van Wyk, *Diamond Res.* 11–14 (1977).
35. A. Gali, Maria Fyta, and Efthimios Kaxiras, *Phys. Rev. B*, **77**, 155206 (2008).
36. J. H. Van Vleck, *Phys. Rev.* **74**, 1168 (1948).
37. F. Dolde1, H. Fedder, M. W. Doherty, *et al.*, *Nat. Phys.*, **7**, 459 (2011).
38. J. R. Maze, J. M. Taylor, and M. D. Lukin, *Phys. Rev. B*, **78**, 094303 (2008).
39. D. D. Awshalom, D. Loss, and N. Samarth, *Semiconductor Spintronics and Quantum Computation* (Spring-Verlag, Berlin, 2002).
40. G. D. Fuchs, V. V. Dobrovitski, D. M. Toyli, *et al.*, *Science*, **326**, 1520 (2009).
41. J. Wrachtrup, F. Jelezko, *J. Phys. Conden. Mat.* **18**. SD807 (2006).
42. K. Ohno, F. J. Heremans, L. C. Bassett, *et al.*, *Appl. Phys. Lett.*, **101**, 082413 (2012).
43. T. Staudacher, F. Ziem, L. Häussler, *et al.*, *Appl. Phys. Lett.*, **101**, 212401 (2012).
44. B. K. Ofori-Okai, S. Pezzagna, K. Chang, *et al.*, *Phys. Rev. B* **86**, 081406(R) (2012).
45. J. Tisler, G. Balasubramanian, B. Naydenov, *et al.*, *ACS Nano*, **3**, 1959 (2009).
46. J. R. Maze, P. L. Stanwix, J. S. Hodges, *et al.*, *Nature* **455**, 644 (2008).
47. G. Balasubramanian, I. Y. Chan, R. Kolesov, *et al.*, *Nature* **455**, 648 (2008).
48. V. M. Acosta, E. Bauch, M. P. Ledbetter, *et al.*, *Phys. Rev. B* **80**, 115202 (2009).
49. J. M. Taylor, P. Cappellaro, L. Childress, *et al.*, *Nat. Phys.*, **4**, 810 (2008).
50. T. Staudacher, F. Shi, S. Pezzagna, *et al.*, *Science,* **339**, 561 (2013).
51. N. Mizuochi, T. Makino, H. Kato, *et al.*, *Nat. Photon.*, **6**, 299–303 (2012).
52. A. Lohrmann, S. Pezzagna, I. Dobrinets, *et al.*, *Appl. Phys. Lett.* **99**, 251106 (2011).
53. Y. Doi, T. Makino, H. Kato, *et al.*, *Phys. Rev. X.* accepted.
54. X. Zhu, S. Saito, A. Kemp, *et al.*, *Nature:* **478**, 221 (2011).
55. Y. Kubo, F. R. Ong, P. Bertet, *et al.*, *Phys. Rev. Lett.*, **105**, 140502 (2010).
56. H. Bernien, B. Hensen, W. Pfaff, *et al.*, *Nature* **497**, 86 (2013).
57. E. Togan, Y. Chu, A. S. Trifonov, *et al.*, *Nature,* **466**, 730 (2010).

13

Ultrafast Light-Induced Spin Reversal in Amorphous Rare Earth-Transition Metal Alloy Films

Arata Tsukamoto[1] *and Theo Rasing*[2]
[1]*Department of Electronic Engineering, College of Science and Technology, Nihon University, Japan*
[2]*Institute for Molecules and Materials, Radboud University Nijmegen, the Netherlands*

13.1 Introduction

The speed limits for magnetization reversal are of vital importance not only for storage media but also for spintronic devices. Conventional pure magnetic recording schemes have a serious and unavoidable problem known as the ferromagnetic resonance (FMR) limit. For ultrafast manipulation of magnetization, optical laser pulses could serve as an alternative stimulus to trigger magnetization reversal. An ultrashort laser pulse allows excitation of magnetic systems at time scales much shorter than fundamental quantities such as spin precession or spin-lattice relaxation times. In particular, the laser excitation brings the magnetic medium into a strong nonequilibrium state [1], where a conventional description of magnetic phenomena in terms of equilibrium thermodynamics and adiabatic approximations is no longer valid. Consequently ultrafast laser-induced magnetization dynamics is a new and rather unexplored topic at the frontier of modern magnetism [2].

In this chapter, we summarize the results of our recent experimental studies of metallic multisublattice magnets. In particular, we focus on magnetization dynamics triggered by an ultrashort laser pulse in ferrimagnetic amorphous rare earth-transition metal alloys. The inequivalency of the magnetic sublattices, on the one hand, and a fine balance of their angular momenta on the other, lead to a very peculiar dynamic behavior. This becomes

Spintronics for Next Generation Innovative Devices, First Edition. Edited by Katsuaki Sato and Eiji Saitoh.
© 2015 John Wiley & Sons, Ltd. Published 2015 by John Wiley & Sons, Ltd.

particularly obvious at short time scales, such as the appearance of a ferromagnetic-like state at time scales below a few picoseconds. This state is followed by an intersublattice relaxation of the angular momentum, leading to a deterministic switching of the magnetization driven by ultrafast laser-induced heating. The role of the light helicity in this process is clarified as well.

13.2 Control of Magnetization Dynamics with Precessional Motion

13.2.1 Magnetization Reversal Time

The interactions of magnetic moments with magnetic fields are basic to the understanding of all magnetic phenomena. According to quantum theory, the magnetic moment $\boldsymbol{m}$ associated with an angular momentum $\boldsymbol{L}$ via the gyromagnetic ratio $\gamma = \boldsymbol{m}/\boldsymbol{L}$. The dynamics of the magnetic moments is described by a simple mechanical relation between the change of angular momentum and the torque applied to the system. The law of the conservation of angular momentum is therefore an important issue to be considered when sudden changes of the magnetization are applied. The precessional motion of a magnetic moment is well described by a torque equation with a dissipative term proportional to the generalized velocity, $(-d\boldsymbol{m}/dt)$. This gives the Landau–Lifshitz–Gilbert (LLG) equation of motion:

$$\frac{d\boldsymbol{m}}{dt} = -|\gamma|(\boldsymbol{m} \times \boldsymbol{H}_{\text{eff}}) + \frac{\alpha}{|m|}\left(\boldsymbol{m} \times \frac{d\boldsymbol{m}}{dt}\right), \tag{13.1}$$

where α and H_{eff} are the dimensionless phenomenological Gilbert damping parameter and the effective magnetic field, respectively. If the spins experience not only the action of the external magnetic field, but are also affected by the magnetocrystalline anisotropy, shape anisotropy, magnetic dipole interaction, etc., the combined action of all these contributions can be considered as an effective magnetic field. It is well known that the magnetization reversal time τ in ferromagnetic material can be calculated based on LLG equation by

$$\tau = \frac{1}{\omega \cdot \alpha} = \frac{1+\alpha^2}{\alpha}\frac{1}{|\gamma|}\frac{1}{H_{\text{ext}}}, \tag{13.2}$$

where ω and H_{ext} are the precessional frequency and the external magnetic field, respectively. Although this formula has been derived from simple assumptions: (a) coherent magnetization rotation (b) constant magnetic field (c) without magnetic anisotropy, it indicates that the τ is governed by α and γ. Essentially, those values are intrinsic properties of ferromagnetic materials. In ferrimagnetic materials, we can characterize those values effectively under some conditions as mentioned below.

13.2.2 Angular Momentum Compensation

Heavy rare earth and 3D transition metal (RE-TM) ferrimagnetic compounds such as GdFeCo have antiparelly coupled sublattice moments. The net magnetic moment and net angular momentum can be continuously tuned, for example, by changing the composition ratio in the amorphous system, which forms a unique playground to study the role of the angular momentum. Depending on their composition, RE-TM ferrimagnets can exhibit a magnetic compensation composition ratio C_M where the magnetizations of the RE and TM sublattice moments cancel each other. Similarly, they exhibit an angular momentum compensation composition ratio C_A where the net angular momentum of the sublattices vanishes; however, the net magnetization remains. It implies that effectively the gamma value drastically increases near C_A. In a ferrimagnetic system, the LLG equation should be written for each ith sublattice (i = RE, TM). These equations are coupled by the presence of the intersublattice exchange fields. By combining the LLG equations of each sublattices, a macro spin model of ferrimagnetic magnetization dynamics can actually be described by a single LLG equation with an effective gyromagnetic ratio $\gamma_{\rm eff}$ and an effective Gilbert damping parameter $\alpha_{\rm eff}$ as follows.

$$\gamma_{\rm eff}(x) = \frac{M_{\rm RE}(x) - M_{\rm TM}(1-x)}{\frac{M_{\rm RE}(x)}{|\gamma_{\rm RE}|} - \frac{M_{\rm TM}(1-x)}{|\gamma_{\rm TM}|}} = \frac{M(x)}{A(x)} \tag{13.3}$$

$$\alpha_{\rm eff}(x) = \frac{\frac{\lambda_{\rm RE}}{|\gamma_{\rm RE}|^2} + \frac{\lambda_{\rm TM}}{|\gamma_{\rm TM}|^2}}{\frac{M_{\rm RE}(x)}{|\gamma_{\rm RE}|} - \frac{M_{\rm TM}(1-x)}{|\gamma_{\rm TM}|}} = \frac{A_0}{A(x)} \tag{13.4}$$

$$\omega_{\rm FMR} = 2\pi f_{\rm FMR} = \gamma_{\rm eff} H_{\rm eff}, \quad \boldsymbol{H}_{\rm eff} = \boldsymbol{H}_{\rm a} + \boldsymbol{H}_{\rm s} + \boldsymbol{H}_{\rm ext}, \tag{13.5}$$

where x is the RE composition ratio; $\lambda_{\rm RE}$ and $\lambda_{\rm TM}$ are the Landau–Lifshitz damping parameters for the RE and TM sublattices, respectively; $M(x)$ and $A(x)$ are the net magnetic moment and net angular momentum, respectively; A_0 is a constant under the assumption that the Landau–Lifshitz damping parameters are independent of composition; $\boldsymbol{H}_{\rm a}$ is anisotropy field, $\boldsymbol{H}_{\rm s}$ is demagnetization field, $\boldsymbol{H}_{\rm ext}$ is the external applied field, respectively. Except for $\boldsymbol{H}_{\rm ext}$, all other contributions will be material dependent. Thus, Equations (13.3) and (13.4) indicate a divergence of both the precession frequency $\omega_{\rm FMR}$ and Gilbert damping parameter of the ferromagnetic resonance (FMR) mode at C_A in the case of $\gamma_{\rm RE} \neq \gamma_{\rm TM}$. Such combination of a high frequency and large damping of the spin precession would provide a means for ultrafast and ringing-free magnetization reversal via precessional motion.

13.2.3 Compositional Dependence of Magnetization Dynamics

Here, we describe a time-resolved study of the composition dependency of ultrafast magnetic response in ferrimagnetic GdFeCo amorphous alloy film [3]. As an advantage of amorphous alloy systems, we can tune the composition ratio continuously and can make C_A appearing at room temperature. SiN (60 nm)/$Gd_x(Fe_{87.5}Co_{12.5})_{100-x}$ (20 nm)/SiN (5 nm)/$Al_{90}Ti_{10}$ (10 nm)/glass substrate was prepared by using a magnetron sputtering system. For the study, we used an all-optical pump and probe method in which an intense (pump) light beam excited a medium due to ultrafast laser (full width half maximum: 90 fs) heating and a less intense (probe) beam monitored this photo-excited state through the magneto-optical Faraday effect. In this way, response times and damping can directly be measured in the time domain with femtosecond resolution. Then, the effective Gilbert damping factor $\alpha_{\rm eff}$ was determined from a direct fitting of the measured results with model calculations.

The ultrafast magnetic response of 20 nm thick GdFeCo films showed precession frequencies of several tens of GHz and relaxation times in the sub nanosecond range under

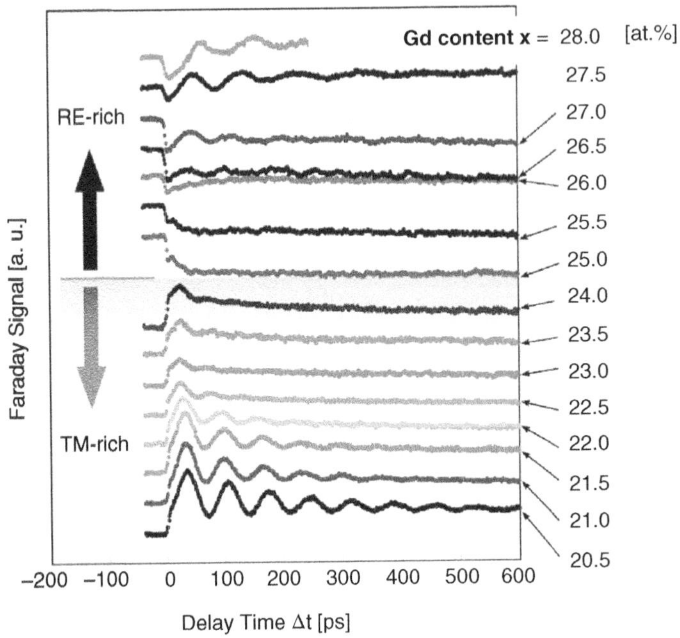

Figure 13.1 *Compositional dependence of magnetization dynamics of $Gd_x(Fe_{87.5}Co_{12.5})_{100-x}$ films measured at room temperature. Applied external field is 0.42 T*

various external magnetic fields. It was determined by eliminating the nonoscillating term caused by the thermal effect from the time-domain measured results. The precession frequency f was determined by nonlinear fitting of the measured result with a damped harmonic equation. The experimental curves were very well fitted by the LLG equation. The dynamic magnetic response of GdFeCo showed strong pump fluence and applied field dependence. Figure 13.1 shows the compositional dependence ($Gd_x(Fe_{87.5}Co_{12.5})_{100-x}$: $x = 20.5–28.0$ at%) of magnetization dynamics measured at room temperature. The compositional dependence of the f and α_{eff}, were derived from those results as shown in Figure 13.2(a) Clearly f and α_{eff} were affected by the composition ratio. Note that only the magnetization of the transition metal subsystem is probed by the linear Faraday effect at laser wavelength of 800 nm. It makes possible to measure the magnetic dynamics near C_M. Figure 13.1 shows that the phase of the oscillations changes around Gd content $x = 24.3$ at%. This happens because of the presence of the magnetization compensation composition C_M. For $x < C_M$ the Gd magnetic moment M_{RE} is smaller than that of TM. Thus the M_{TM} is aligned along the applied field direction. At $x > C_M$, M_{RE} becomes larger than M_{TM} and the ferrimagnetic system flips, allowing the RE component to be aligned along the direction of the applied field H_{ext}. It is clear that this will lead to a sign change of the observed component of the magnetization, M_{TM}. At $x \sim 23.5$ at% a significant increase is observed in both α_{eff} and ω_{FMR}. As expected from Equations (13.3) and (13.4), the fact that both α_{eff} and f peak at the same composition, clearly indicates the existence of angular momentum compensation near this composition of $x \sim 23.5$ at%. The strong compositional dependence of γ_{eff} demonstrates the nonequivalent character of the gyromagnetic ratios

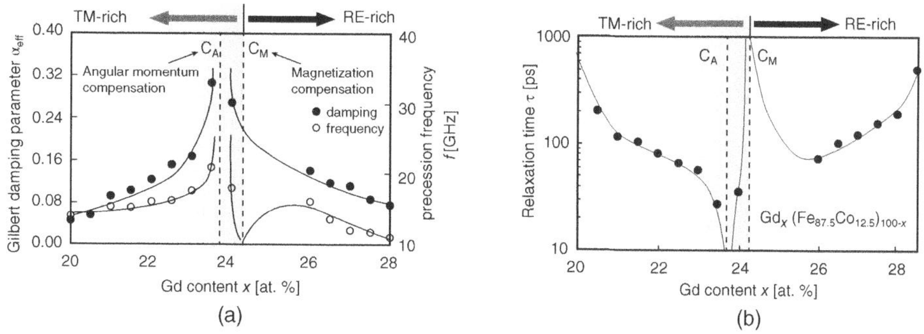

Figure 13.2 *Compositional dependence of (a) effective Gilbert damping factor α_{eff} and precession frequency f of $Gd_x(Fe_{87.5}Co_{12.5})_{100-x}$ derived from the experimental results shown in Figure 13.1 (b) relaxation time τ derived from the results in (a) with Equation (13.2). The dotted lines are a qualitative representation of the expected trend from Equations (13.3) and (13.4)*

of the two magnetic sublattices in GdFeCo. This inequivalence also leads to a difference between the composition of magnetization compensation C_M and angular momentum compensation C_A.

The observed high-speed and strongly damped magnetization dynamics (f: 21 GHz, α_{eff}: 0.31) near C_A is expected as an ideal situation for an ultrafast ringing-free precessional switching that can be triggered with short pulse laser irradiation. This value of α_{eff} is much higher than that of conventional magnetic recording media CoCrPt ($\alpha_{eff} = 0.038$[4]). Figure 13.2(b) shows the compositional dependency of relaxation time τ derived by Equation (13.2) with experimental values of Figure 13.2(a). The minimum value of τ was near C_A, and the value of τ varied more than 10 times in this composition range. This result suggests magnetization reversal time can be controlled vastly by the RE-TM composition ratio.

Similar results are obtained for the temperature dependence of M_{RE} and M_{TM}. We reported the dynamic behavior of ferrimagnetic GdFeCo, when the temperature approaches the angular momentum compensation point T_A, both γ_{eff} and α_{eff} increase significantly [5]. A small difference between the gyromagnetic ratio of the two sublattices leads to T_A slightly (typically ~50K) above magnetization compensation point T_M.

13.2.4 Precessional Switching by Ultrashort Pulse Laser

The fastest conventional way to reverse magnetization is based on a precessional motion when a magnetic field $\boldsymbol{H}$ is applied orthogonal to $\boldsymbol{m}$, so that the created torque $\boldsymbol{m} \times \boldsymbol{H}$ leads to a rapid change of the angular momentum and a possible switching of the magnetization. A realistic switching time is determined by the strength and duration of the magnetic field pulse. However, such precessional switching requires a magnetic field pulse precisely tuned to half of the precession period [6, 7].

High-speed and strongly damped precessional switching was triggered by the laser irradiation with ultrafast heating of a GdFeCo across its T_M toward T_A, under a static applied magnetic field. A dc magnetic field can be applied to the ferrimagnet parallel to the original

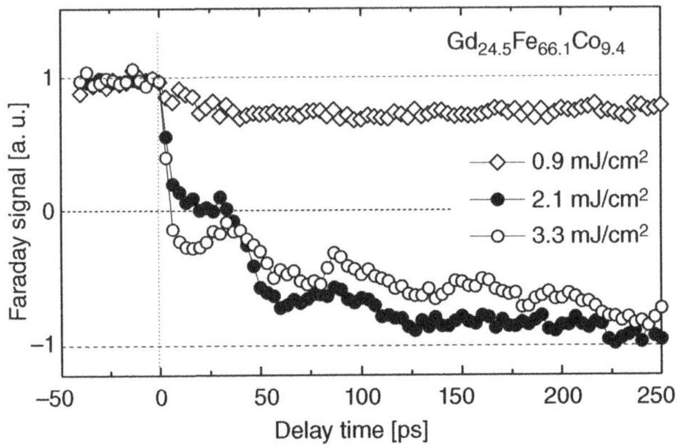

Figure 13.3 *Transient magnetization reversal dynamics of* $Gd_{24.5}Fe_{66.1}Co_{9.4}$ *(*T_M*: 345 K) measured with a pump fluences of 0.9, 2.1, and 3.3 mJ cm*$^{-2}$*. Applied external field is 0.42 T*

magnetization direction, at a temperature $T < T_M$. By crossing the T_M, ultrafast switching from parallel to antiparallel configuration between magnetic field and net magnetization allows excitation of magnetic systems at time scales much shorter than thermal cooling process. Therefore, one may expect, for the net magnetization, the above process effectively acts as an ultrafast magnetic field triggering the magnetization reversal. We have measured hysteresis loops at different time delays after the intense pump pulses [8]. It is clearly observed that in this field range, the measured Faraday signal changes its sign after about 0.7 ps. The sign change reflects the change of the TM sublattice direction toward the applied magnetic field, as the spin temperature of the ferrimagnetic system increases over T_M in the probed area. This observation unambiguously demonstrates that the magnetization reversal takes place on a subpicosecond time scale. To initiate and investigate the precessional switching, we have used an all-optical pump–probe technique employing an amplified Ti:Sapphire laser system with 90 fs pulses [3]. The dynamic magnetic response of $Gd_{24.5}Fe_{66.1}Co_{9.4}$ (T_M: 345 K) showed strong pump fluence and applied field dependence. In particular, following the laser excitation with low pump fluence (0.9 mJ cm^{-2}) the magnetic system relaxes back toward the initial state while the rather high pump fluence (3.3 mJ cm^{-2}) excitation induces a metastable opposite magnetic state as shown in Figure 13.3. After a sudden heating with 3.3 mJ cm^{-2} which causes just 30% reduction of the magnetization at the arrival of the pump pulse, after which the magnetization started to rotate across $M_z = 0$ within 6 ps during the first precession (around 30 GHz) and finished the high damped precessional motion within a few cycles into the opposite direction.

13.3 Ultrafast Distinct Dynamics of Sublattices and Transient Ferromagnetic State

Laser-induced heating of a magnetic medium in the presence of an external field is used in thermomagnetic writing to record a bit of information via magnetization reversal. Revealing the ultimate speed at which such a recording event can be realized is a fundamentally interesting issue with possible consequences for the future development of magnetic recording and information processing. As shown in Section 13.2.4, high-speed and strongly damped precessional switching was triggered by femtosecond laser irradiation, resulting in an ultrafast heating of a GdFeCo thin film across its T_M toward T_A, under a static applied magnetic field. And, the whole process of demagnetization and subsequent switching of a two-sublattice ferrimagnet can happen on a subpicosecond time scale, i.e., before the system approaches thermodynamic equilibrium. These experiments raised several questions. The most important one is actually why such a reversal occurs at all. First, looking at the data from a thermodynamical point of view, one may conclude that the reversal takes place just after the sample crosses the magnetization compensation point and approaches the angular momentum compensation by ultrafast heating. However, the laser-induced ultrafast demagnetization of ferromagnets, already demonstrated in 1996 [9] to occur at a subpicosecond time scale, is still a subject of hot debate [10]. Whether the angular momentum is dissipated into the lattice via phonons and defects [11, 12], or whether it is carried away by hot electrons [13] or the photons [14] – are still questioned at the forefront of ultrafast magnetism.

To study this transient regime of spin dynamics, one faces the challenge of how to probe the response of one magnetic sublattice relative to another. For this, we can exploit an element-specific detection technique like X-ray magnetic circular dichroism (XMCD) [15, 16]. This will allow us to "color" spins in the magnet and to probe the response of one sublattice relative to another. Second, as mentioned in Section 13.2.4, ultrafast heating of a ferrimagnet over its compensation point in an external magnetic field allows the initiation of the fastest spin reversal. For simplicity, in the following we focus on the behavior of the Gd and Fe moments and ignore the small percentage of Co in the TM sublattice. The XMCD measurements were performed in transmission for a fixed X-ray light helicity and opposite orientations of the external magnetic field of 0.5 T. The photon energy of the X-ray light was tuned at the FeL_3 and at the GdM_5 absorption edges in order to measure their magnetic response separately. In order to trigger ultrafast spin dynamics in this alloy, the reversal of the magnetizations of the two sublattices is initiated by ultrafast heating of the sample using a linearly polarized 60 fs laser pulse with photon energy of 1.55 eV. The measurements were performed with an incident laser fluence of 4.4 mJ cm^{-2} and a sample temperature of 83 K, which is well below $T_M = 250$ K. First, the net magnetizations of both sublattices rapidly decrease. However, whereas the net magnetization of Fe has collapsed within 300 fs, the demagnetization of Gd takes as long as 1.5 ps. Remarkably, in spite of the strong antiferromagnetic (AFM) exchange coupling between the RE and TM sublattices, they apparently lose their net magnetizations independently. We find a time constant $\tau_{Fe} = 100 \pm 25$ fs for the Fe and $\tau_{Gd} = 430 \pm 100$ fs for the Gd magnetic moment, characterizing the fast initial drops in the transient dichroic signal. Second, the magnetizations of both sublattices switch their directions by crossing the zero signal level and rebuilding their net magnetic moments. Up to 10 ps, Gd and Fe sublattices show distinctly different switching dynamics. Even more surprisingly, within the time scale between the zero crossings of the Fe and Gd moments (that is, between 300 fs and 1.5 ps), the net Fe and Gd moments are aligned parallel along the z-axis despite the AFM coupling of their spins in the ground state. This, together with the substantial laser-induced increase in temperature, indicates a rather strong transient parallel alignment of the Fe and Gd moments. Thus, we have entered an unexplored regime of magnetization dynamics, where two exchange-coupled magnetic sublattices are not in equilibrium with each other.

13.4 All Optical Magnetization Switching Phenomena with an Ultrashort Pulsed Laser

13.4.1 Ultrafast Heating as a Sufficient Stimulus for Magnetization Reversal in a Multisublattice Ferrimagnet

It has been unexpectedly found that the ultrafast laser-induced spin reversal in GdFeCo, where spins are coupled antiferromagnetically, occurs by way of a transient ferromagnetic-like state [17]. Such a novel strongly nonequilibrium spin dynamics may lead to yet unexplored magnetization reversal. This raises the question of what the actual magnetic susceptibility of this novel nonequilibrium state is and what is the minimum external magnetic field required to trigger magnetization reversal? We found that magnetization reversal could be achieved without any magnetic field, using an ultrafast thermal energy load alone [18]. Until now it has been generally assumed that heating alone, not represented as a vector at all, cannot result in a deterministic reversal of magnetization, although it may assist this process. We found deterministic magnetization reversal in a ferrimagnetic GdFeCo driven by an ultrafast heating of the medium resulting from the absorption of a subpicosecond laser pulse without the presence of a magnetic field. First, To verify the feasibility of such a hypothetical magnetization reversal scenario, atomistic scale simulations were used for modeling of laser-induced spin dynamics in a Heisenberg GdFe ferrimagnet [18]. The numerical results support the idea of purely heat-induced switching. Specifically, it is demonstrated that magnetization can be reversed after the application of femtosecond laser pulses that increase the temperature of the thermal bath very rapidly in the absence of an applied field. The rapid transfer of thermal energy into the spin system leads to switching of the magnetization within a few picoseconds. The simulations show that such a switching process proceeds with such impetus that even an opposing 40 T field is not able to prevent the magnetization from reversing on a short time scale.

Next, this novel heat-induced switching mechanism [18] was experimentally verified in patterned GdFeCo structures with both in-plane and out-of-plane magnetic anisotropy. The magnetization direction in these microstructures was measured with a photoemission electron microscope (PEEM) employing the XMCD effect. Starting from the initial state, we applied a series of linearly polarized femtosecond laser pulses. After the first single pulse, the contrast is simultaneously reversed in both structures, meaning that the magnetization has switched relative to the initial state. This magnetization reversal occurs after every subsequent single laser pulse. As typical experimental results in continuous GdFeCo, Figure 13.4 shows magneto-optical images of magnetic domains after single pulse laser irradiations. In case of a multidomain structure, the magnetic pattern was reversed. A further set of experiments shows that this switching occurs independently of polarization and initial state in thin films of GdFeCo. Importantly for technological applications, we experimentally show that this type of switching can occur when starting at room temperature.

From the theoretical discussion [19], the reversal happens because of the interplay of these different demagnetization rates with the exchange interaction coupling the sublattices. A general theoretical framework for ultrafast spin dynamics in multisublattice magnets was given in [19], which contains longitudinal relaxation of both relativistic

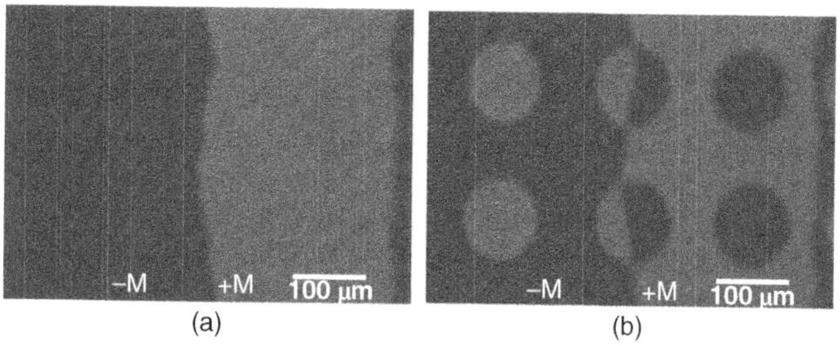

Figure 13.4 *Magneto-optical images of magnetic domains (a) before and (b) after single pulse laser irradiations. White and black areas correspond to up (+M) and down (−M) magnetic domains, respectively*

(spin-orbit) and exchange origin. The latter is the key new ingredient and is only present in magnets with more than one sublattice. The reason is simply that the exchange interaction conserves the total angular momentum and therefore longitudinal exchange relaxation in magnets with only one sublattice is not possible. From here one could thus conclude that the same reversal mechanism is valid for a much broader range of RE and TM concentrations, and should, e.g., work if the starting point is both below and above Tcomp. Confirmation of this can actually be found already in [20] where, even though interpreted differently, such switching was observed for samples of various compositions and at a variety of temperatures. This observation makes the observed phenomenon truly universal for a wide range of ferrimagnetic materials. The only question that is left is how this heat-driven reversal relates to the helicity-dependent process discussed in next section.

13.4.2 The Contribution of MCD in All-Optical Light Helicity-Dependent Magnetic Switching

Direct demonstration of all-optical light helicity-dependent magnetic switching (AO-HDS) was observed in ferrimagnetic GdFeCo alloys in the absence of an external magnetic field [21], which became subject of intense discussion in modern magnetism. A very straightforward experiment is performed by placing a GdFeCo sample under a polarizing microscope, where domains with magnetization "up" and "down" could be observed as white and black regions, respectively. To excite the material, amplified pulses from a Ti:sapphire laser were used at a wavelength of $\lambda = 800$ nm, repetition rate of 1 kHz and a pulse width of 40 fs. The laser pulses were incident normal to the sample surface, so that an effective optically generated magnetic field would be directed along the magnetization, similar to a conventional recording scheme. The direction of this opto-magnetic switching is determined only by the helicity of light. This finding reveals an ultrafast and efficient pathway for all-optically writing magnetic bits at record-breaking speeds. The most obvious explanation via the inverse Faraday effect (IFE) [22] could only very qualitatively account for the previously observed features. The equation describing it was derived in a nondissipative approximation and thus the microscopical origin of the IFE-field in metallic GdFeCo is still a subject of research. How does it relate to the heat-induced switching discussed above? Hence, the microscopic mechanisms of the light–matter interaction responsible for AO-HDS have to be looked at more carefully. In general, the effects of light-matter interaction can be divided into two groups: those that result in an absorption of light and those that lead to a change of the phase of the radiation. Consequently, helicity-dependent effects in optics of magnetic media can be seen as magnetic circular dichroism (MCD) or magnetic circular birefringence (MCB), respectively. It is the aim of this section to reveal which one of these two magneto-optical effects is responsible for AO-HDS. To achieve this goal, we performed spectral studies and explored the correlation with the sublattice magnetizations.

First, we performed spectral studies of AO-HDS as a function of light polarization and intensity [23]. The switching was studied by exciting the GdFeCo with a single 60 fs laser pulse at different wavelengths ($\lambda = 500$–800 nm) using magneto-optical microscopy to determine the final magnetic state of the exposed area. It was hypothesized that circular polarized (CP) light acts as a strong effective magnetic field pulse H_{OM} on the magnetization of the medium through the IFE. The direction of H_{OM} is then defined by the helicity of the light. From spectral measurements of the Faraday rotation angle θ_{F} and n, we calculated H_{OM} as a function of the wavelength. The calculated H_{OM} has a strong wavelength dependence, while the switching window Δ is almost constant in this spectral range. We defined the window of AO-HDS, as $\Delta = 2(F_{\mathrm{RC}} - F_{\mathrm{LC}})/(F_{\mathrm{RC}} + F_{\mathrm{LC}})$, where F_{RC} and F_{LC} denote the switching threshold of GdFeCo for right- (RC) and left-handed circularly polarized (LC) excitation pulses, respectively. These values show that the switching window in GdFeCo, expressed in terms of relative fluence Δ, exactly corresponds to the relative absorption of RC and LC light, given by the MCD values. Note that this correlation shows that the switching thresholds for LC and RC light are different by exactly the same amount as the difference between the total absorption in the GdFeCo layer for LC and RC light, respectively. Thus the effective switching threshold (the actual absorbed energy density in the GdFeCo layer) is also independent of the wavelength and is equal to 2.6 ± 0.2 mJ cm^{-2} [23].

Secondly, we performed experiments to explore the correlation between the sublattice magnetization and AO-HDS. Samples were grown by magnetron sputtering in a multilayer structure: glass/AlTi(10 nm)/SiN(5 nm)/GdFeCo(20 nm)/SiN(60 nm). Net magnetization of GdFeCo is attributed to the difference of magnetization in anti-parallelly coupled rare-earth Gd and transition-metal FeCo. Increasing the amount of Gd, the configuration of sublattice (transition metal) magnetization and net magnetization is changed parallel to antiparallel by across the magnetization compensation composition ratio C_M. Using magneto-optical microscopy in combination with femtosecond pulsed light, we measured the compositional dependency ($x = 22$–27 at%) of AO-HDS in $\mathrm{Gd}_x(\mathrm{Fe}_{87.5}\mathrm{Co}_{12.5})_{100-x}$. To excite the material, we used amplified pulses from a Ti:sapphire laser at a wavelength of 800 nm, pulse width 90 fs and a repetition rate of 1 kHz. The direction of this switching is determined by the helicity of the light pulse. The relation between the direction of reversed net magnetization and helicity of the light changed sign by going across the magnetization compensation composition ratio C_M ($x \sim 24.5$ at%). We also found that the sign of MCD originated from the transition metal contribution with photon energies in the visible range changed at C_M.

These results demonstrate that all-optical switching depends only on the amount of energy absorbed by the magnetic system, independent of the wavelength or helicity of the laser pulse. A natural consequence is that switching can be achieved with any polarization as long as the absorbed intensity in the GdFeCo layer is sufficiently strong. Furthermore, we present a quantitative explanation of the intensity window in which all-optical helicity-dependent switching (AO-HDS) occurs, based on magnetic circular dichroism. Because of different absorption coefficients for RC and LC light in GdFeCo, the switching threshold is helicity dependent. This explanation is consistent with all the experimental findings on AO-HDS so far, varying from single- to multiple-shot experiments. We can conclude unambiguously that AO-HDS originates from MCD. The presented results give a solid understanding of the origin of AO-HDS, and give novel insights into the physics of ultrafast, laser-controlled magnetism.

13.5 Conclusions

In this chapter, we briefly summarized the results of our recent experimental studies on ultrafast light-induced spin reversal in amorphous rare earth-transition metal alloy films. The work directed to the understanding of the exact mechanism led to many interesting discoveries, that fueled a lot of interest to this area. As a result, optical manipulation of magnetic order by femtosecond laser pulses has developed into an exciting and still expanding research field that keeps producing a continuous stream of new and sometimes counterintuitive results.

Acknowledgments

The work reviewed in this chapter was a result of an intense collaboration between our groups, with important contributions from the BESSY (I. Radu) and LCLS (H. Duerr) teams. This work was supported by the PRESTO Japan Science and Technology Agency, the Netherlands Organization for Scientific Resarch (NWO), the Stichting Fundamenteel Onderzoek der Materie (FOM), the EU program Femtospin, the Nihon University Strategic Project for Academic Research and the MEXT-Supported Program for the Strategic Research Foundation at Private Universities 2013-2017.

References

1. K. Vahaplar, A. M. Kalashnikova, A. V. Kimel, *et al.*, Ultrafast oath for optical magnetization reversal via a strongly nonequilibrium state, *Phys. Rev. Lett.*, **103**, 117201 (2009).
2. J. Stöhr and H. C. Siegmann, Magnetism: From Fundamentals to Nanoscale Dynamics, Berlin: Springer (2006).
3. A. Tsukamoto, T. Sato, S. Toriumi, and A. Itoh, Precessional switching by ultrashort pulse laser: Beyond room temperature ferromagnetic resonance limit, *J. Appl. Phys.*, **109**, 07D302 (2011).
4. N. Inaba, Y. Uesaka, A. Nakamura, *et al.*, Damping constant of Co–Cr–Ta and Co–Cr–Pt thin films, *IEEE Trans. Magn.* **33**, 2989 (1997).
5. C. D. Stanciu, A. V. Kimel, F. Hansteen, *et al.*, Ultrafast spin dynamics across compensation points in ferrimagnetic GdFeCo: The role of angular momentum compensation, *Phys. Rev. B*, **73**, 220402(R) (2006).

6. Th. Gerrits, H. A. M. van den Berg, J. Hohlfeld, *et al.*, Ultrafast precessional magnetization reversal by picosecond magnetic field pulse shaping, *Nature*, **418**, 509–512 (2002).
7. H. W. Schumacher, C. Chappert, R. C. Sousa, *et al.*, Quasiballistic magnetization reversal, *Phys. Rev. Lett.*, **90**, 017204 (2003).
8. C. D. Stanciu, A. Tsukamoto, F. Hansteen, *et al.*, Subpicosecond magnetization reversal across ferrimagnetic compensation points, *Phys. Rev. Lett.*, **99**, 217204 (2007).
9. E. Beaurepaire, J.-C. Merle, A. Daunois, and J.-Y. Bigot, Ultrafast spin dynamics in ferromagnetic nickel, *Phys. Rev. Lett.*, **76**, 4250–4253 (1996).
10. Andrei Kirilyuk, Alexey V. Kimel, and Theo Rasing, Ultrafast optical manipulation of magnetic order, *Rev. Mod. Phys.*, **82**, 2731–2784 (2010).
11. B. Koopmans, G. Malinowski, F. Dalla Longa, *et al.*, Explaining the paradoxical diversity of ultrafast laser-induced demagnetization, *Nature Mater.*, **9**, 259–265 (2010).
12. M. Sultan, U. Atxitia, A. Melnikov, *et al.*, Electron- and phonon-mediated ultrafast magnetization dynamics of Gd(0001), *Phys. Rev. B*, **85**, 184407 (2012).
13. M. Battiato, K. Carva, and P. M. Oppeneer, Superdiffusive spin transport as a mechanism of ultrafast demagnetization, *Phys. Rev. Lett.*, **105**, 027203 (2010).
14. G. P. Zhang and T. F. George, Total angular momentum conservation in laser-induced femtosecond magnetism, *Phys. Rev. B*, **78**, 052407 (2008).
15. G. Schütz, W. Wagner, W. Wilhelm, *et al.*, Absorption of circularly polarized x rays in iron, *Phys. Rev. Lett.*, **58**, 737–740 (1987).
16. J. Stöhr, Y. Wu, B. D. Hermsmeier, *et al.*, Element-specific magnetic microscopy with circularly polarized X-rays, *Science*, **259**, 658–661 (1993).
17. I. Radu, K. Vahaplar, C. Stamm, *et al.*, Transient ferromagnetic-like state mediating ultrafast reversal of antiferromagnetically coupled spins, *Nature*, **472**, 205–208 (2011).
18. T. A. Ostler, J. Barker, R. F. L. Evans, *et al.*, Ultrafast heating as a sufficient stimulus for magnetization reversal in a ferrimagnet, *Nature Communications*, **3**, 666 (2012).
19. J. H. Mentink, J. Hellsvik, D. V. Afanasiev, *et al.*, Ultrafast spin dynamics in multisublattice magnets, *Phys. Rev. Lett.*, **108**, 057202 (2012).
20. K. Vahaplar, A. M. Kalashnikova, A. V. Kimel, *et al.*, All-optical magnetization reversal by circularly polarized laser pulses: Experiment and multiscale modeling, *Phys. Rev. B*, **85**, 104402 (2012).
21. C. D. Stanciu, F. Hansteen, A. V. Kimel, *et al.*, All-optical magnetic recording with circularly polarized light, *Phys. Rev. Lett.*, **99**, 047601 (2007).
22. A. V. Kimel, A. Kirilyuk, P. A. Usachev, *et al.*, Ultrafast non-thermal control of magnetization by instantaneous photomagnetic pulses, *Nature*, **435**, 655–657 (2005).
23. A. R. Khorsand, M. Savoini, A. Kirilyuk, *et al.*, Role of magnetic circular dichroism in all-optical magnetic recording, *Phys. Rev. Lett.*, **108**, 127205 (2012).

Index

Page numbers in italics refer to *Figures* and those in bold refer to **Tables**.

Spintronics for Next Generation Innovative Devices, First Edition. Edited by Katsuaki Sato and Eiji Saitoh.
© 2015 John Wiley & Sons, Ltd. Published 2015 by John Wiley & Sons, Ltd.

Printed and bound by CPI Group (UK) Ltd, Croydon, CR0 4YY
17/09/2026
14964302-0002